Prep for Success

Florida's PERT
Math Study Guide

Stacey L Francis

President: Track 2 Success, Inc.
Kissimmee, Florida

Editor

Travis L Jackson

CEO: Track 2 Success, Inc.

This study guide is intended for informational purposes only. The material herein represents study techniques for Florida's Postsecondary Education Readiness Test (PERT) mathematics section as summarized by the authors. The information herein is meant to supplement materials obtained from your testing institution and can only be used as a general study guide. Be sure to contact your institution to obtain their specific requirements and guidelines before taking the PERT.

Contact us at *info@track2success.com* to obtain more information about this guide or for additional assistance.

Prep for Success:
Florida's PERT Math Study Guide

Copyright © 2014 Track 2 Success, Inc.

All Rights Reserved

Printed in the United States of America

ISBN: 978-0-9830558-4-6

Cover Design and Graphics by Stacey Francis
Illustrations by Stacey Francis

Copy Edited by Travis L. Jackson

Track 2 Success, Inc.

Website: www.track2success.com
Email: info@track2success.com

Preface

Track 2 Success, Inc. was founded in 2000 as a small math tutoring business, catering to students who learn more effectively outside of a large classroom setting. We have always been dedicated to helping our students reach their full potential, firmly believing that everyone has the capacity to learn given the right tools. Now, with over 15 years of experience and seeing hundreds of students successfully attain their goals, we have continued to strive to provide the best in test preparation materials.

The following study guide is the successor to the very well received first edition. I recall taking a college placement test ten years ago and feeling utterly dejected because there were no study materials available, only a small booklet that listed the topics that the test would cover. With no real clue as to what questions to expect, I was forced to study as much as I could from an array of textbooks from basic arithmetic to college algebra. In writing this study guide, my goal was to spare others of my experience so that they could go into the test feeling confident and prepared.

I believe that this guide is indispensable to anyone preparing to take the math portion of Florida's Postsecondary Education Readiness Test. As always, I would like to thank my husband for enduring all my complaints, giving me the strength to continue when I wanted to quit and for taking the time to edit this study guide. I would also like to thank my sister and mother for always being my inspiration and reminding me why I chose to become an educator.

Good Luck to you,

President, Track 2 Success, Inc.

Test Information

For the past few years, Community Colleges and Universities in Florida have been using the Postsecondary Education Readiness Test (PERT) as a tool to determine the appropriate course level for incoming freshmen. This test is mandatory to all first-time-in-college students who have earned no college credits. In 2011, the Florida Department of Education indicated that there would be a shift in policy so that the PERT would now be administered to high school students depending on their score on the FCAT. There may be some cases in which a student can be exempt from taking a placement exam (see your advisor to determine your exemption eligibility). Students having obtained a minimum score on either the ACT or SAT may not have to take the PERT but may opt to take the test solely to be placed into a higher level college course. Depending on your score, you can receive credit for and avoid having to take remedial classes, saving time and money. Therefore, it is worthwhile to spend some time preparing for the test.

Make sure to obtain as much information as possible from your institution. The PERT is not a pass/fail exam; institutions set point levels that will determine class placement. It is important to do your best since a better score may mean entrance into a higher level class. The PERT Math is an untimed exam consisting of 30 questions and the maximum scaled score that can be obtained is 150. Of the 30 questions, 25 are scored while 5 are field test items used to improve the quality of future exams. You will not be notified which questions are field test questions so it is important to treat each question as equally important. The average time needed to complete the exam is 30 minutes. Make sure to ask your academic advisor and testing center any questions before or during registration for the test. This guide is meant to cover all the pertinent topics on the PERT Math as defined in the list of Postsecondary Readiness Competencies (PRCs) provided by the Florida Department of Education. This ensures that you are well prepared no matter what questions you are given. However, we do not attempt to provide details about the way the test will be administered because this may differ by institution.

Typical questions to ask your advisor/testing center:

- What are the classes that I can possibly be assigned to and what are the corresponding score levels?

- Can I take the test more than once and if yes, how long do I have to wait between attempts?

- What do I need to bring on the day of the test?

The PERT is a computerized adaptive test (CAT) which adjusts as the test progresses based on whether previous questions were answered correctly. Since the test is a computer adaptive test and questions are graded as you progress through the exam, you cannot return to a previously visited question to change your answer and your final score will be displayed onscreen once you complete the exam. This test usually begins with a question of average difficulty and adjusts based on your responses. You should also receive a printout which you should keep for your records since your institution may charge a fee for duplicates.

This guide does not provide information about scores needed to place into a particular course because they may vary by institution but the table below displays the ranges provided by the Florida Department of Education.

Math Course	Score Range
Lower Level Development Education	50 – 95
Higher Level Development Education	96 – 113
Intermediate Algebra (MAT 1033)	114* – 122
College Algebra or higher	123 - 150

*114 is the college-ready cut score for math

The course number and title may differ by institution. A college-level cut score of 113 indicates a student's readiness for college as determined by the Florida Department of Education. If you score lower than 114 on the PERT, you will probably be assigned to a Lower Level or Higher Level Development Education course which may be remedial math classes such as Basic Algebra.

This second edition revises the first edition to more closely adhere to the Postsecondary Readiness Competencies (PRCs) tested on the exam. The study guide portion has been rearranged to provide a better flow from Arithmetic concepts to Advanced Algebra. We have added more chapter questions and examples throughout.

To learn more about the Postsecondary Education Readiness Test (PERT), please visit:

http://www.fldoe.org/fcs/pert.asp

http://www.fldoe.org/fcs/pdf/pertfaq.pdf

http://www.fldoe.org/fcs/OSAS/Evaluations/pdf/Zoom2010-03.pdf

http://www.fldoe.org/fcs/collegecareerreadiness.asp

Table of Contents

Part One – Study Guide 1

Chapter 1 – The Real Number System 1

Classification of Numbers 1

Properties of Real Numbers 2

    Associative Property 2

    Commutative Property 2

    Distributive Property 2

    Inverse Property 2

Operations on Integers 3

    Adding Integers 3

    Subtracting Integers 4

    Multiplying and Dividing Integers 4

    Absolute Value 5

    Exponents 5

    Order of Operations 6

    Prime and Composite Numbers 7

    Factors and Divisibility 7

    Least Common Multiple (LCM) 8

    Greatest Common Factor (GCF) 9

Radicals and Roots 9

    Addition and Subtraction of Radicals 11

    Multiplication and Division of Radicals 12

Chapter 1 Exercises 13

Chapter 2 – Conversions 17

Fractions, Decimals & Percentages 17

    Reducing Fractions 17

    Mixed Number to an Improper Fraction 18

    Improper Fraction to a Mixed Number 18

    Decimal to Percentage 18

Decimal to Fraction 19

Fraction to Decimal.. 19

Fraction to Percentage 20

Percentage to Fraction 20

Percentage to Decimal 21

Chapter 2 Exercises 22

Chapter 3 – Fractions & Decimals 25

Comparing Fractions 25

Operations on Fractions 26

Adding and Subtracting Fractions 26

Multiplying and Dividing Fractions 27

Operations on Decimals 28

Adding and Subtracting Decimals 28

Multiplying and Dividing Decimals 29

Applications 30

Chapter 3 Exercises 31

Chapter 4 – Ratios, Proportions & Percentages 37

Percentages 37

Percentage Increase or Decrease 38

Percentage Error 39

Simple Interest 40

Arithmetic Mean 42

Ratios 44

Proportions 44

Rates 46

Important Conversions 46

Chapter 4 Exercises 48

Chapter 5 – The Decimal System 53

The Decimal System 53

Rounding 54

Estimation 55

Scientific Notation 56

Converting Standard Form to Scientific Notation 56

Comparisons 58

Chapter 5 Exercises 60

Chapter 6 – Linear Equations of One Variable 63

Basic Geometry 63

Perimeter of Plane Figures 63

Area of Plane Figures 63

Other Geometric Shapes 64

Algebraic Equations 66

Formulate and Solve First Degree Equations 67

Solving Equations involving Absolute Values 69

Relevant vs. Irrelevant Information 69

Other Word Problems 70

Chapter 6 Exercises 73

Chapter 7 – Linear Equations of Two Variables 77

Linear Equations 77

The Coordinate Plane 77

Equations of a Line 78

Intercepts 79

Graphs of Linear Equations 80

Graphs of Horizontal and Vertical Lines 83

Parallel Lines 84

Perpendicular Lines 85

Distance and Midpoint 85

Determine Equation of Line Given Graph 86

Determine Equation of Line Given Table of Values 87

Applications 87

Chapter 7 Exercises 89

Chapter 8 – Inequalities 93

Linear Inequalities 93

Solving Linear Inequalities 93

Solving Linear Inequalities Graphically 95

vi   Prep for Success: Florida's PERT Math Study Guide

Graphs of Linear Inequalities 95

Absolute Values and the Less Than Inequality 99

Absolute Values and the Greater Than Inequality 100

Word Problems Involving Inequalities 100

Chapter 8 Exercises 102

Chapter 9 – Terms, Factors & Expressions 107

Factors 107

Coefficients, Variables, Terms & Factors 107

Expressions and Equations 107

Solving Literal Equations 107

Evaluating Expressions 108

Order of Operations with Algebraic Terms 109

Exponential Expressions 110

Solving Exponential Equations 111

Basic Factoring 111

Chapter 9 Exercises 113

Chapter 10 – Polynomials 117

Adding and Subtracting Polynomials..117

Standard Form 117

Like Terms 117

Factoring Polynomials 118

Factoring Quadratic Expressions 118

Difference and Sum of Cubes 121

Factoring 4th Degree Polynomials 121

Multiplying and Dividing Polynomials 122

FOIL..122

Multiplying Polynomials...123

Dividing Polynomials 124

Expanding Polynomials 126

Chapter 10 Exercises 127

Chapter 11 – Radicals & Exponents 129

Complex Roots and Exponents 129

Complex Exponents 129

Complex Roots 132

Complex Roots with Variables 133

Rationalizing the Denominator 133

Solving Equations involving Radicals 135

Chapter 11 Exercises 137

Chapter 12 – Quadratics 141

Complex Numbers 141

Algebraic Operations involving Complex Numbers 141

Powers of i 143

Quadratic Equations 143

Solving Quadratic Equations by Factoring 143

The Quadratic Formula 144

Graphs of Quadratic Equations 147

Domain and Range 148

Chapter 12 Exercises 151

Chapter 13 – Rational Expressions 155

Adding and Subtracting Rational Expressions 155

Multiplying and Dividing Rational Expressions 156

Multiplying Rational Expressions 156

Dividing Rational Expressions 157

Solving Equations involving Rational Expressions 158

Inequalities and Rational Expressions 160

Chapter 13 Exercises 162

Chapter 14 – Systems of Equations 165

Solving Systems of Equations 165

Solving by Inspection 165

Solving Algebraically 166

Elimination 166

Substitution 168

Solving Systems of Equations Graphically 170

Applications..170

Chapter 14 Exercises 172

Chapter 15 – Inverse & Composite Functions 177

Inverse Functions 177

Composite Functions 179

Rules of Inverses 180

Chapter 15 Exercises 181

Chapter 16 – Geometric Reasoning 185

Geometric Reasoning 185

Types of Triangles 187

Pythagorean Theorem 188

Types of Angles 189

Polygons and Their Properties 190

Chapter 16 Exercises 191

Part Two – Workbook 197

Worksheet #1 Bubble Sheet 199

Worksheet #1 201

Worksheet #1 Answer Sheet 211

Worksheet #1 Solutions 213

Worksheet #2 Bubble Sheet 233

Worksheet #2 235

Worksheet #2 Answer Sheet 245

Worksheet #2 Solutions 247

Worksheet #3 Bubble Sheet 269

Worksheet #3 271

Worksheet #3 Answer Sheet 281

Worksheet #3 Solutions 283

Worksheet #4 Bubble Sheet 303

Worksheet #4 305

Worksheet #4 Answer Sheet 317

Worksheet #4 Solutions 319

Worksheet #5 Bubble Sheet 339

Worksheet #5 341

Worksheet #5 Answer Sheet 353

Worksheet #5 Solutions 355

Extra Bubble Sheets 377

Appendix – Formulas 383
Index 389

Chapter 1 — The Real Number System

This first chapter provides a very basic review of the arithmetic concepts used on the exam. If you are already familiar with topics such as the number line, addition and subtraction of integers, absolute value and order of operations, skip ahead to the next chapter.

Classification of Numbers

The word "numbers" is used loosely to define the points on the number line depicted below. The point 0, often called the origin, is the midpoint. Numbers to the left of the origin are negative and numbers to the right are positive. Positive and negative numbers are called **signed numbers**.

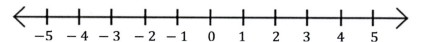

An important thing to remember is that on the number line, numbers get larger as we move to the right and smaller as we move to the left. You will use this on questions that ask you to order a set of values from largest to smallest or smallest to largest.

Some of the most common types of numbers are as follows:

- The set of **natural numbers**, also known as counting numbers, are the numbers used to count or quantify objects.
  $N = \{1, 2, 3, 4, 5, 6, \ldots\}$
- The set of **whole numbers** is the set of natural numbers plus the number zero.
  $W = \{0, 1, 2, 3, 4, 5, 6, \ldots\}$
- The set of **integers** include the set of whole numbers and their opposites (also called **additive inverses**). The integers less than zero (to the left on the number line) are called **negative integers** and those greater than zero are called **positive integers**. Zero is also an integer but is neither positive nor negative.
  $I = \{\ldots, -4, -3, -2, -1, 0, 1, 2, 3, 4, \ldots\}$
- The set of **rational numbers** include any number that can be expressed as a fraction therefore rational numbers include fractions, the set of integers and many decimals. We say many because not all decimals are rational. We take a closer look at fractions and decimals later but here are some examples of rational numbers:

  4 can be expressed as an improper fraction such as $\frac{12}{3}$ therefore it is rational.

  0.6 can be written as the fraction $\frac{6}{10}$ and so it is rational as well.

- The set of **irrational numbers** include those that cannot be written in fractional form. Irrational numbers include radicals like $\sqrt{2}$ and $\sqrt{3}$ and non-terminating decimals such as π or $0.1121231234\ldots$
- The set of **even integers** are those divisible by 2 leaving no remainder and ending in 0, 2, 4, 6 or 8. Note that 0 is considered to be an even integer.
- The set of **odd integers** are the integers that are not even and ending in 1, 3, 5, 7 or 9.
- The set of **real numbers** encompasses all of the above numbers.
- A **prime number** is a positive integer that is only divisible by 1 and itself. Note that 1 is not considered to be a member of the set of prime numbers. The prime numbers between 1 and 100 are 2, 3, 5, 7, 11, 13, 17, 19, 23, 29, 31, 37, 41, 43, 47, 53, 59, 61, 67, 71, 73, 79, 83, 89 and 97.

Properties of Real Numbers

The following properties of real numbers will come in handy as we work through future chapters and are therefore worth listing.

Associative Property

Addition and multiplication are associative and can be done in any order (parentheses indicate calculations that need to be done first).
$$a + (b + c) = (a + b) + c$$
$$a \times (b \times c) = (a \times b) \times c$$

Commutative Property

Addition and multiplication are also commutative so we can reverse the numbers and get the same result. Like the associative property, this does not apply to subtraction and division.
$$a + b = b + a$$
$$a \times b = b \times a$$
Note that subtraction and division are not commutative because $a - b \neq b - a$ and $a \div b \neq b \div a$.

Distributive Property

When multiplying the sum or difference of two numbers by a third number, we can use the distributive property as follows:
$$a(b + c) = ab + ac$$

Inverse Property

The inverse of multiplication is division and the inverse of addition is subtraction.

$a + (-a) = 0$ *Additive Inverse*

$a \times \dfrac{1}{a} = 1$ *Multiplicative Inverse*

Note that 0 has no multiplicative inverse because division by 0 is undefined and not allowed.

Example 1.1:

Choose the expression that is equivalent to $4 \times 2(b + 5)$

A. $6(b + 5)$　　　　B. $(4 + 2)5b$　　　　C. $2 \times 4(b + 5)$　　　　D. $4 \times 2b + 5$

Choice A gets eliminated because $4 \times 2 = 8$ not 6

Choice B gets eliminated because the multiplication and addition signs were switched and that is not allowed.

Choice C looks acceptable because by the Commutative Property 4×2 is equal to 2×4.

Choice D gets eliminated because the 2 was only distributed to b and not to 5 which violates the Distributive Property.

Therefore, the answer is C.

Operations on Integers

Before we begin reviewing addition, subtraction, multiplication and division of integers, we present a few guidelines. Since all integers are either even or odd, the following rules can act as a quick check helping you to identify if you've calculated incorrectly.

$even \pm even = even$
$odd \pm odd = even$

$even \pm odd = odd$
$odd \pm even = odd$

$even \times even = even$
$even \times odd = even$
$odd \times odd = odd$

Adding Integers

Remember that signed integers are simply negative and positive numbers. Below are the rules for adding signed integers.

Rule	Example
When adding two positive numbers, the resulting sum is also positive.	$11 + 13 = 24$
When adding two negative numbers, add the numbers and keep the sign so the result is also negative.	$-4 + (-8) = -12$
When adding two numbers of opposite signs, ignore the signs and subtract the smaller from the larger. The sign of the result will be the sign of the larger number.	$-15 + 6$ $= -(15 - 6)$ $= -9$

Subtracting Integers

Subtraction can be treated as one of the last two addition rules on the previous page.

Example 1.2:

$9 - 21 =$

This can be treated as addition between numbers of opposite signs.

$9 + (-21)$

Ignore the signs and subtract 9 from 21.

$21 - 9 = 12$

Keep the sign of 21 since it is larger so the result will be negative.

Therefore $9 - 21 = -12$

Example 1.3:

$-3 - 14 =$

This can be treated as addition between numbers of the same sign.

$-3 + (-14)$

We add the numbers 3 and 14.

$3 + 14 = 17$

Since both numbers were negative, the result is also negative.

$-3 - 14 = -17$

Multiplying and Dividing Integers

Multiplication and division of signed numbers follow these rules for determining the sign of the result:

$(+)$ *multiplied or divided by* $(+) = (+)$
$(+)$ *multiplied or divided by* $(-) = (-)$
$(-)$ *multiplied or divided by* $(+) = (-)$
$(-)$ *multiplied or divided by* $(-) = (+)$

When two negative signs come together, it is treated as multiplication of two negatives.
The expression $-(-4)$ can actually be written as $-1(-4)$.
Multiplication of two negatives yields a positive therefore $-(-4) = +4$.

Example 1.4:

$(-20) \div (-4) =$

Division of two negatives will result in a positive.

$(-20) \div (-4) = 5$

Example 1.5:

$-11 \times 3 =$

Multiplication of a negative and a positive will result in a negative.

$-11 \times 3 = -33$

Chapter 1 – The Real Number System **5**

Absolute Value

The set of **integers** include the number zero, positive whole numbers and negative whole numbers. Decimals and fractions can also be negative or positive. The absolute value of a number gives the magnitude of the number and so the result is always positive. Think of the positive/negative sign as the direction travelled and the number as the distance travelled. Since distance is always positive, so is the absolute value. Therefore, the absolute value of a number is its distance from the origin and the distance between two points on the number line is the absolute value of their difference.

Example 1.6:

$|-7| =$

> The result of an absolute value is always positive because we only care about the distance travelled and not the direction. Therefore:
> $|-7| = 7$

Example 1.7:

$|-15| - |-4| =$

> $|-15| = 15$
> $|-4| = 4$
> Therefore, $|-15| - |-4| = 15 - 4 = 11$

Example 1.8:

What is the distance between -9 and 5 on the number line?

> To find the distance between the two numbers, subtract then take the absolute value of the result. Note that from the properties of real numbers we know that subtraction is not commutative and typically cannot be reversed i.e. $a - b \neq b - a$. However, since we are finding the distance and will be taking the absolute value of the difference, we can subtract either one from the other.
> $distance = |-9 - 5| = |-14| = 14$

Exponents

Exponents are used to represent repeated multiplication of the same number. We call the number being multiplied the **base** and number of times it is to be multiplied the **exponent**.

Base $\Longrightarrow 2^5 \Longleftarrow$ Exponent

When the base is positive, the result will always be positive. A negative number will yield a negative result when raised to an odd exponent but will yield a positive result when raised to an even exponent.

$(-2)^5 = (-2)(-2)(-2)(-2)(-2) = -32$
$(-3)^4 = (-3)(-3)(-3)(-3) = 81$

Note that $(-2)^4 \neq -2^4$

$(-2)^4 = (-2)(-2)(-2)(-2) = 16$ while $-2^4 = -[(2)(2)(2)(2)] = -16$

Here are a couple basic laws of exponents of which you need to be aware.

$0^x = 0$ 0 raised to any power equals 0.

$x^0 = 1$ Any number raised to the power 0 equals 1.

The only exception to the above two rules occurs when $x = 0$ because 0^0 is undefined.

$$a^{-x} = \frac{1}{a^x} \qquad \frac{1}{a^{-x}} = a^x$$

We will look at more laws of exponents in Chapter 9.

Example 1.9:

$(-2)^3 - 3^4 =$
$$[(-2)(-2)(-2)] - [(3)(3)(3)(3)]$$
$$= -8 - 81 = -89$$

Example 1.10:

$5^2 \cdot |-1|^3 =$

Remember that an absolute value results in a positive number.
$$5^2 \cdot |-1|^3 = 5^2 \cdot 1^3 = (25)(1) = 25$$

Order of Operations

When a question involves multiple operators $(+, -, \times$ and $\div)$ the order in which you complete the operations is very important. The acronym PEMDAS serves as a good reminder of the order in which the operations should be conducted. Many students also use phrases such as "Please Excuse My Dear Aunt Sally." The operations that match the letters of the acronym are as follows:

Parentheses

Exponents

Multiplication

Division

Addition

Subtraction

Work though the operations from top to bottom so anything in parentheses would be solved before any exponents and so on.

There are two important things to note here:

1. Even though multiplication appears before division, they are actually done together so do them *as they appear from left to right* in the expression. The same rule applies to addition and subtraction.

2. The word "parentheses" is slowly being replaced by the word "groupings" to make it clear that all types of groupings including brackets, expressions within an absolute value or under a radical are to be done first. Hence the acronym GEMDAS is gaining popularity. Also, when the parentheses contain other operators, they also need to be done in the order of PEMDAS.

Example 1.11:

$6 - 18 \div 2 + 7 - 4^2 =$

There are no parentheses so solve the exponent first[1].
$= 6 - 18 \div 2 + 7 - \mathbf{16}$
Now complete any multiplication or division moving from left to right.
$= 6 - \mathbf{9} + 7 - 16$
Finally, complete the addition and subtraction, also moving from left to right.
$= \mathbf{-3} + 7 - 16$
$= \mathbf{4} - 16$
$= -12$

Prime and Composite Numbers

A **prime number** is a number that is only divisible by 1 and itself. Note that 1 is not considered to be a prime number.

Prime numbers: 2, 3, 5, 7, 11, 13, 17, 19, 23, 29, 31, 37, 41, 43, 47, 53, 59, 61, 67, 71, 73, 79, 83, 89, 97...

A composite number is the opposite of prime and is a number that can be broken up into prime factors (see the next section).
Examples of composite numbers:

$24 = 2 \times 12 \ or \ 4 \times 6 \ or \ 2 \times 2 \times 2 \times 3$
$35 = 5 \times 7$

Example 1.12:

Which of the following is NOT a prime number?

A. 31   B. 43   C. 57   D. 79

From the list of the prime numbers between 1 and 100 above, we see that 57 is not prime. It is actually divisible by 3 ($57 = 3 \times 19$).

If you do not memorize the list, you can easily use the divisibility rules found in the table in the next section to test the answer choices. For 57, $5 + 7 = 12$ which is divisible by 3, therefore 57 is divisible by 3.

Factors and Divisibility

A **factor** of a number leaves no remainder when divided into the number. Therefore 2 is a factor of 6. A **prime factor** is a factor that is also a prime number.

Prime factors: 2, 3, 5, 7, 11, 13, 17...

---

[1] The numbers in bold represent the result of a completed operation

8 Prep for Success: Florida's PERT Math Study Guide

One easy method of breaking a number into its prime factors is demonstrated below. Remember to go through the list of prime factors starting with 2, dividing each result until left with 1.

The table below gives some basic rules that can be used to determine divisibility.

Divisible by:	Rule
2	Number ends in an even digit
3	Sum of digits is divisible by 3
4	Last two digits are divisible by 4
5	Number ends in 0 or 5
6	Combine first two rules
9	Sum of digits is divisible by 9
10	Number ends in 0

Example 1.13:

What is the prime factorization of 420?

We can begin by dividing by 2 because 420 is even.

$420 \div 2 = 210$

$210 \div 2 = 105$

$105 \div 3 = 35$

$35 \div 5 = 7$

$7 \div 7 = 1$

The prime factors of 420 are therefore 2, 2, 3, 5, and 7 and so the prime factorization is:

$2^2 \cdot 3 \cdot 5 \cdot 7$

Example 1.14:

Which of the following does not divide into 1155?

A. 3   B. 6   C. 7   D. 11

Using the rules of divisibility:

1155 is divisible by 3 because $1 + 1 + 5 + 5 = 12$ which is divisible by 3.

To be divisible by 6, a number must be divisible by both 2 and 3. Since 1155 does not end in an even number, it is not divisible by 2 and therefore is not divisible by 6.

Least Common Multiple (LCM)

A **multiple** of a number is the result of multiplication by another number, as we learned in the multiplication tables.

The multiples of 2 are 2, 4, 6, 8, 10, 12 …

The multiples of 5 are 5, 10, 15, 20, 25 …

A **common multiple** is a multiple that appears in the list of multiples for 2 or more numbers.

Multiples of 3: 3, 6, 9, 12, 15, 18, 21 …

Multiples of 6: 6, 12, 18, 24, 30, 36 …

Therefore 6, 12 and 18 are all common multiples of 3 and 6.

Chapter 1 – The Real Number System   **9**

The **Least Common Multiple** is the smallest possible common multiple for 2 or more numbers i.e. it is the smallest number that is divisible by all the numbers. So 6 is the LCM of 3 and 6 because it is the smallest number divisible by both 3 and 6. To find the least common multiple, we list the prime factors of all the numbers. All the factors of each number become factors of the LCM, without including unnecessary repeated factors.

Example 1.15:
What is the least common multiple of 3, 4, 5 and 12?

First list the prime factors of all 4 numbers
$3 = \mathbf{3}$
$4 = \mathbf{2 \times 2}$
$5 = \mathbf{5}$
$12 = 2 \times 2 \times 3$
The LCM needs to include a pair of 2's, one 3 and one 5 (highlighted bold above). Note that the factors that are not in bold do not need to be included because they already appear in the LCM.
Therefore, the LCM $= 2 \times 2 \times 3 \times 5 = 60$

Greatest Common Factor (GCF)

Also called the Greatest Common Divisor, the GCF of a set of numbers is the largest factor common to all the number in the set. To find the GCF, list the prime factors of each of the numbers. Then extract and multiply all the common prime factors.

Example 1.16:
Find the greatest common factor of the numbers 30, 36 and 48.

List the prime factors of each
$30 = 2 \times 3 \times 5$
$36 = 2 \times 2 \times 3 \times 3$
$48 = 2 \times 2 \times 2 \times 2 \times 3$
There are two prime factors common to all 3 numbers: 2 and 3
So the GCF is $2 \times 3 = 6$ i.e. 6 is the largest factor by which all 3 numbers are divisible.

Radicals and Roots

Consider the equation:
$6^2 = 36$
We say that 6 is the square root of 36 because 6 multiplied by itself gives 36. Therefore a **root** is a repeated factor. Squared indicates that the factor is repeated twice, cubed indicates that the factor is repeated three times and so on.

A **radical**, denoted by the $\sqrt{}$ symbol, indicates that we need to find the specified root of the number under the radical. The number at the top left of the radical symbol specifies what root needs to be found. No number at the top left of the radical symbol indicates that we should take the square root. The number or expression under the radical is called the **radicand**.

$\sqrt[3]{}$ means take the cube root and $\sqrt[5]{}$ means take the fifth root.

Note that we cannot find the even root of a negative number without the use of complex numbers as we will see in later chapters.

Find $\sqrt[3]{-8}$

$$-8 = (-2)(-2)(-2)$$
$$\sqrt[3]{-8} = \sqrt[3]{(-2)(-2)(-2)} = -2$$

But ...

Find $\sqrt[4]{-16}$

$$(2)(2)(2)(2) = 16$$
$$(-2)(-2)(-2)(-2) = 16$$

A negative number multiplied by itself an even number of times will always yield a positive result. Therefore, $\sqrt[4]{-16}$ does not exist.

The table below lists the first 25 squares and the first 10 cubes. Knowing some or all of these will make calculations on the test easier and faster but is not mandatory.

	Squares		Cubes	
$1^2 = 1$	$10^2 = 100$	$19^2 = 361$	$1^3 = 1$	$6^3 = 216$
$2^2 = 4$	$11^2 = 121$	$20^2 = 400$	$2^3 = 8$	$7^3 = 343$
$3^2 = 9$	$12^2 = 144$	$21^2 = 441$	$3^3 = 27$	$8^3 = 512$
$4^2 = 16$	$13^2 = 169$	$22^2 = 484$	$4^3 = 64$	$9^3 = 729$
$5^2 = 25$	$14^2 = 196$	$23^2 = 529$	$5^3 = 125$	$10^3 = 1000$
$6^2 = 36$	$15^2 = 225$	$24^2 = 576$		
$7^2 = 49$	$16^2 = 256$	$25^2 = 625$		
$8^2 = 64$	$17^2 = 289$			
$9^2 = 81$	$18^2 = 324$			

Numbers do not always have a perfect root and so will not simplify to a whole number. We can however reduce the radical by breaking the number into its prime factors and simplifying.

Example 1.17:

$\sqrt{48} =$

First break 48 into its prime factors.

$48 = 2 \times 2 \times 2 \times 2 \times 3$

Since we are taking the square root, look for factors that appear in pairs.

$\sqrt{48} = \sqrt{2 \times 2} \times \sqrt{2 \times 2} \times \sqrt{3}$

Finally, simplify by taking the root of the paired factors, leaving the unpaired factors under the radical symbol.

$\sqrt{48} = 2 \times 2 \times \sqrt{3} = 4\sqrt{3}$

Chapter 1 – The Real Number System   **11**

Example 1.18:

$\sqrt[3]{270} =$

Again break the number into its prime factors.

$270 = 2 \times 3 \times 3 \times 3 \times 5$

Since we are taking the cube root, look for factors that appear in groups of three.

$\sqrt[3]{270} = \sqrt[3]{3 \times 3 \times 3} \times \sqrt[3]{2 \times 5}$

Finally, simplify by taking the cube root of the grouped factors leaving the other factors under the radical symbol.

$\sqrt[3]{270} = 3\sqrt[3]{10}$

Addition and Subtraction of Radicals

Before attempting to add or subtract radicals, ensure that the number under the radical is the same throughout the entire problem. Some radicals may need to be simplified first using the procedure demonstrated above.

Example 1.19:

Simplify: $5\sqrt{3} - 2\sqrt{75} + \sqrt{3}$

In order to add or subtract the terms of the expression, we need to ensure that all the terms contain $\sqrt{3}$. Therefore $\sqrt{75}$ needs to be simplified. Since we know that 3 is most likely a factor, start by dividing 75 by 3.

$75 = 25 \times 3 = 5 \times 5 \times 3$

Therefore:

$5\sqrt{3} - 2\sqrt{75} + \sqrt{3}$

$= 5\sqrt{3} - 2\sqrt{5 \times 5 \times 3} + \sqrt{3}$

$= 5\sqrt{3} - 2\sqrt{5 \times 5} \times \sqrt{3} + \sqrt{3}$

$= 5\sqrt{3} - 2(5)\sqrt{3} + \sqrt{3}$

$= 5\sqrt{3} - 10\sqrt{3} + \sqrt{3}$

Now combine the numbers in front of the radicals. Remember that if there appears to be no coefficient, the coefficient is actually 1.

$(5 - 10 + 1)\sqrt{3} = -4\sqrt{3}$

Therefore $5\sqrt{3} - 2\sqrt{75} + \sqrt{3} = -4\sqrt{3}$

Example 1.20:

$2\sqrt{63} + 3\sqrt{28} =$

$2\sqrt{63} + 3\sqrt{28}$

$= 2\sqrt{9 \times 7} + 3\sqrt{4 \times 7}$

$= 2\sqrt{9}\sqrt{7} + 3\sqrt{4}\sqrt{7}$

$= 2 \cdot 3\sqrt{7} + 3 \cdot 2\sqrt{7}$

$= 6\sqrt{7} + 6\sqrt{7}$

$= 12\sqrt{7}$

12   Prep for Success: Florida's PERT Math Study Guide

Multiplication and Division of Radicals

First check to determine whether each radical can be simplified. Multiplication of radicals follows the rule:

$$\sqrt[n]{a} \cdot \sqrt[n]{b} = \sqrt[n]{a \cdot b}$$

To multiply the radicals, multiply any numbers outside the radical and then multiply the radicands. Simplify the new radical further if possible.

Example 1.21:

$3\sqrt{48} \times 5\sqrt{60} =$

We first simplify each of the radicals.

$3\sqrt{48} \times 5\sqrt{60}$

$= 3\sqrt{16 \times 3} \times 5\sqrt{4 \times 15}$

$= 3(4\sqrt{3}) \times 5(2\sqrt{15})$

$= 12\sqrt{3} \times 10\sqrt{15}$

Now that the radicals have been simplified, multiply the numbers outside the radical and then multiply the radicands.

$= 120\sqrt{45}$

Further simplify the radical if possible.

$= 120\sqrt{3 \times 3 \times 5}$

$= 120(3\sqrt{5}) = 360\sqrt{5}$

To divide two radicals, we again first simplify each radical if possible. Division of radicals follows the rule:

$$\frac{\sqrt[n]{a}}{\sqrt[n]{b}} = \sqrt[n]{\frac{a}{b}}$$

If the radicand in the numerator and the denominator share a common factor, combine the two radicals as shown in the rule above and simplify.

Example 1.22:

$\dfrac{\sqrt{120}}{\sqrt{54}} =$

Simplify the numerator and denominator if possible.

$\dfrac{\sqrt{120}}{\sqrt{54}}$

$= \dfrac{\sqrt{4 \times 30}}{\sqrt{9 \times 6}} = \dfrac{2\sqrt{30}}{3\sqrt{6}}$

Since 30 is divisible by 6, we combine the two radicals and simplify.

$= \dfrac{2}{3}\sqrt{\dfrac{30}{6}} = \dfrac{2}{3}\sqrt{5}$

We will cover rationalizing the denominator and more complex radicals in Chapter 11.

Chapter 1 – The Real Number System   **13**

Chapter 1 Exercises

1. Which of the following does not divide into 462?
   A. 2   B. 3   C. 7   D. 9

2. Which of the following numbers is rational?
   A. $\sqrt{7}$ B. $\sqrt{9}$ C. π D. $\sqrt{11}$

3. The prime numbers between 1 and 12 inclusive are:
   A. $\{2, 3, 5, 7, 9, 11\}$
   B. $\{1, 2, 3, 5, 7, 11\}$
   C. $\{2, 3, 5, 7, 11\}$
   D. $\{3, 5, 7, 9\}$

4. $7 - 6^2 \div 4 - (2^3 - 5) \cdot 3 =$
   A. -11 B. -5 C. 1   D. 7

5. $-18 - |-4| + 3 =$
   A. 17   B. 12   C. -11 D. -19

6. $5\sqrt{27} - 2\sqrt{3} =$
   A. 13   B. $43\sqrt{3}$ C. $13\sqrt{3}$ D. $6\sqrt{3}$

7. Which of the following is NOT a prime number?
   A. 19   B. 31   C. 43   D. 57

8. $13 + (5 - |2|)^3 - 18 \div 2 =$
   A. 347   B. 121   C. 31   D. 13

9. The minimum required temperature for the survival a particular house plant is 52°F. If the current room temperature is 4°F below 0, by how many degrees Fahrenheit must the room temperature be increased so that the house plant will not die?
   A. 4°   B. 13°   C. 52°   D. 56°

10. What is the least common multiple of 15, 36 and 42?

   ☐

Solutions

1. **Answer: [D]**   Using the rules of divisibility, 462 is even so it is divisible by 2. The sum of the digits in 462 ($4 + 6 + 2 = 12$) is divisible by 3, so 462 is divisible by 3. The divisibility rule

for 7 takes longer than actually dividing by 7 so just divide. $462 \div 7 = 66$ so 462 is divisible by 7. The sum of the digits in 462 is not divisible by 9 ($12 \div 9 = 1R3$) therefore 9 does not divide into 462 and the answer is D.

2. **Answer: [B]** At first glance, all the above numbers appear to be irrational and that is because the number 9 is often mistaken as a prime number because it is an odd number.

$$\sqrt{9} = \sqrt{3 \times 3} = 3$$

The number 3 is rational and therefore the answer is B.

3. **Answer: [C]** Recall that a prime number is a number that is divisible by itself and 1 only and that 1 is not considered a prime number. Therefore, eliminate choices B and D. Again, 9 is often mistaken as being prime because it is odd but $9 = 3 \times 3$ and is therefore not prime. All but one of the answer choices includes 1 and/or 9 so the correct answer is C.

4. **Answer: [A]** Applying the order of operations, we simplify the expression within the parentheses first by first simplifying the exponent and then subtracting.
$$7 - 6^2 \div 4 - (2^3 - 5) \cdot 3$$
$$= 7 - 6^2 \div 4 - (\mathbf{8} - 5) \cdot 3$$
$$= 7 - 6^2 \div 4 - (\mathbf{3}) \cdot 3$$
Next we simplify the other exponent in the expression
$$= 7 - \mathbf{36} \div 4 - 3 \cdot 3$$
Now simplify multiplication and division working from left to right
$$= 7 - \mathbf{9} - 3 \cdot 3$$
$$= 7 - 9 - \mathbf{9}$$
Finally, simplify addition and subtraction from left to right
$$= \mathbf{-2} - 9$$
$$= -11$$

5. **Answer: [D]** The absolute value of a number is always positive so:
$$|-4| = 4$$
The expression therefore becomes:
$$-18 - 4 + 3$$
Now we follow the order of operations and simplify addition and subtraction from left to right
$$-18 - 4 + 3$$
$$= \mathbf{-22} + 3$$
$$= -19$$

6. **Answer: [C]** Since neither 27 nor 3 is a perfect square, we use the techniques taught in the section on addition and subtraction of radicals.
$$27 = 3 \times 3 \times 3$$
$$\sqrt{27} = \sqrt{3 \times 3 \times 3} = \sqrt{3 \times 3} \times \sqrt{3} = 3\sqrt{3}$$
Therefore:
$$5\sqrt{27} - 2\sqrt{3} = 5(3\sqrt{3}) - 2\sqrt{3} = 15\sqrt{3} - 2\sqrt{3} = 13\sqrt{3}$$

7. **Answer: [D]**   From the list of prime numbers between 1 and 100 listed on page 7 we see that 57 is not prime. It is actually divisible by 3 ($57 = 3 \times 19$). If you decide not to memorize the list of prime numbers, you can easily use the divisibility rules found in the table on page 8 to test the answer choices. For 57, $5 + 7 = 12$ which is divisible by 3 therefore 57 is divisible by 3 and cannot be prime.

8. **Answer: [C]**   Simplify the expression within the parentheses first.
$13 + (5 - |2|)^3 - 18 \div 2 =$
$= 13 + (5 - \mathbf{2})^3 - 18 \div 2$
$= 13 + (\mathbf{3})^3 - 18 \div 2$
Next simplify the exponent
$= 13 + \mathbf{27} - 18 \div 2$
Calculate the division
$= 13 + 27 - \mathbf{9}$
Lastly, calculate the addition and subtraction, working from left to right
$= \mathbf{40} - 9$
$= 31$

9. **Answer: [D]**   The current temperature is $-4°F$ because we were told that the temperature is 4°F below 0 and numbers less than 0 are negative. The number of degrees by which we must increase the room temperature is the same as the distance between the two temperatures.

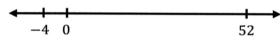

So we find the absolute value of the difference.
$Temperature\ increase = |52° - (-4°)| = |52° + 4| = |56°| = 56°$
The room temperature must be increased by 56°F.

10. **Answer: [1,260]**   List the prime factors of each:
$15 = 3 \times 5$
$36 = 2 \times 2 \times 3 \times 3$
$42 = 2 \times 3 \times 7$
The LCM must contain all the prime factors of each without including unnecessary repeated factors.
$LCM = 2 \times 2 \times 3 \times 3 \times 5 \times 7 = 1260$
Therefore, 1260 is the lowest possible multiple that is divisible by 15, 36 and 42.

Prep for Success: Florida's PERT Math Study Guide

Chapter 2 – Conversions **17**

Chapter 2

Conversions

In order to properly solve some of the problems on the test, it is important to know how to complete some simple conversions: switch between fractional, decimal and percentage forms. If you are already comfortable with these types of conversions, skip ahead to the next chapter.

Fractions, Decimals & Percentages

A **fraction** is a ratio in which the bottom number, called the **denominator**, represents the whole and the top number, called the **numerator**, represents the number of parts of the whole.

A **decimal** is another type of number that uses the decimal system where every place value can be written in terms of a power of 10. We take closer look at decimals later on but all fractions can be expressed as decimals. The reverse is not always true as non-terminating (irrational) numbers such as pi cannot be expressed as fractions.

Note

π is often approximated by $\dfrac{22}{7}$ but it is just an estimation, not the exact value.

The word **percent** is derived from the words per (part of) and centum (hundred). Therefore, percentage is the number of parts per one hundred. 20% therefore means 20 parts of 100 and 100% means 100 parts of 100.

Now that we have a basic definition of fractions, decimals and percentages, let's look at converting between the three.

Reducing Fractions

If a fraction contains composite numbers in both the numerator and the denominator, we check to see if the fraction can be reduced by finding a factor common to both.

Example 2.1:

Reduce the fraction $\dfrac{30}{42}$ to its lowest terms.

We begin by identifying the prime factors of both the numerator and the denominator.

$30 = 2 \times 3 \times 5$ and $42 = 2 \times 3 \times 7$

Therefore:

$$\frac{30}{42} = \frac{2 \times 3 \times 5}{2 \times 3 \times 7}$$

Both 2 and 3 are factors common to both so they can be eliminated.

$$\frac{30}{42} = \frac{5}{7}$$

18   Prep for Success: Florida's PERT Math Study Guide

Mixed Number to an Improper Fraction

A **mixed number** is the combination of a whole number and a fraction. An **improper fraction** is a fraction in which the numerator is larger than the denominator. To convert a mixed number to an improper fraction, follow these 3 steps:

1. Multiply the whole number and the denominator.
2. Add the result of step 1 to the numerator.
3. Place the result of step 2 over the denominator to form the improper fraction.

Example 2.2:

Convert $4\frac{7}{9}$ to an improper fraction.

1. *Multiply the whole number and the denominator*
   $4 \times 9 = 36$
2. *Add the result of step 1 to the numerator*
   $36 + 7 = 43$
3. *Place the result of step 2 over the denominator to form the improper fraction*
   $\frac{43}{9}$

   Therefore, $4\frac{7}{9}$ as an improper fraction is $\frac{43}{9}$.

Improper Fraction to a Mixed Number

To convert an improper fraction back to a mixed number, divide the numerator by the denominator to find the whole number and place the remainder over the denominator to form the fractional part.

Example 2.3:

Convert $\frac{44}{7}$ to a mixed number.

We can divide 44 by 7 by using long division or any other method that you may already be comfortable with.

$$\begin{array}{r} 6 \\ 7\overline{\smash{)}44} \\ -42 \\ \hline 2 \end{array}$$

The whole number is 6 and the remainder 2 is placed over the denominator 7.

Therefore $\frac{44}{7}$ as a mixed number is $6\frac{2}{7}$.

Decimal to Percentage

This is the simplest conversion to do. To convert from a decimal to a percentage, simply move the decimal point 2 places to the right.

Example 2.4:

Convert 0.0624 to a percentage.

$$0.06\overset{\curvearrowright}{24}$$

After moving the decimal point 2 places to the right we get 6.24%.

Decimal to Fraction

To change from a decimal to a fraction:
1. Count the number of decimal places.
2. Ignore the decimal point and place the number over 1.
3. Add the same number of zeros to the denominator as there were decimal places.
4. If possible, reduce the fraction formed.

Example 2.5:

Convert 0.45 to a fraction

1. *Count the number of decimal places*
   0.45 has 2 decimal places
2. *Ignore the decimal point and place the number over 1*
   $$\frac{45}{1}$$
3. *Add the same number of zeros to the denominator as there were decimal places*
   Add 2 zeros to the denominator
   $$\frac{45}{100}$$
4. *Reduce the fraction if possible*
   $$\frac{45}{100} = \frac{9}{20}$$
   Therefore, 0.45 as a fraction is $\frac{9}{20}$.

Fraction to Decimal

To convert a fraction to a decimal, simply divide the numerator by the denominator using long division, adding padding zeros as necessary.

Example 2.6:

Convert $\dfrac{7}{8}$ to a decimal

We need to divide 7 (**dividend**) by 8 (**divisor**) by using long division or any other method that you may already be comfortable with.

$$
\begin{array}{r}
0.875 \\
8\,\overline{\smash{)}\,7.0} \\
-\ 64 \\
\hline
60 \\
-\ 56 \\
\hline
40 \\
-\ 40 \\
\hline
0 \\
\hline
\end{array}
$$

Add a 0 to the 7 because 7 is not divisible by 8
$70 \div 8 = 8\,R\,6$ *and* $8 \times 8 = 64$
Add a 0 to the 6 because 6 is not divisible by 8
$60 \div 8 = 7\,R\,4$ *and* $7 \times 8 = 56$
Add a 0 to the 4 because 4 is not divisible by 8
$40 \div 8 = 5\,R\,0$ *and* $5 \times 8 = 40$

Therefore $\dfrac{7}{8}$ as a decimal is 0.875.

Below is a table of commonly used fractions and their decimal counterparts. Committing these to memory can save you time on the exam.

Fraction	Decimal	Fraction	Decimal	Fraction	Decimal	Fraction	Decimal
$\frac{1}{2}$	0.5	$\frac{1}{5}$	0.2	$\frac{1}{8}$	0.125	$\frac{1}{10}$	0.10
$\frac{1}{3}$	0.333...	$\frac{2}{5}$	0.4	$\frac{3}{8}$	0.375	$\frac{1}{11}$	$0.\overline{09}$
$\frac{2}{3}$	0.666...	$\frac{3}{5}$	0.6	$\frac{5}{8}$	0.625	$\frac{1}{12}$	0.0833...
$\frac{1}{4}$	0.25	$\frac{4}{5}$	0.8	$\frac{7}{8}$	0.875	$\frac{1}{16}$	0.0625
$\frac{3}{4}$	0.75	$\frac{1}{6}$	0.166...	$\frac{1}{9}$	0.111...	$\frac{1}{20}$	0.05

Fraction to Percentage

Two steps are involved in changing a fraction to a percent:
1. Convert from fraction to decimal.
2. Convert from decimal to percent.

We already learned how to complete the two steps so let's look at an example.

Example 2.7:

Convert $\frac{3}{16}$ to a percentage

1. *Convert from fraction to decimal*

    This can be done by long division or by using the table above.

    $$\frac{3}{16} = 3 \times \frac{1}{16} = 3 \times 0.0625 = 0.1875$$

2. *Convert from decimal to percent*

    Remember, to convert from a decimal to a percent we move the decimal place two places to the right.

    $$0.1875$$

    Therefore $\frac{3}{16}$ as a percentage is 18.75%.

Percentage to Fraction

To convert a percentage to a fraction, follow these 4 steps:
1. Place the percentage over 100.
2. If the percent was a decimal, move the decimal point to the right until the number becomes a whole number.
3. Add the same number of zeros to the denominator as the number of decimal places moved in the numerator.
4. Reduce the fraction to its lowest terms.

Chapter 2 – Conversions 21

Example 2.8:
Convert 87.5% to a fraction
1. *Place the percentage over 100*
   $$\frac{87.5}{100}$$
2. *If the percentage was a decimal, move the decimal point to the right until the number becomes a whole number*
   $$\frac{875}{100}$$ Decimal point needs to be moved 1 place to the right
3. *Add the same number of zeros to the denominator as the number of decimal places moved in the numerator*
   $$\frac{875}{1000}$$ Add 1 zero to the denominator
4. *Reduce the fraction to its lowest terms*
   $$\frac{7}{8}$$
   Therefore 87.5% as a fraction is $\frac{7}{8}$.

Example 2.9:
Convert 45% to fraction

1. *Place the percentage over 100*
   $$\frac{45}{100}$$

2. *If the percent was a decimal, move the decimal point to the right until the number becomes a whole number*
   45 is not a decimal so move on to step 4

4. *Reduce the fraction to its lowest terms*
   $$\frac{9}{20}$$
   Therefore 45% as a fraction is $\frac{9}{20}$.

Percentage to Decimal
The steps are again really simple. Move the decimal point two places to the left.

Example 2.10:
Convert 325.8% to a decimal

After moving the decimal point 2 places to the left we get 3.258.

22   Prep for Success: Florida's PERT Math Study Guide

Chapter 2 Exercises

1. Convert 0.85 to a fraction.

2. Convert $\dfrac{4}{11}$ to a decimal.

3. Convert 7.25% to a fraction.

4. Convert $15\dfrac{2}{5}$ to a percentage.

5. Which of the following fractions is equivalent to 0.0375?

   A. $\dfrac{3}{8}$ B. $\dfrac{5}{16}$ C. $\dfrac{3}{80}$ D. $\dfrac{11}{20}$

6. $4\dfrac{3}{5} =$

   A. 4.6%   B. $\dfrac{17}{5}$ C. 0.046%   D. 4.6

7. Write the decimal 0.56 as a fraction in its lowest terms.

   A. $\dfrac{7}{25}$ B. $\dfrac{14}{25}$ C. $\dfrac{7}{125}$ D. $\dfrac{28}{50}$

8. $\dfrac{9}{20} =$

   A. 48%   B. 0.45%   C. 4.8   D. 0.45

9. Which of the following is equivalent to 265%?

   A. 26.5   B. 2.65   C. 0.265   D. 0.0265

10. Write the mixed number $5\dfrac{4}{9}$ as an improper fraction.

   A. $\dfrac{29}{4}$ B. $\dfrac{56}{9}$ C. $\dfrac{49}{9}$ D. $\dfrac{45}{4}$

Solutions

1. This is a decimal to fraction conversion. 0.85 has 2 decimal places. Ignore the decimal point, place 85 over 1 to create a fraction, and then add 2 zeros to the denominator.

   $$0.85 = \frac{0.85}{1} = \frac{85}{100}$$

   Reduce the resulting fraction if possible.

   $$\frac{85}{100} = \frac{17}{20}$$

Chapter 2 – Conversions

2. We could divide using long division but we know from the table on page 20 that fractions with an 11 in the denominator will have 2 repeating digits to the right of the decimal point. If you have memorized the decimal values in the table, the problem becomes really simple.

$$\frac{1}{11} = 0.\overline{09}$$

$$\frac{4}{11} = 0.\overline{09} \times 4 = 0.\overline{36}$$

If you choose to do long division:

$$
\begin{array}{r}
0.3636 \\
11 \,\overline{\smash{)}\, 4.0} \\
-\ 33 \\
\hline
70 \\
-\ 66 \\
\hline
40 \\
-\ 33 \\
\hline
70 \\
-\ 66 \\
\hline
4
\end{array}
$$

Add a 0 to the 4 because 4 is not divisible by 11
$40 \div 11 = 3\,R\,7$ *and* $3 \times 11 = 33$
Add a 0 to the 7 because 7 is not divisible by 11
$70 \div 11 = 6\,R\,4$ *and* $6 \times 11 = 66$
Same as above

Same as above

$$\frac{4}{11} = 0.3636\ldots = 0.\overline{36}$$

3. To convert a percentage to a fraction, first convert to a decimal by moving the decimal 2 places to the left.

$7.25\% = 0.0725$

0.0725 has 4 decimal places. Ignore the decimal point, place 725 over 1 to create a fraction, and then add 4 zeros to the denominator.

$$0.0725 = \frac{0.0725}{1} = \frac{725}{10000}$$

Reduce the resulting fraction if possible.

$$\frac{725}{10000} = \frac{29}{400}$$

4. We first convert to a decimal by long division or by using the table of common fractions.

$$\frac{1}{5} = 0.2 \ \text{ so } \ \frac{2}{5} = (0.2)(2) = 0.4$$

Therefore:

$$15\frac{2}{5} = 15.4$$

Next, convert from the decimal to a percentage by moving the decimal point two places to the right.

$15.4 = 1540\%$

Recall that numbers greater than 1 result in a percentage greater than 100% since $1 = 100\%$

5. **Answer: [C]**   0.0375 has 4 decimal places so there will be 4 zeros added to the denominator.

24   Prep for Success: Florida's PERT Math Study Guide

$$0.0375 = \frac{0.0375}{1} = \frac{375}{10000}$$

Reduce the fraction if possible.

$$\frac{375}{10000} = \frac{15}{400} = \frac{3}{80}$$

6. **Answer: [D]**   There are more decimal and percentage choices than fractional so convert to a decimal first and then to a percentage if necessary.

$$\frac{1}{5} = 0.2 \ \text{ so } \ \frac{3}{5} = (0.2)(3) = 0.6$$

Therefore:

$$4\frac{3}{5} = 4.6$$

We don't need to convert any further because 4.6 is one of the choices so the answer is D.

7. **Answer: [B]**   0.56 has 2 decimal places so we place 2 zeros in the denominator.

$$0.56 = \frac{56}{100} = \frac{28}{50} = \frac{14}{25}$$

Note that even though choice D is also equal to 0.56, it is not in the lowest terms.

8. **Answer: [D]**   First, convert to a decimal either by using the table on page 20 or by long division.

$$
\begin{array}{r}
0.45 \\
20 \overline{\smash{)}\ 9.0} \\
-\ 80 \\
\hline
100 \\
-\ 100 \\
\hline
0
\end{array}
$$

Add a 0 to the 9 because 9 is not divisible by 20
$90 \div 20 = 4\ R\ 10$ *and* $4 \times 20 = 80$
Add a 0 to the 10 because 10 is not divisible by 20
$100 \div 20 = 5\ R\ 0$ *and* $5 \times 20 = 100$

Therefore, the answer is D.

9. **Answer: [B]**   All the responses are in decimal notation and to convert from a percentage to a decimal, we move the decimal point 2 places to the left.

265.0%

After moving the decimal point 2 places to the left we get 2.65 and the answer is B.

10. **Answer: [C]**   Multiply the whole number by the denominator, add the result to the numerator, then place over the denominator.

$$(5 \times 9) + 4 = 45 + 4 = 49$$

So:

$$5\frac{4}{9} = \frac{49}{9}$$

Therefore, the answer is C.

Chapter 3 – Fractions & Decimals **25**

Chapter 3

Fractions & Decimals

This chapter reviews basic arithmetic operations on fractions and decimals. Before we go into arithmetic operations, we review how to determine whether one fraction is larger than the other fraction.

Comparing Fractions

* If two positive fractions have the same denominator, the larger fraction will have the larger numerator because the further right we move on the number line, the larger the numbers become.

Therefore $\dfrac{7}{11} > \dfrac{2}{11}$

* If two negative fractions have the same denominator, the larger fraction will have the smaller numerator. Why? Remember the further left you move on the number line, the smaller the numbers become.

So $-5 < -3$ and $-\dfrac{5}{7} < -\dfrac{3}{7}$

* If two positive fractions have the same numerator, the larger fraction will have the smaller denominator. This is because the more pieces the whole is cut into, the smaller each piece will become.

Therefore $\dfrac{3}{8} > \dfrac{3}{64}$

Your share of 3 slices of pizza is larger when we cut the pizza into 8 slices than if we cut the pizza into 64 slices!

* If the fractions share neither a common numerator nor a common denominator, we need to find the **lowest common denominator** (LCD) before comparing. The LCD is the smallest number that is divisible by all the current denominators therefore the procedure is the same as for finding the LCM. If the fractions need to be converted, first figure out what number you need to multiply the denominator by in order to get the LCD. Remember to multiply both the numerator and the denominator by this number.

Example 3.1:

Which of the following is smallest?

A. $\dfrac{5}{9}$ 　　　B. $\dfrac{8}{45}$ 　　　C. $\dfrac{2}{3}$ 　　　D. $\dfrac{11}{27}$

26 Prep for Success: Florida's PERT Math Study Guide

We first find the LCD and then use this to convert all the fractions. Note that we could have simply multiplied all the denominators which would have given us a common denominator albeit not the smallest one possible.

$9 = 3 \times 3$

$27 = 3 \times 3 \times 3$

$45 = 3 \times 3 \times 5$

The LCD must contain at least all the factors of each denominator without adding unnecessary factors.

$\text{LCD} = 3 \times 3 \times 3 \times 5 = 135$

Now convert all the fractions so that they have the LCD in their denominator.

$$\frac{5}{9} = \frac{(15)5}{(15)9} = \frac{75}{135} \qquad \frac{8}{45} = \frac{(3)8}{(3)45} = \frac{24}{135} \qquad \frac{2}{3} = \frac{(45)2}{(45)3} = \frac{90}{135} \qquad \frac{11}{27} = \frac{(5)11}{(5)27} = \frac{55}{135}$$

Therefore $\dfrac{8}{45}$ is the smallest.

Operations on Fractions

Adding and Subtracting Fractions

There are a few easy-to-follow steps for adding and subtracting fractions.

1. Change any mixed numbers to improper fractions.
2. Change any whole numbers to fractions.
3. Make sure that the fractions have the same denominator.
4. Add or subtract the fractions and reduce the result if possible.

Just as with integers, the sign of the answer depends on the sign of the larger fraction. Remember that if there is no sign in front, the fraction is positive. To determine which fraction is larger, ignore the signs and compare their numerators. Note that this can only be done after making sure that the fractions have the same denominator.

The rules for deciding the sign of the answer are:
❑ When the signs are the same, add and keep the sign.
❑ When the signs are different, subtract and keep the sign of the larger fraction.

Example 3.2:

$-5 + 3\dfrac{2}{3} =$

First, we change the mixed number to an improper fraction. Review Chapter 2 if you need help doing this.

$-5 + \dfrac{11}{3} =$

Next we make the whole number into a fraction by placing it over 1.

$-\dfrac{5}{1} + \dfrac{11}{3} =$

Now find the LCD and convert the fractions so that they both have the same denominator. In this case, the LCD is 3 since 3 is divisible by both 1 and 3. The second fraction already has the LCD in the denominator so we do not need to do anything to this fraction. The denominator of the first fraction needs to be multiplied by 3 to get the LCD. Remember to multiply the numerator by 3 as well.

$$-\frac{(3) \cdot 5}{(3) \cdot 1} + \frac{11}{3} =$$

$$-\frac{15}{3} + \frac{11}{3} =$$

Since the signs are different, we subtract the smaller fraction from the larger and keep the sign of the larger. In this case, the first fraction is larger because it has the larger numerator and it is negative so the answer will be negative as well.

$$\frac{15}{3} - \frac{11}{3} = \frac{4}{3}$$

Therefore, the final answer is $-\frac{4}{3}$.

Remember, there is more than one way to approach a problem. We implement the method that has had the most success with our students but that does not mean you should abandon your own method. If you are comfortable with the way you do a problem and you achieve the same results, you should stick with your method as it is going to be the easiest to remember and execute on the test.

Multiplying and Dividing Fractions

The first two steps are the same as for adding & subtracting fractions. The main difference is that we do not need to find a common denominator.

- ❏ For multiplication, we multiply straight across (multiply the numerators then multiply the denominators), and reduce the resulting fraction.
- ❏ For division, we flip the second fraction, multiply straight across, and then reduce the resulting fraction.

The rules to determine the sign of the final answer are the same as for integers.

Example 3.3:

$$-7 \times 1\frac{4}{5} =$$

Change the mixed number to an improper fraction.

$$-7 \times \frac{9}{5} =$$

Change the whole number to a fraction by placing it over 1.

$$-\frac{7}{1} \times \frac{9}{5} =$$

Multiply straight across and reduce the resulting fraction if possible. The answer will be negative because we are multiplying a negative fraction with a positive fraction.

$$-\frac{63}{5}$$

28   Prep for Success: Florida's PERT Math Study Guide

Our answer is an improper fraction. If the answers all contain mixed numbers, divide the numerator by the denominator and place the remainder over the denominator.

$$-12\frac{3}{5}$$

Example 3.4:

$$-5\frac{1}{3} \div -\frac{3}{4} =$$

Again we change the mixed number to an improper fraction.

$$-\frac{16}{3} \div -\frac{3}{4} =$$

Since there are no whole numbers, we move on to flipping the second fraction and changing to multiplication.

$$-\frac{16}{3} \times -\frac{4}{3} =$$

If nothing can be reduced, multiply straight across and reduce the resulting fraction if possible. If the answer is an improper fraction, convert to a mixed number if necessary. Note that our answer will be positive because both fractions were negative.

$$\frac{64}{9} = 7\frac{1}{9}$$

Therefore, the final answer is $7\frac{1}{9}$.

Operations on Decimals

Adding and Subtracting Decimals

To add or subtract decimals you need to line up the decimal points, add padding zeros if necessary and work from right to left. The rules for deciding the sign of the result is the same as for adding and subtracting fractions. It is a good idea to always place the larger decimal above the smaller decimal because it is necessary for subtraction and is therefore a good habit to form.

Example 3.5:

$0.037 + 3.9821 =$

We place 3.9821 above, making sure to line up the decimal points, and we add an extra zero to the end of the first decimal to make everything line up and to reduce possibility of error. Since both decimals are positive, we add and keep the sign so the answer will also be positive. Do not forget to add from right to left.

$$
\begin{array}{r}
3.9821 \\
+\ 0.0370 \\
\hline
4.0191 \\
\hline
\end{array}
$$

Therefore, $0.037 + 3.9821 = 4.0191$

Example 3.6:

$-7.043 + 2.651 =$

Chapter 3 – Fractions & Decimals   **29**

Again we place the larger decimal above, line up the decimal points and subtract from right to left. If the digit below is larger than the digit above, take 1 (regroup) from the digit to the left.

$$7.043$$
$$- \ 2.651$$
$$4.392$$

Since the larger decimal is negative, the answer will also be negative. The answer is therefore -4.392.

Multiplying and Dividing Decimals

To multiply two decimals, ignore the decimal points and multiply as you would with two whole numbers. Next count the number of decimal places in the original two decimals and make sure that the answer has the same number of decimal places by moving the decimal point to the left. Use the same rules as multiplying integers to determine the sign of the answer.

Example 3.7:

$-5.23 \times 2.7 =$

Ignore the decimal points and multiply 523 by 27 instead.

$$523$$
$$\times \ \ 27$$
$$10460$$
$$+ \ \ 3661$$
$$14121$$

The two decimals had 3 decimal places combined (5.23 had 2 decimal places and 2.7 had 1 decimal place) so we move the decimal point 3 places to the left in the final answer. The answer will be negative because we multiplied a negative decimal and a positive decimal.

14.121   Therefore the final answer is -14.121.

To divide two decimals:
1. Count the number of decimal places in both decimals.
2. Whichever decimal had the larger number of decimal places, move the decimal point this number of places to the right in both decimals to form two whole numbers.
3. Divide the whole numbers using long division as demonstrated in the previous chapter (for a decimal answer) or use reduction (for a fractional answer).

Example 3.8:

Divide $- 0.36$ by $- 0.048$

We are dividing two negative decimals so the answer will be positive. The first decimal has 2 decimal places and the second decimal has 3 decimal places so we need to move both decimal points 3 places to the right.

$-0.36 \div -0.048$

$= -360 \div -48$

Now we divide 360 by 48 using long division for a decimal answer.

$$\begin{array}{r} 7.5 \\ 48\,\overline{\smash{)}\,360} \\ -\,336 \\ \hline 240 \\ -240 \\ \hline 0 \end{array}$$

Add a 0 to 24 because 24 is not divisible by 48

$240 \div 48 = 5\,R\,0$

Therefore -0.36 divided by -0.048 is 7.5.

If we wanted a fractional answer we would have set up the divisor and dividend as a fraction and reduced.

$$\frac{-360}{-48}$$

We easily recognize that both the numerator and denominator are divisible by 12.

$$\frac{30}{4}$$

We can now tell that they are further divisible by 2.

$$\frac{15}{2}$$

Finally, we can convert to a mixed number.

$$7\frac{1}{2}$$

Applications

Some of the more basic word problems may contain fractions and decimals or a mix of the two. These questions typically mimic real life situations.

Example 3.9:

A group of friends go out to dinner. Sally, Trina and Paul are each about to order the 2 slice pizza combo for $4.50. The waitress suggests that they order a medium 8 slice pizza combo for $15.00 instead. If they can convince Steven to eat pizza and split the bill 4 ways, how much would Sally, Trina and Paul each save?

A. $0.50   B. $0.60   C. $0.75   D. $1.13

If they get the 8 slice pizza combo and split it 4 ways, the cost per person is:

$$\frac{\$15}{4} = \$3.75$$

Since Sally, Trina and Paul were each going to pay $4.50 for the 2 slice combo, they would each save $4.50 – $3.75 = $0.75

Therefore, the answer is C.

Chapter 3 – Fractions & Decimals **31**

Chapter 3 Exercises

1. $-3 + \dfrac{4}{5} - \dfrac{7}{15} =$

   A. $-\dfrac{2}{5}$　　　B. $-2\dfrac{2}{3}$　　　C. $-2\dfrac{2}{15}$　　　D. $-1\dfrac{11}{15}$

2. $-2\dfrac{5}{6} \div \dfrac{2}{3} =$

   A. $-4\dfrac{1}{3}$　　　B. $-4\dfrac{1}{4}$　　　C. $-2\dfrac{1}{6}$　　　D. $-1\dfrac{8}{9}$

3. $2.8594 - (-8.193) =$

   A. -5.3336　　　B. -6.2436　　　C. 10.0514　　　D. 11.0524

4. $25.2 \div 0.84 =$

   A. 30　　　B. 3.0　　　C. 0.3　　　D. 0.03

5. What number when multiplied by $1\dfrac{1}{6}$ gives $1\dfrac{1}{4}$?

   A. $\dfrac{1}{12}$　　　B. $\dfrac{24}{35}$　　　C. $1\dfrac{1}{14}$　　　D. $1\dfrac{11}{24}$

6. A phone call from Capital City to Steelsburg costs \$0.20 to connect, \$0.47 for the first 3 minutes and \$0.23 for each additional minute. How many full minutes can you get for \$2.55?

   A. 8　　　B. 9　　　C. 11　　　D. 12

7. The Bronson family collects rainwater in a barrel for watering their plants and measures the height of the water in the barrel at the end of every day. If the barrel held 10.25 inches of water on Sunday and the table below shows the change in the number of inches of water in the barrel each day, how many inches of water were there in the barrel at the start of Wednesday?

Day	Change in water level (inches)
Monday	$-2\dfrac{3}{8}$
Tuesday	$3\dfrac{5}{16}$
Wednesday	$1\dfrac{2}{5}$
Thursday	$2\dfrac{1}{4}$
Friday	$-3\dfrac{11}{25}$

   A. 15.9375　　　B. 11.1875　　　C. 9.3125　　　D. 4.5625

32   Prep for Success: Florida's PERT Math Study Guide

8. A typist charges $1.05 for every 15 words or $22.50 for every complete page. If one complete page typically contains 550 words, what is an estimate of the charge for typing 2,350 words?

    A. $100.50   B. $112.50   C. $164.50   D. $247.50

9. $\left(\frac{4}{3}-\frac{1}{2}\right)-\left(\frac{3}{4}-\frac{2}{3}\right)$

If the above expression is reduced to its simplest form, what is the denominator of the resulting fraction?

    A.   3   B.   4   C.   6   D.   12

10. Sandra weighs 235.5 lbs and would like to reach her target weight of 190 lbs in 3 months. She loses 12.57 lbs during the first month of her diet. The next month she loses another 15¾ lbs. How much weight must she lose during the third month of her diet in order to reach her target?

    A.   38.32   B.   28.32   C.   17.18   D.   16.78

Solutions

1. **Answer: [B]**   We need to use the LCD to convert all to the same denominator.

$5 = 5$

$15 = 3 \times 5$

Since 5 is a factor of 15, 15 is the LCD. Convert the fractions so they all share the same LCD in their denominator.

$$-\frac{3}{1}+\frac{4}{5}-\frac{7}{15}$$

$$= -\frac{3}{1}\cdot\frac{(15)}{(15)}+\frac{4}{5}\cdot\frac{(3)}{(3)}-\frac{7}{15}$$

$$= -\frac{45}{15}+\frac{12}{15}-\frac{7}{15}$$

$$= \frac{-45+12-7}{15} = \frac{-40}{15} = -2\frac{10}{15} = -2\frac{2}{3}$$

2. **Answer: [B]**   First convert the mixed number to an improper fraction (multiply the whole number by the denominator, add the result to the numerator and place the result over the denominator).

$$2\frac{5}{6} = \frac{12+5}{6} = \frac{17}{6}$$

Therefore:

$$-2\frac{5}{6}\div\frac{2}{3} = -\frac{17}{6}\div\frac{2}{3}$$

Flip the second fraction and multiply instead.

$$-\frac{17}{6}\times\frac{3}{2} = -\frac{(17)(3)}{12} = -\frac{17}{4} = -4\frac{1}{4}$$

3. **Answer: [D]** $2.8594 - (-8.193) = 2.8594 + 8.193$

Place the decimals one above the other, lining up the decimal points and adding padding zeroes where necessary. Since both decimals are positive, we add and keep the sign so the answer will be positive. Do not forget to add from right to left.

$$\begin{array}{r} 2.8594 \\ +\ 8.1930 \\ \hline 11.0524 \end{array}$$

4. **Answer: [A]**   To divide decimals, it's easiest to first convert to whole numbers. There is 1 decimal place in the dividend and 2 decimal places in the divisor so we move the decimal point 2 places to the right in both decimals.

$25.2 \div 0.84 = 2520 \div 84$

Since all the answers are decimals we can use long division. However, by rounding both to the nearest hundred, we can easily identify the answer.

$2520 \div 84 \approx 2500 \div 100 = 25$

Therefore, the answer is clearly A.

5. **Answer: [C]**   We use inverse operations here.

If a number multiplied by $1\frac{1}{6}$ gives $1\frac{1}{4}$ then dividing $1\frac{1}{4}$ by $1\frac{1}{6}$ would give the number.

We demonstrate with a bit of algebra:

Let the number $= n$

$$(n)\left(1\frac{1}{6}\right) = 1\frac{1}{4}$$

$$n = \frac{1\frac{1}{4}}{1\frac{1}{6}}$$

To divide the two fractions we first convert them to improper fractions.

$$1\frac{1}{4} \div 1\frac{1}{6} = \frac{5}{4} \div \frac{7}{6}$$

Flip the second fraction and multiply instead.

$$\frac{5}{4} \times \frac{6}{7} = -\frac{30}{28} = \frac{15}{14} = 1\frac{1}{14}$$

6. **Answer: [C]**   Start by subtracting the cost to connect.

$$\begin{array}{r} 2.55 \\ -\ 0.20 \\ \hline 2.35 \end{array}$$

Next, subtract the cost of the first 3 minutes.

$$\begin{array}{r} 2.35 \\ -\ 0.47 \\ \hline 1.88 \end{array}$$

34   Prep for Success: Florida's PERT Math Study Guide

Finally, to determine how many full minutes we can get for $1.88, divide $1.88 by the cost of each min ($0.23). They each have 2 decimal places, so move the decimal point 2 places to the right in both then divide the whole numbers.

$1.88 \div 0.23 = 188 \div 23$

$$
\begin{array}{r}
8 \\
23\overline{\smash{)}188} \\
-184 \\
\hline
4
\end{array}
$$

We stop here because we are finding the number of full minutes so we don't need anything after the decimal point. Therefore, you get 8 mins plus the initial 3 mins for a total of 11 minutes.

7. **Answer: [B]**   Since the table provides the change in the number of inches of water each day, and we were given height of the water in the barrel on Sunday, we need the change in inches for Monday and Tuesday.

We start with $10\frac{1}{4}$, subtract $2\frac{3}{8}$ inches for Monday then add $3\frac{5}{16}$ inches for Tuesday.

$$10\frac{1}{4} - 2\frac{3}{8} + 3\frac{5}{16}$$

Convert the mixed numbers to improper fractions.

$$\frac{41}{4} - \frac{19}{8} + \frac{53}{16}$$

The LCD is clearly 16 so we convert all the fractions to the same denominator using the LCD.

$$\frac{41}{4}\cdot\frac{(4)}{(4)} - \frac{19}{8}\cdot\frac{(2)}{(2)} + \frac{53}{16} = \frac{164}{16} - \frac{38}{16} + \frac{53}{16} = \frac{164-38+53}{16} = \frac{179}{16}$$

Since the answer are all decimals, divide using long division.

$$
\begin{array}{r}
11 \\
16\overline{\smash{)}179} \\
-16 \\
\hline
19 \\
-16 \\
\hline
3
\end{array}
$$

We can stop here because 3 is not divisible by 16 and continuing would give the decimal portion of the answer. Since there is only one answer that begins with 11, we can save time and move on. The answer is B.

8. **Answer: [A]**   From the word "or" we can tell that you are either charged by the page or by the word. If a page is completely filled you get charged per page, otherwise you get charged per 15 words. First determine how many complete pages there are by dividing the total number of words by the number of words per page.

$$\frac{2350}{550} = \frac{235}{55} = \frac{47}{11} = 4\frac{3}{11}$$ So we know there will be 4 complete pages.

$4\ pages = 4 \times 550 = 2200\ words$

$Number\ of\ words\ on\ 5th\ page = 2,350 - 2,200 = 150\ words$

These words will be charged at $1.05 for every 15 words.

$$\frac{150\ words}{15\ words} = 10\ groups\ of\ 15\ words$$

$$Total\ charge = (\#complete\ pages \times \$22.50) + (\#\ groups\ of\ 15\ words \times \$1.05)$$
$$= (4 \times \$22.50) + (10 \times \$1.05)$$
$$= \$90 + \$10.50 = \$100.50$$

9. **Answer: [B]**   We can convert all the fractions to the same denominator using an LCD of 12 since 2, 3 and 4 are all factors of 12.

$$\left(\frac{4}{3} \cdot \frac{(4)}{(4)} - \frac{1}{2} \cdot \frac{(6)}{(6)}\right) - \left(\frac{3}{4} \cdot \frac{(3)}{(3)} - \frac{2}{3} \cdot \frac{(4)}{(4)}\right) = \left(\frac{16}{12} - \frac{6}{12}\right) - \left(\frac{9}{12} - \frac{8}{12}\right)$$

Using the order of operations, simplify the expressions within the parentheses first, then subtract the two resulting fractions and reduce.

$$= \frac{10}{12} - \frac{1}{12} = \frac{9}{12} = \frac{3}{4}$$

Therefore, the denominator of the resulting fraction is 4.

10. **Answer: [C]**   Her weight loss in the second month is fractional and since the answers are all decimals, we first convert the fraction to a decimal.

$$15\frac{3}{4} = 15.75$$

Next, find out how many pounds Sandra needs to lose in total to achieve her goal.

$$Weight\ loss\ goal = initial\ weight - target\ weight$$
$$= 235.5\ lbs - 190\ lbs = 45.5\ lbs$$

We were given her weight loss during the first 2 months. To find the number of pounds she needs to lose in the 3rd month, subtract her weight loss for the first couple months from her weight loss goal.

$$3rd\ month = weight\ loss\ goal - (1st\ month\ loss + 2nd\ month\ loss)$$
$$= 45.5\ lbs - (12.57\ lbs + 15.75\ lbs)$$
$$= 45.5\ lbs - 28.32\ lbs$$
$$= 17.18\ lbs$$

Therefore, she needs to lose another 17.18 lbs and the answer is C.

36 Prep for Success: Florida's PERT Math Study Guide

Chapter 4 – Ratios, Proportions & Percentages   **37**

Chapter 4 — Ratios, Proportions & Percentages

Here we review percentages in detail as they relate to problems involving simple interest and percentage increase or decrease. We also introduce the concepts of ratios, direct and indirect proportions, rates and arithmetic mean.

Percentages

Recall that percentage represents the number of parts per 100. For most problems involving percentages, the following formula can be used to solve for the missing variable.

$$\frac{is}{of} = \frac{\%}{100}$$

Notice that the above formula has three (3) variables. The keywords "is" and "of" will indicate which numbers go where. Read the problem carefully to identify the given values. In simple percentage problems, you will be given two of the values. Use the variable x to represent the unknown quantity and then solve for x.

Example 4.1:

What is 15% of 60?

> We are given the values of two variables:
>
> of $= 60$
>
> % $= 15$
>
> Substitute these values and use x to represent "is".
>
> $$\frac{x}{60} = \frac{15}{100}$$
>
> Next, cross multiply and solve.
>
> $(x)(100) = (60)(15)$
>
> $100x = 900$
>
> $$x = \frac{900}{100} = 9$$
>
> Therefore 9 is 15% of 60.

Example 4.2:

16 is 32% of what number?

> We are given the values of two variables:
>
> is $= 16$
>
> % $= 32$
>
> Substitute these values and use x to represent "of".

$$\frac{16}{x} = \frac{32}{100}$$

Next, cross multiply and solve.

$(x)(32) = (16)(100)$

$32x = 1600$

$x = \dfrac{1600}{32} = 50$

Therefore 16 is 32% of 50.

If you simply need to find the percentage of a number, change the percentage to a decimal and multiply it by the number.

Example 4.3:

80% of 20% of 40 is?

80% of $(0.20 \times 40) = 80\%$ of $8 = 0.8 \times 8 = 6.4$

Alternatively, we could have used the percentage formula twice which would have given the same result. Try it for yourself.

Problems involving percentages may not be as simply laid out as the three previous examples. You may find word problems involving calculating tips, taxes, interest and percent increase/decrease which build on the basic concepts of percentage.

Percentage Increase or Decrease

Finding percent increase or decrease is easy if we remember that the original price or number before any changes can be represented by 100%. A percent increase, as in the case of adding taxes or interest, results in a percentage larger than 100%. A percent decrease, as in discounted merchandise, results in a percentage smaller than 100%.

Examples:

A 30% increase = original + increase

$= 100\% + 30\% = 130\% = 1.30$

A 55% decrease = original − decrease

$= 100\% - 55\% = 45\% = 0.45$

A 9% increase = original + increase

$= 100\% + 9\% = 109\% = 1.09$

Therefore, we can find the total after a percentage increase or decrease by finding the decimal representation of the percentage then multiplying as in Example 4.3 above.

Example 4.4:

The Great Electronics store is having a sale on televisions. If one of their televisions costs $195 and the store is offering a 20% discount, how much will the television cost after the reduction in price?

Chapter 4 – Ratios, Proportions & Percentages **39**

The quickest method of solving this problem involves recognizing that if the television is being offered at a 20% discount, the total paid will be 80% of the price.
Discount = 20%
Remaining = 100% – 20% = 80%
Now we only have to figure out what 80% of $195 will be.
80% of $195 = 0.80 × 195 = $156

The longer method involves finding the amount of money to be discounted and then subtracting this from the original price to find the new price of the television. To do this, we can use the percentage formula or simply multiply $195 by 0.20.
$0.20 \times \$195 = \$39$
Since the television is being offered on discount, subtract the discount from the original amount to find the price of the television after the discount.
$\$195 - \$39 = \$156$
Therefore the price of the television after the discount is $156.

Example 4.5:
The cost of a blouse is increased by 50% then decreased by 50%. The final price is _____?
A. The same
B. A 25% decrease from the original price
C. A 25% increase from the original price
D. A 75% increase from the original price

Let price $= x$
Increase of 50% $= 150\%$ of price $= 150\%$ of $x = 1.5x$
Decrease of 50% $= 50\%$ of new price $= 50\%$ of $1.5x = 0.50 \times 1.5x = 0.75x$
$0.75x = 75\%$ of x
This means an overall 25% decrease in price.

Example 4.6:
At Pete's Discount Store, electronic items and appliances are billed at 8% tax and everything else is billed at 6%. What would be the total bill if a television is purchased for $195, a toaster for $18.50 and 4 chairs at $22.75 each?

The toaster and television will get 8% tax added while the chairs will be billed at 6% tax.
Pre-tax total for TV and toaster $= \$195 + \$18.50 = \$213.50$
Total after 8% tax $= \$213.50 \times 1.08 = \$230.58$

Pre-tax total for 4 chairs $= \$22.75 \times 4 = \$91.00$
Total after 6% tax $= \$91.00 \times 1.06 = \$96.46$
Therefore:
Total bill $= \$230.58 + \$96.46 = \$327.04$

Percentage Error

A **margin of error** represents the difference between what was estimated and the value that actually occurred. Therefore, percentage error is the ratio of this difference to the actual value.

$$Percentage\ Error = \frac{Estimated\ value - Actual\ value}{Actual\ value} \times 100\%$$

A negative percentage error indicates that the actual was higher than the estimate (we underestimated the error).

A positive percentage error indicates that the actual value was lower than the estimate (we overestimated the value).

Sometimes, we take the absolute value of the percentage error if we are only concerned with how wrong our guess was but not whether our guess was higher or lower.

Example 4.7:

A concert was expected to have a turnout of 1400 people but only 1050 people showed up. What was the percentage error?

A. Overestimate by 12.5%
C. Underestimate by 12.5%
B. Underestimate by 33.3%
D. Overestimate by 33.3%

Using the percentage error equation given above:

$$Percentage\ Error = \frac{Estimated\ value - Actual\ value}{Actual\ value} \times 100\%$$

$$= \frac{1400 - 1050}{1050} \times 100\%$$

$$= \frac{350}{1050} \times 100\% = \frac{35000}{1050}\% = \frac{3500}{105}\% = 33.3\%$$

The error is positive because the estimate was higher than the actual value and we overestimated how many people were going to attend the concert. The answer is D.

Simple Interest

The simple interest formula is given by:

$$I = PRT = Principle \times Rate \times Time$$

where: I is the interest earned

P is the principle or original amount of money held

R is the interest rate per period as a decimal

T is the number of time periods

If for example the interest is added per month, then T is the number of months in the given period.

Example 4.8:

How much interest would Karen receive on $500 after 18 months if the interest rate is 3.6% per year?

Since the interest rate is applied annually, we need the time in years. There are 12 months per year so:

$$T = 18\ mths = \frac{18\ mths}{12\ mths/yr} = 1.5\ yrs$$

$P = \$500$

$R = 3.6\% = 0.036$

Therefore:

$$Interest\ Earned = PRT = 500 \times 0.036 \times 1.5 = \$27.00$$

You may also encounter percentage problems that involve diagrams or graphs.

Example 4.9:
What percentage of the grid is shaded?

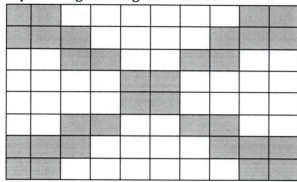

A. 30%   B. 40%   C. 50%   D. 60%

There are 8 rows and 10 columns so there are $8 \times 10 = 80$ squares. There are 32 shaded squares. Therefore the percentage shaded is:

$$\frac{\# \text{ shaded squares}}{\text{total } \# \text{ squares}} \times 100\% = \frac{32}{80} \times 100\% = \frac{2}{5} \times 100\% = \frac{200}{5}\% = 40\%$$

Example 4.10:
There are 30,000 people in Splitsville. From the graph below, what percentage of homes have cable television?

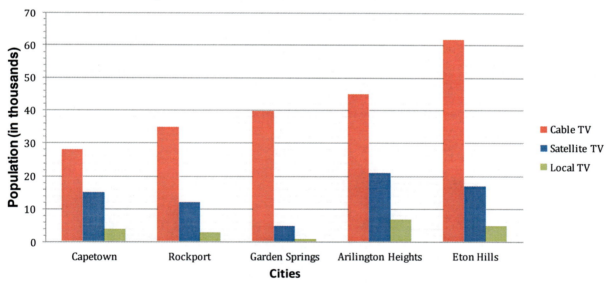

The red levels indicate the cable subscribers. Read off the values for each of the 5 bars and add to find the total number of cable subscribers.

$28 + 35 + 40 + 45 + 62 = 210$

The vertical axis represents the population in hundreds so 21,000 people subscribe to cable. Therefore the percentage of homes that subscribe to cable is:

$$\frac{\# \text{ cable subscribers}}{\text{total population}} \times 100 = \frac{21000}{30000} \times 100 = \frac{7}{10} \times 100 = \frac{700}{10} = 70\%$$

42   Prep for Success: Florida's PERT Math Study Guide

Arithmetic Mean (Average)

The **arithmetic mean** or **average** is found by adding all the items in the list and dividing by the number of items. This is a pretty straightforward question but be certain to read the question carefully as sometimes you are given the average and asked to work backwards.

Example 4.11:

Find the arithmetic mean of the following:

$\frac{3}{4}, \frac{1}{3}$ and $\frac{5}{12}$

A. $\frac{1}{2}$ B. $\frac{3}{2}$ C. $\frac{1}{6}$ D. $\frac{1}{4}$

There are 3 fractions in the list so we add them together and divide by 3 to find the arithmetic mean.

$$\frac{\frac{3}{4} + \frac{1}{3} + \frac{5}{12}}{3}$$

Find the common denominator of the fractions and convert all the fractions so they have the same denominator.

$$\frac{\frac{9}{12} + \frac{4}{12} + \frac{5}{12}}{3}$$

Add the fractions in the numerator together and convert the denominator to a fraction by placing the 3 over 1.

$$\frac{\frac{18}{12}}{\frac{3}{1}}$$

Flip the bottom fraction, multiply then reduce (or reduce then multiply like we do).

$$\frac{18}{12} \times \frac{1}{3} = \frac{3}{2} \times \frac{1}{3} = \frac{1}{2}$$

Therefore the answer is A.

Example 4.12:

Johnny took four math exams and scored an 82, 90 and 84 on the first three. What score does he need on the fourth exam in order to receive an average grade of 80?

A. 54 B. 64 C. 74 D. 84

The average score is already given. Use a variable to represent Johnny's score on the fourth test then set up the formula for calculating the average.

$$\frac{82 + 84 + 90 + x}{4} = 80$$

Set the 80 over 1 and then cross multiply to make the equation easier to solve.

$$\frac{82 + 84 + 90 + x}{4} = \frac{80}{1}$$

$$82 + 84 + 90 + x = (4)(80)$$

Now that the equation has been formed, solve for x.

$$256 + x = 320$$
$$x = 360 - 256 = 64$$

Therefore the answer is B.

Some other concepts to be aware of are the range, the median and the mode.
The **range** of a list of numbers is the difference between the largest and smallest numbers in the list. The **mode** of a list of numbers is the number that appears most frequently in the list. The **median** of a list of numbers is the number that appears in the middle of the list after the list has been ordered in either ascending or descending order. If the list contains an even number of values, then the median will be the average of the middle two numbers.

Example 4.13:
Find the range and the mode of the following numbers: 8.9, 8.1, 8.8, 7.4, 8.9, 8.8, 6.2, 9.8, 7.3 and 8.8.

Rearrange the numbers from smallest to largest so that the mode is easier to see.
6.2, 7.3, 7.4, 8.1, 8.8, 8.8, 8.8, 8.9, 8.9, 9.8

Therefore the mode = 8.8
Range = 9.8 − 6.2 = 3.6

Example 4.14:
Find the median score of the data in the following table.

# Students	Score
12	55
8	63
30	76
24	82
17	87
9	92

Since there are 100 students and this is an even number, we need to find the average of the 50th and 51st scores. The scores are already ordered from least to greatest.
Figure out in what row the 50th and 51st students lie by adding the number of students in each row.

$$12 + 8 = 20$$
$$12 + 8 + 30 = 50$$

Therefore, the 50th student lies in the third row and the 51st student lies in the fourth row.

$$\text{Median} = \frac{76 + 82}{2} = \frac{158}{2} = 79$$

Ratios

Ratios are particularly useful in solving word problems when given the relationship between two variables and asked to use this relationship to solve a measurement problem.

The ratio of 2 to 7 can be written as 2 : 7 or as the fraction $\frac{2}{7}$.

Ratios can be used in distribution problems where the whole is the sum of the parts of the ratio and the ratio determines how much of each part is used to form the whole.

Example 4.15:

In Ms. Miller's English class, the ratio of boys to girls is 2 to 3. If the class has 35 students, how many girls are in the class?

$$2 + 3 = 5$$

This means that for every 5 children, 3 are girls. Therefore:

$$\text{Number of girls in the class} = \frac{3}{5} \times 35 = \frac{3}{5} \times \frac{35}{1} = \frac{(3)(35)}{5} = 3 \times 7 = 21$$

Example 4.16:

A basketball team has played and won 11 games and lost 7. What fraction of games did they lose?

A. $\frac{7}{11}$　　　B. $\frac{11}{7}$　　　C. $\frac{11}{18}$　　　D. $\frac{7}{18}$

It is important to remember that the denominator represents the total, in this case, the total number of games. So we must first add the number of games won and lost.

Total games played $= 11 + 7 = 18$

Therefore:

$$\text{fraction of games lost} = \frac{\#games\ lost}{total\ \#\ games\ played} = \frac{7}{18}$$

Proportions

By setting two ratios equal to each other, we form a **proportion** which can be used to solve for a missing quantity. Before setting up the proportion, determine whether the relationship/variation is direct or indirect. In a **direct relationship**, as one variable increases so does the other variable. Let's look at an example.

Example 4.17:

How many mini cupcakes can be made with 45 cups of flour if 8 cups make 64 mini cupcakes?

The relationship between cupcakes and flour has been established in the problem so we know that for every 8 cups of flour, 64 cupcakes can be made. The ratio of cups to cupcakes:

$$\frac{8\ cups}{64\ cupcakes}$$

As the number of cups of flour increases, so does the number of cupcakes that can be made. Therefore the relationship is direct. Use the ratio of cups to cupcakes to set up a proportion that can be used to solve for the number of cupcakes that can be made with 45 cups of flour. Use x to represent the unknown number of cupcakes.

Chapter 4 – Ratios, Proportions & Percentages   **45**

$$\frac{8\ cups}{64\ cupcakes} = \frac{45\ cups}{x\ cupcakes}$$

To solve the proportion for x, first cross multiply.

$$\frac{8}{64} = \frac{45}{x}$$

$$(8)(x) = (64)(45)$$

$$x = \frac{(64)(45)}{8} = (8)(45) = 360\ cupcakes$$

Therefore 45 cups of flour can make 360 mini cupcakes.

With **indirect variation**, as one variable increases the other variable decreases. This is often true with rate and time. If distance is kept constant, an increase in speed means a shorter travel time.

Example 4.18:

Peter is driving to visit his mother next weekend. The last time he visited, the trip took 5 hours driving at 60 mph. How long will the trip take this time if he drives at 40 mph?

This relationship is indirect because as speed decreases, the length of time to reach the destination increases. Instead of forming two ratios and setting them equal to form the proportion, we go top left to bottom right and bottom left to top right.

5 hours   at x hours

40 mph   60 mph

$$\frac{5}{x} = \frac{40}{60}$$

Again cross multiply and solve for x

$$40x = 5(60) = 300$$

$$x = \frac{300}{40} = 7.5\ hrs$$

Though it is easier to use the above method, we can create a formula that can be used to solve questions involving proportions. A direct relationship between two variables x and y can be represented by $y = kx$ where k is the constant of proportionality. Solving for k:

$$k = \frac{y}{x}$$

Therefore, k represents the ratio of the two variables which is always constant and so we call it the **constant of proportionality**. An indirect relationship between two variables x and y, can be represented by:

$$y = \frac{k}{x}$$

Let's apply this concept to Example 4.17 to show that the result is the same.

Example 4.19:

How many mini cupcakes can be made with 45 cups of flour if 8 cups make 64 mini cupcakes?

Let $x = \#cups\ of\ flour$

Let $y = \#cupcakes$

Since the relationship is direct:

$y = kx$

8 cups makes 64 cupcakes so $x = 8$ and $y = 64$

$64 = k(8)$

$k = \dfrac{64}{8} = 8$

So the function becomes $y = 8x$

We were asked to find the number of cupcakes (y) that can be made using 45 cups of flour (x).

Substituting $x = 45$ into the equation:

$y = 8(45) = 360 \; cupcakes$

Rates

Problems involving rate are often word problems. Read the problem carefully to identify what formula you should use and what variables were given. It is important to make sure that the measurement units of the variables match. Therefore, if the problem contains variables in both feet and inches, convert so that they are either all in feet or all in inches.

Important formulas

$$\text{rate} = \frac{\text{distance}}{\text{time}} \qquad \text{time} = \frac{\text{distance}}{\text{rate}}$$

$$\text{distance} = \text{rate} \times \text{time}$$

$$\text{average rate} = \frac{\text{total distance}}{\text{total time}}$$

Important conversions

$1 \; foot = 12 \; inches$	$1 \; pound \; (lb) = 16 \; ounces \; (oz)$
$3 \; feet = 1 \; yard$	
$1 \; mile = 5,280 \; feet$	$1 \; cup = 8 \; fluid \; ounces \; (fl \; oz)$
	$1 \; pint \; (pt) = 2 \; cups$
$60 \; seconds = 1 \; minute$	$1 \; quart \; (qt) = 2 \; pints$
$60 \; minutes = 1 \; hour$	$1 \; gallon \; (gal) = 4 \; quarts$
$24 \; hours = 1 \; day$	
$7 \; days = 1 \; week$	$1 \; gal = 4 \; qt = 8 \; pt = 16 \; cups = 128 \; fl \; oz$
$52 \; weeks = 1 \; year$	

Example 4.20:

Susan is running late for work and only has 15 minutes to get there. If her job is 12 miles away from her house, how fast would she need to drive, in miles per hour, to make it to work on time?

We are given time in minutes and distance in miles. Notice that the answer needs to be in miles per hour. This means that the time needs to be converted to hours before substituting into the formula to find rate. We use the fact that there are 60 minutes in 1 hour to set up the following ratio to convert 15 minutes to hours.

$$\frac{60 \; minutes}{1 \; hour} = \frac{15 \; minutes}{x \; hours}$$

Cross multiply and solve for x.

$(60)(x) = (15)(1)$

$$60x = 15$$

$$x = \frac{15}{60} = \frac{1}{4} \text{ hour}$$

We now have time in hours and can proceed to substituting into the rate formula.

$$\text{rate} = \frac{distance}{time}$$

$$\text{rate} = \frac{12 \; miles}{1/4 \; hour}$$

Flip the fraction in the denominator and multiply.

$$\text{rate} = \frac{12}{1} \times \frac{4}{1} \; miles \; per \; hour$$

$$\text{rate} = 48 \; miles \; per \; hour$$

Therefore Susan would have to drive at a rate of 48 mph in order to make it to work on time.

Some questions like the example given below require you to calculate the average rate which is not simply the average of the given rates.

Example 4.21:

A boat travels up a river from city A to city B at an average of 16 mph. It makes the return trip at an average of 48 mph. What is the boat's average speed for the entire trip?

A. 20 mph B. 24 mph C. 32 mph D. 36 mph

You may be tempted to add the speeds and divide by 2 but this would be incorrect. The average speed is given by:

$$\text{Average Speed } s = \frac{\text{Total Distance } d}{\text{Total Time } t}$$

The distance both ways will be the same so we can choose a distance and then find the time. Choose a distance that is divisible by both 48 and 16. A distance of 48 miles works well because it is divisible by 16 and itself.

$$Let \; d = 48 \; miles$$

$$Time \; up \; river = \frac{\text{Distance } d}{\text{Speed } s} = \frac{48 \; miles}{16 \; mph} = 3 \; hrs$$

$$Time \; down \; river = \frac{\text{Distance } d}{\text{Speed } s} = \frac{48 \; miles}{48 \; mph} = 1 \; hr$$

$$\text{Average Speed } s = \frac{\text{Total Distance } d}{\text{Total Time } t}$$

$$= \frac{48 + 48 \; miles}{3 + 1 \; hours} = \frac{96}{4} = 24 \; mph$$

48 Prep for Success: Florida's PERT Math Study Guide

Chapter 4 Exercises

1. A diamond bracelet has been reduced by 15%. If the current price is $544.00, what was the original price?
   A. $535.84   B. $552.16   C. $625.60   D. $640.00

2. On average, every shipment of bulbs contains 12.5% broken or non-working bulbs. If a particular shipment contains 4,200 bulbs, how many bulbs will be broken?
   A. 420   B. 525   C. 630   D. 840

3. Mary is late for class and drives at 36 mph on the highway. On the return trip she drives at 60 mph. What was her average speed?
   A. 40 *mph*   B. 45 *mph*   C. 48 *mph*   D. 52 *mph*

4. Sarah divided the shares of her company equally between her son and daughter. Her son sold $\frac{3}{8}$ of his share to a friend at market value for $4,800. How much money is the company worth?
   A. $7,680   B. $12,800   C. $25,600   D. $38,400

5. Water flows from two pipes into a swimming pool at a rate of 18 gallons per hour and 30 gallons per hour respectively. How many days will it take both pipes working together to fill the pool if it holds 2,880 gallons?
   A. 2.5   B. 5   C. 30   D. 60

6. Two workers can complete a job alone in 3 and 6 hours respectively. How long will it take them to complete the job together?
   A. 2 hrs   B. 3 hrs   C. 4.5 hrs   D. 6 hrs

7. If Patsy can type 120 pages in 5 days, how many weeks will it take her to type 1,680 pages?
   A. 2   B. 7   C. 10   D. 14

8. A map is drawn to scale so that every quarter inch represents 12 miles. If the distance between two towns is $2\frac{5}{8}$ inches on the map, approximately how many miles are there between the towns?
   A. 31.5 miles   B. 63 miles   C. 94.5 miles   D. 126 miles

9. In one particular English test, 30 students scored 77 on the test, 20 students scored 89 and 10 students scored 95. What is the sum of the mean, median and mode scores?

10. How many pints are there in 4 gallons?
    A. 16 pints   B. 24 pints   C. 32 pints   D. 48 pints

Chapter 4 – Ratios, Proportions & Percentages   **49**

Solutions

1. **Answer: [D]**   Let the original price $= x$

   Since the price is reduced by 15%, the current price is 85% of the original price. Therefore, we can say that $544 is 85% of x.

   $$\frac{is}{of} = \frac{\%}{100}$$

   $$\frac{544}{x} = \frac{85}{100}$$

   Cross multiply and solve.

   $$85x = (544)(100) = 54{,}400$$

   $$x = \frac{54{,}400}{85} = \$640.00$$

2. **Answer: [B]**   The questions can be simply written as "what is 12.5% of 4,200?" Since we are finding the percentage of a number, we change the percentage to a decimal and multiply.

   $$12.5\% = 0.125$$

   $$4{,}200 \times 0.125 = 525 \text{ bulbs}$$

3. **Answer: [B]**   The distance both ways will be the same so we choose a distance that is divisible by both 36 and 60. Find the LCM.

   $$36 = 2 \times 2 \times 3 \times 3$$

   $$60 = 2 \times 2 \times 3 \times 5$$

   $$\text{LCM} = 2 \times 2 \times 3 \times 3 \times 5 = 180$$

   Let $d = 180$ miles

   $$Time\ from\ home\ to\ class = \frac{distance\ d}{speed\ s} = \frac{180}{36} = \frac{20}{4} = 5\ hrs$$

   $$Time\ from\ class\ to\ home = \frac{distance\ d}{speed\ s} = \frac{180}{60} = 3\ hrs$$

   $$Average\ Speed = \frac{Total\ Distance}{Total\ Time} = \frac{180 + 180}{5 + 3} = \frac{360\ miles}{8\ hrs} = 45\ mph$$

4. **Answer: [C]**   Her son owned half of the company.

   He sold $\frac{3}{8}$ of his $\frac{1}{2}$ therefore he sold $\frac{3}{8} \times \frac{1}{2} = \frac{3}{16}$ of the company

   Let $x =$ total company worth

   $$\frac{3}{16}x = \$4{,}800$$

   $$x = \$4{,}800 \div \frac{3}{16}$$

   $$= \$4{,}800 \times \frac{16}{3}$$

   $$= \$25{,}600$$

5. **Answer: [A]**   The two rates are given in gallons per hour but the question asks how many days it will take to fill the pool. We first find how many hours it will take to fill the pool. Since

50   Prep for Success: Florida's PERT Math Study Guide

both pipes are working together, simply add the rates to find how many gallons of water per hour enter the pool.

$18 + 30 = 48 \; gallons \; per \; hour$

Since each hour, the pipes add 48 gallons of water to the pool and the pool holds a total of 2,880 gallons, we can use the rate formula to find out how many hours are needed to fill the pool.

$gallons = gallons \; per \; hour \times time$

$2,880 = 48 \times time$

$time = \dfrac{2,880}{48} = \dfrac{240}{4} = 60 \; hours$

We now know that it takes 60 hours to fill the pool but we still need to convert this to days. There are 24 hours in 1 day so we can set up a ratio to find the number of days in 60 hours (or just divide 60 by 24).

$\dfrac{24 \; hours}{60 \; hours} = \dfrac{1 \; day}{x \; days}$

Cross multiply and solve for x

$(24)(x) = (60)(1)$

$24x = 60$

$x = \dfrac{60}{24} = \dfrac{5}{2} = 2.5 \; days$

Therefore it takes 2.5 days to fill the pool using the two pipes together and the answer is A.

6. **Answer: [A]**   This is a popular question in algebra but often gives students a tough time because it has to be set up properly. It is different from question 5 where we were given rates and asked for time. Here we are given times and asked for time. Given the time for n people, use the format below:

$$\dfrac{1}{person \; 1} + \dfrac{1}{person \; 2} + \cdots + \dfrac{1}{person \; n} = \dfrac{1}{together}$$

$\dfrac{1}{3} + \dfrac{1}{6} = \dfrac{1}{x}$

LCD of left side $= 6$

$\dfrac{(2)1}{(2)3} + \dfrac{1}{6} = \dfrac{1}{x}$

$\dfrac{2}{6} + \dfrac{1}{6} = \dfrac{1}{x}$

$\dfrac{3}{6} = \dfrac{1}{x}$

Cross multiply and solve for x

$3x = 6$

$x = \dfrac{6}{3} = 2 \; hrs$

Chapter 4 – Ratios, Proportions & Percentages   **51**

7. **Answer: [C]**   We can set up a direct proportion of pages to days because as the number of pages increase so does the time needed to complete the typing. Use x to represent the number of days it will take her to type 1,680 pages.

$$\frac{5 \; days}{120 \; pages} = \frac{x \; days}{1680 \; pages}$$

Cross multiply and solve for x

$$120x = (5)(1680)$$

$$x = \frac{(5)(1680)}{120} = \frac{(5)(168)}{12} = \frac{(5)(42)}{3} = (5)(14) = 70 \; days$$

Note however that the question asks for the answer in weeks.

$$1 \; week = 7 \; days$$

$$70 \; days = \frac{70}{7} weeks = 10 \; weeks$$

8. **Answer: [D]**   We can set up a direct proportion of inches to miles. Use x to represent the number of miles between the two towns.

$$\frac{\frac{1}{4} \; inch}{12 \; miles} = \frac{2\frac{5}{8} \; inches}{x \; miles}$$

Cross multiply and solve for x

$$\frac{1}{4}x = (12)\left(2\frac{5}{8}\right)$$

$$\frac{1}{4}x = \left(\frac{12}{1}\right)\left(\frac{21}{8}\right)$$

$$x = \left(\frac{12}{1}\right)\left(\frac{21}{8}\right)\left(\frac{4}{1}\right) = \left(\frac{12}{1}\right)\left(\frac{21}{2}\right) = (6)(21) = 126 \; miles$$

9. **Answer: [244]**   First find the total number of students in the class because we'll need it to find the mean.

Total # of students $= 30 + 20 + 10 = 60$

The mode of all the scores is the score that appears the most i.e. the score that was obtained by the largest number of students. Half the students scored a 77 on the test so the mode of the scores $= 77$.

When there is an even number of values in the list, the median is the average of the two middle scores. In this case, we need to find the scores obtained by the 30th and 31st students after the scores are arranged either in ascending or descending order. The scores are already arranged in ascending order (77, 89, 95) so the 30th student scored a 77 while the 31st student scored an 89.

Median of scores = average of scores obtained by 30*th* and 31*st* students $= \dfrac{77 + 89}{2} = 83$

To find the mean, we need to find the sum of all the scores. For each score, simply multiply the score by the number of students who earned that score.

Sum of scores $= 30(77) + 20(89) + 10(95) = 2310 + 1780 + 950 = 5040$

Mean score $= \dfrac{Sum \; of \; scores}{Total \; \# \; students} = \dfrac{5040}{60} = 84$

Therefore, the sum of the mode, median and mean is:

Sum $= 77 + 83 + 84 = 244$

10. **Answer: [C]**   Look up the conversion from gallons to pints in the given conversion table on page 46 and formulate a proportion.

$$\frac{1 \: gallon}{8 \: pints} = \frac{4 \: gallons}{x \: pints}$$

Cross multiply and solve for x

$x = (8)(4) = 32 \: pints$

Chapter 5 – The Decimal System   **53**

Chapter 5

The Decimal System

This chapter introduces the concepts of rounding, estimation and scientific notation.

The Decimal System

The decimal system uses the ten digits 0 through 9 and defines the numbers preceding and trailing the decimal point in terms of powers of ten. The farther right of the decimal point a digit falls, the smaller its place value. Similarly, the farther left of the decimal point that a digit falls, the higher its place value.

The following chart shows the place values relative to the decimal point.

Millions	Hundred Thousands	Ten Thousands	Thousands	Hundreds	Tens	Ones	.	Tenths	Hundredths	Thousandths	Ten Thousandths	Hundred Thousandths	Millionths
10^6	10^5	10^4	10^3	10^2	10^1	10^0	.	$\dfrac{1}{10^1}$	$\dfrac{1}{10^2}$	$\dfrac{1}{10^3}$	$\dfrac{1}{10^4}$	$\dfrac{1}{10^5}$	$\dfrac{1}{10^6}$

If a decimal is written in words, you need to be able to extract the information and write in decimal notation.

Example 5.1:

Write nine hundred and two and forty-five thousandths in decimal notation.

A.   902.45   B.   902.0045   C.   902.045   D.   900.2045

> We separate into the part before the decimal point and the part after.
> Before the decimal point ⇒ Nine hundred and two can be written as nine hundreds plus two ones or 902.
> After the decimal point ⇒ Forty-five thousandths. The thing to remember is that with digits after the decimal point, we write the digits from right to left.
> So with 45, the 5 goes under the place value indicated in the problem (thousandths) and the 4 goes under the place value to the left (hundredths). So 45 thousandths is written as 0.045 in decimal notation.
> We see why if we use the base ten representations in the chart above.

$$1 \text{ thousandth} = \frac{1}{10^3} = \frac{1}{1000}$$
$$45 \text{ thousandths} = \frac{45}{1000} = 0.045$$

Therefore, nine hundred and two and forty-five thousandths can be written as 902.045 in decimal notation and the answer is C.

Example 5.2:
Write the decimal in the following statement in words:
"The wedding cake weighed an astonishing 127.39lbs!"

On the left side of the decimal point, there is a 1 in the hundreds place (one hundred), a 2 in the tens place (twenty) and a 7 in the ones place (seven). On the right side of the decimal point, the rightmost digit 9 is in the hundredths place so 0.39 equals $\frac{39}{100}$, written as thirty-nine hundredths.

So 127.39 can be written as "one hundred and twenty-seven and thirty-nine hundredths."

Rounding

It is important to learn rounding techniques because the on-screen calculator is usually only provided for select questions. To round a decimal to the nearest digit, determine whether the place value to the right of the digit is greater than or equal to 5. If it is, then round the digit up before dropping any digits to the right. Otherwise, simply drop all the digits to the right. Consider the next few examples of rounding. If rounding to a whole number, add padding zeros as necessary.

Example 5.3:
Round 695.8726 to the nearest hundredth.

First identify the digit that represents the hundredths place value. Comparing the number to the chart on the previous page, we see that the associated digit is 7. The digit to the right of 7 is 2 which is less than 5 so we drop all the digits to the right of 7.
Therefore, 695.8726 to the nearest hundredth is 695.87

Example 5.4:
Round 1,685.209 to the nearest ten.

Again, first identify the digit that represents the tens place value. Comparing the number to the chart, we see that the associated digit is 8. The digit to the right is 8 is 5 so we round up to 9. Replace any digits between the tens place value and the decimal point with zeros and drop all the digits to the right of the decimal point.
Therefore, 1,685.209 to the nearest ten is 1, 690

To round a fraction or mixed number to the nearest whole number, determine whether the fractional part is greater than or equal to ½.

Chapter 5 – The Decimal System   **55**

Example 5.5:

Round $13\frac{7}{8}$ to the nearest whole number.

> The denominator is 8 and half of 8 is 4. Since 7 is greater than 4, we know that the fractional part is greater than ½ so we round the whole number up to 14.

Estimation

When asked to solve a problem using estimation, use the rounding techniques above to round all the numbers in the problem to a place value that makes sense. For instance, if most of the numbers in the problem have two digits, it would be appropriate to round to the nearest ten. Once the numbers have been rounded, solve the problem and round the answer if necessary. Some estimation problems can be tricky especially when they involve money. In cases such as those, you should always round up whether or not the digit to the right is greater than or equal to 5.

Example 5.6:

One gallon of milk costs $2.35 at the farmers market. If Sally needs to purchase 6.5 gallons of milk, what is a reasonable estimate of the amount of money she should take with her to the market?

A. $12.00   B. $13.00   C. $14.00   D. $16.00

> The exact amount of money needed is $2.35 × 6.5 = $15.28 ≈ $16. On the test it will be easier (and faster) to round before multiplying.
>
> If the numbers are rounded too low then Sally will not take enough money with her and will not be able to purchase the milk she needs. The first thing to be done is round the two numbers in the problem. It would be easier to round the two decimals to whole numbers because they would be easier to multiply.
>
> $2.35 would get rounded to $3
> 6.5 gallons would get rounded to 7 gallons
> $3 × 7 gallons = $21
> The closest answer would therefore be D.
>
> Alternatively, the numbers could have been rounded to the nearest tenth and so $2.35 could be rounded to $2.40 and 6.5 kept as is.
> $2.40 × 6.5 gallons = $15.60
>
> This rounds to $16 which works but multiplying the decimals is harder and takes more time than multiplying whole numbers.

** Note **

If we had used the regular rules when rounding the decimals to whole numbers, we would have ended rounding $2.35 to $2 and 6.5 gallons to 7 gallons.

$2 × 7 gallons = $14

This would not have been enough money to purchase the 6.5 gallons of milk needed.

56   Prep for Success: Florida's PERT Math Study Guide

Example 5.7:
Which of the following is closest to 1048 ÷ 21.5?

A.   43   B.   50   C.   55   D.   60

This problem does not involve money so the regular rounding rules apply. It would be appropriate to round to the nearest ten in this example so 21.5 would be rounded to 20 and 1048 would be rounded to 1050.

$1050 \div 20 = 52.5$

We round this down to 50, therefore the correct answer would be B.

Scientific Notation
- -

Scientific notation is used to condense very large or very small numbers. The components of a number written in scientific notation are given below.

Coefficient $\Longrightarrow$ 5.83×10^{-6} $\Longleftarrow$ Exponent

$\Uparrow$ Base

Note

The base is always ten in scientific notation because each decimal place moved represents a power of ten. A negative exponent indicates that the number is small while a positive exponent indicates a large number. There is always one digit to the left of the decimal point and it cannot be 0.

Converting Standard Form to Scientific Notation

We convert standard form to scientific notation by moving the decimal point to the right of the first significant (non-zero) digit. The number of decimal places moved becomes the exponent and is negative if moved right (the number was small) and positive if moved left (the number was large).

Example 5.8:
Convert 0.0000008271 to scientific notation.

We can tell the number is small because it contains a decimal point followed by a number of zeros. Move the decimal point to the right of the 8 which is the first non-zero digit from the left. Count the number of decimal places moved.

0.0000008271

The decimal point was moved seven places to the right so the exponent will be –7.

Scientific notation: 8.271×10^{-7}

When there are more than 4 or 5 significant figures we often round to 3 significant figures to condense the number.

Chapter 5 – The Decimal System **57**

Example 5.9:

Convert 5, 738, 924 to scientific notation using 3 significant figures.

> When dealing with whole numbers, the decimal point is located after the last digit. Move the decimal place immediately right of the 5 which is the first non-zero digit from the left. Count the number of decimal places moved.
>
> $$5, 738, 924$$
>
> The decimal point was moved six places to the left so the exponent will be 6.
>
> Scientific notation: 5.738924×10^6
>
> Notice that this did not really condense the number but we still need to round to 3 significant figures. The third significant figure is 3 but the fourth significant figure is 8 which is greater than 5 so we round 3 up to 4.
>
> Scientific notation using 3 significant figures: 5.74×10^6

When multiplying or dividing in scientific notation, the following laws of exponents may become necessary:

1. $(x^a)^b = x^{ab}$
2. $(xy)^a = x^a y^a$
3. $\left(\dfrac{x}{y}\right)^a = \dfrac{x^a}{y^a}$
4. $x^a x^b = x^{a+b}$
5. $\dfrac{x^a}{x^b} = x^{a-b}$
6. $x^{-a} = \dfrac{1}{x^a}$ and $\dfrac{1}{x^{-a}} = x^a$

Note that $x^a + x^b \neq x^{a+b}$

Example 5.10:

Simplify the following and write the answer in scientific notation:

$$\frac{200000 \times 0.7}{100 \times 0.0002}$$

> Write each piece in scientific notation
>
> $$\frac{2 \times 10^5 \times 7 \times 10^{-1}}{1 \times 10^2 \times 2 \times 10^{-4}}$$
>
> Gather coefficients together and power of tens together.
> Combine powers of 10 using law #4 of exponents.
>
> $$\frac{(2 \times 7) \times (10^5 \times 10^{-1})}{(1 \times 2) \times (10^2 \times 10^{-4})} = \frac{14 \times 10^{5+(-1)}}{2 \times 10^{2+(-4)}} = \frac{14 \times 10^4}{2 \times 10^{-2}}$$
>
> Next, divide and apply law #5 of exponents.
>
> $$= \frac{14}{2} \times \frac{10^4}{10^{-2}} = 7 \times 10^{4-(-2)} = 7 \times 10^6$$

58   Prep for Success: Florida's PERT Math Study Guide

Example 5.11:
Simplify the following and write in scientific notation: $(0.0003 \times 500)(200000 \times 0.04)$

Write each piece in scientific notation.

$= (3 \times 10^{-4} \times 5 \times 10^2)(2 \times 10^5 \times 4 \times 10^{-2})$

Gather coefficients together and power of tens together.

$= (3 \times 5 \times 2 \times 4)(10^{-4} \times 10^2 \times 10^5 \times 10^{-2})$

Combine the powers of 10 using law number 4 of exponents.

$= 120 \times 10^{-4+2+5-2}$

$= 120 \times 10^1$

Check that the final answer is written in scientific notation by ensuring that the decimal point is after the first significant figure. In this case, we need to move the decimal point 2 places to the left while increasing the power of ten by 2.

$= 1.20 \times 10^3$

Comparisons

Questions involving comparisons may provide a range of answers and require you to choose the least or greatest of these answers. The answer choices can be a mix of fractions, decimals, percentages, recurring numbers, radicals and numbers in scientific notation. If possible, first eliminate any choices that are obviously wrong. Convert the remaining answers to one format before comparing. Choose the format with which you feel most comfortable comparing and then convert all the remaining answer choices to this format. Many students prefer comparing decimals. This is a great tactic because not only are decimals easy to compare, but by doing so you avoid having to convert recurring decimals and working with fractions. Fractions may seem easier but you will have to spend some time making sure that they all share the same common denominator.

The table below gives the symbols used when comparing numbers and their meanings.

Symbol	Meaning
=	Equal to
≠	Not equal to
<	Less than
≤	Less than or equal to
>	Greater than
≥	Greater than or equal to

Remember that on a number line, the numbers farther left are less than the numbers farther right. Therefore $-3 < 5$ and $-1 > -6$. Also, if the relationship $x > y$ is true then $y < x$. Knowing the decimal equivalent of $\sqrt{2}$ and $\sqrt{3}$ may also prove useful.

$\sqrt{2} \approx 1.41$ and $\sqrt{3} \approx 1.73$

Example 5.12:
Which of the following is greatest?

A.   5.37%   B. $5.\overline{35}$ C. $-5\dfrac{2}{5}$ D.   5.352

Since one of the answers is a recurring decimal, it would be easier to convert all the answers to decimals. Before attempting to convert the fraction to a decimal, make sure to read what the question is asking very carefully. The question asks to find the answer that is greatest but the fraction is the only negative number therefore it is actually the smallest and can be eliminated immediately. In order to compare choices A, B, and D we convert the percentage to a decimal and write out a few more decimal places for the recurring decimal.

$5.37\% = 0.0537$
$5.\overline{35} = 5.3535\ldots$

Comparing the place values of 0.0537, 5.3535…, and 5.352 we see that the largest decimal is the recurring one so the answer is B.

** Note **

To compare decimals, it is easier to write them one above the other, lining up their decimal points. Compare place values from left to right and eliminate decimals with the smallest number in that place value until you are left with only one answer.

0.0537
5.3535 …
5.352

Comparing the ones place value, answer choice A gets eliminated so we are left with:
5.3535 …
5.352

These two decimals have the same place values until comparing the thousandth position. The top decimal has a 3 in that place value while the other has a 2. Therefore the largest decimal is 5.3535… and the answer is B.

Example 5.13:
Given that $5 < \sqrt{x} < 7$, which of the following could be the value of x?
  A. 6.5   B. 25   C. 36.7   D. 49

The inequality indicates that the square root of x lies between 5 and 7. Since 5 and 7 are both positive integers, we can square the entire inequality without changing the validity of the relationship. Therefore, we see that $25 < x < 49$. This means that x lies between 25 and 49, not inclusive. The only value that fits in this range is 36.7.

60 Prep for Success: Florida's PERT Math Study Guide

Chapter 5 Exercises

1. Which of the following is closest to $\sqrt{63} - \sqrt{31}$?
   A. 2   B. 3   C. 5   D. 6

2. Which of the following is smallest?
   A. $0.33\overline{6}$ B. $\dfrac{\sqrt{2}}{7}$ C. $\dfrac{4}{11}$ D. $|-5|$

3. Round $37,851$ to the nearest hundred.

4. In 2009, there were 3578 thrift stores across the United States. If the stores employed $72,496$ people, what is a good estimate of the number of people employed per store?

5. Samantha can buy a flat panel TV either for \$895 cash or \$50 down with monthly payments of \$71.42 for a year. Which option is cheaper?

6. Gordon, Amy, Lil and Nathaniel each has 150 chocolate bars to sell for their school fundraiser. Gordon sold 46% of his bars, Amy sold $\dfrac{7}{25}$ of her bars, Lily sold $\dfrac{2}{5}$ of hers and then gave 15 bars away to her family and Nathaniel sold 32% of his and then lost a box of 25 bars. Which of them had the fewest number of bars left over?
   A. Amy   B. Gordon   C. Lily   D. Nathaniel

7. What is the value of $-0.000056 + 0.000374$ written in scientific notation?
   A. 4.3×10^{-5} B. 3.18×10^{-4} C. 4.3×10^{5} D. 3.18×10^{4}

8. A class of 50 students took a math test. The lowest score was a 64 while the highest score was a 97. Which of the following could be a reasonable estimate of the students' average test score?
   A. 56   B. 63   C. 78   D. 98

9. All of the following are ways to write $\dfrac{3}{4}$ percent of N except?
   A. $\dfrac{3}{400}N$ B. $0.0075N$ C. $\dfrac{3}{4}N$ D. $7.5 \times 10^{-3}N$

10. If $12 \le a \le 16$ and $4 \le b \le 8$, then the smallest possible value of $\dfrac{a+b}{b-a}$ is?
    A. -6 B. -5 C. -2 D. Cannot be determined

Solutions

1. **Answer: [A]**   According to the list of squares on page 10, the first few squares are:
   $2^2 = 4$ $3^2 = 9$ $4^2 = 16$ $5^2 = 25$ $6^2 = 36$ $7^2 = 49$ $8^2 = 64$
   63 is closest to 64 so $\sqrt{63}$ is closest to $\sqrt{64}$

31 is closest to 36 so $\sqrt{31}$ is closest to $\sqrt{36}$

We can approximate as follows:

$\sqrt{63} - \sqrt{31} \approx \sqrt{64} - \sqrt{36} = 8 - 6 = 2$

2. **Answer: [B]**   Writing each of the answers as decimals makes it easier to compare.

$0.33\overline{6} = 0.3366666\ldots$

We have an estimate of the value of $\sqrt{2}$ so we use that

$\dfrac{\sqrt{2}}{7} \approx \dfrac{1.4}{7} = 0.2$

$\dfrac{4}{11} = 4 \times \dfrac{1}{11} = 4 \times 0.090909\ldots = 0.36363636\ldots$

$|-5| = 5$

We can clearly see that B is the smallest.

3. **Answer: [37, 900]**   The hundreds place digit is 8. The digit to the right of 8 is 5 so we round 8 up to 9 and add padding zeros to all the other place values to the right. Therefore the answer is 37, 900.

4. **Answer: [20]**   Round both numbers. The level of rounding depends on you and the supplied answers but rounding to the nearest hundred or thousand would work well here.

$3,578 \approx 3,600$ $72,496 \approx 72,000$

Now divide to find the number of people per store.

$72,000 \div 3,600 = 720 \div 36 = 20$ people

5. **Answer: [Cash Option]**   We need to calculate or estimate how much the TV would cost if she chose the monthly payment option. First round her monthly payment of \$71.42 to \$71. Multiply this by 12 months to determine how much she would pay in monthly payments for a year.

$71 \times 12 \; months = \$852$

Add the total monthly payments to the down payment to find the total paid using option 2.

Total paid with monthly payment option $= \$852 + \$50 = \$902$

Therefore, the cash option is better because she would only pay \$895.

6. **Answer: [C]**   We need to calculate how many bars each person sold, lost or gave away in order to figure out how many bars they each had left over. The person who got rid of the most bars has the fewest remaining since they all started with the same number of bars.

$Gordon = 46\% \; of \; 150 = 0.46 \times 150 = 69 \; bars$

$Amy = \dfrac{7}{25} \times 150 = \dfrac{1050}{25} = 42 \; bars$

$Lily = \left(\dfrac{2}{5} \times 150 \; sold\right) + (15 \; gave \; away) = \dfrac{300}{5} + 15 = 60 + 15 = 75 \; bars$

$Nathaniel = (32\% \; of \; 150 \; sold) + (25 \; lost) = (0.32 \times 150) + 25 = 48 + 25 = 73 \; bars$

Since Lily got rid of the largest number of chocolate bars, she has the fewest left over.

7. **Answer: [B]**   Ignoring the sign of both decimals, we compare to see which is larger. The first non-zero digit in 0.000056 is 5, which is in the hundred thousandths place. The first non-

zero digit in 0.000374 is 3, which is in the ten thousandths place. Since 0.000374 has fewer leading zeros, it is the larger of the two decimals and goes above in the subtraction.

$$
\begin{array}{r}
0.000374 \\
-\ \ 0.000056 \\
\hline
0.000318
\end{array}
$$

Now move the decimal place immediately right of the first non-zero digit and count the number of places moved. Since we are moving the decimal point to the right, the exponent will be negative. The answer is positive because the larger decimal was positive.

$$0.000318 = 3.18 \times 10^{-4}$$

8. **Answer: [C]** Since we weren't given the actual scores, we use a bit of reasoning. If every person had scored a 64, the average would be 64 so we know that an average of 56 is not possible. At least one person had to score 97 because it was the highest score. Even if all the other students scored a 64, the average would be really close to but slightly higher than 64 (the one score of 97 raises the average above 64). Similarly, if all but one person scored a 97, the average will be close to 97 but slightly lower (the one score of 64 drops the average below 97). Since these two cases are extreme with a very small likelihood of occurring, the average score is most likely somewhere in between 64 and 97. Hence the answer is C, 78.

9. **Answer: [C]** The obviously wrong answer is C because no conversion was done and it is missing the percentage sign. But to prove the others are correct:

$$\frac{3}{4}\% \text{ of N} = 0.75\% \times N$$

To convert the 0.75% to a fraction, place over 100, move the decimal point 2 places to the right in the numerator and the denominator, and then reduce the resulting fraction.

$$0.75\% \times N = \frac{0.75}{100}N = \frac{75}{10000}N = \frac{3}{400}N$$

To convert 0.75% to a decimal, move the decimal point 2 places to the left.

$$0.75\% \times N = 0.0075N$$

Converting the decimal to scientific notation:

Move the decimal point 3 places to the right (after the first non-zero digit). The exponent will be negative because we moved to the right.

$$0.0075N = 7.5 \times 10^{-3}N$$

10. **Answer: [B]** Based on the given intervals, which do not overlap, we know for sure that b is less than a and that they are both positive. Since b is smaller than a, the numerator $(a + b)$ will be positive while the denominator $(b - a)$ will be negative. This means that the entire fraction will be negative. For negative fractions, a larger numerator and a smaller denominator will result in an overall smaller fraction.

For example, on a number line $-\frac{5}{4}$ is further left and is therefore actually smaller than $-\frac{1}{4}$

So we need to ensure that the absolute value of the difference between b and a is as small as possible. Choosing $b = 8$ and $a = 12$ gives us the smallest difference of 4. Therefore:

$$\frac{a + b}{b - a} = \frac{12 + 8}{8 - 12} = \frac{20}{-4} = -5$$

Chapter 6 — Linear Equations of One Variable

Some of the word problems may require knowledge of the most basic geometric formulas such as the area of a rectangle or square. In this chapter, you will learn to translate word problems to first degree equations by extracting relevant information from the given question.

Basic Geometry

Perimeter of Plane Figures
The perimeter of a plane figure is the sum of the length of all the sides or, in the case of a circle, the total distance around the figure. A list of formulas for the perimeter of the most popular plane figures is given below. These formulas can also be found in the comprehensive formula list in Appendix A at end of the book.

Area of Plane Figures
The area of a plane figure is the total amount of surface occupied by the figure and is measured in square units (e.g. ft² or square cm). The formulas to calculate the surface area of the most popular plane figures are also found in the list of formulas below.

Formulas

		Area	Perimeter
Square		$A = s^2$ $A = \dfrac{1}{2}d^2$	$P = 4s$
Rectangle	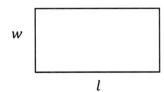	$A = l \times w$	$P = 2l + 2w$

Triangle

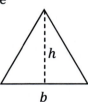

$A = \frac{1}{2}bh$ Add lengths of sides

Circle

$A = \pi r^2$ $C = 2\pi r$
$C = \pi d$

Other Geometric Shapes

Trapezoid

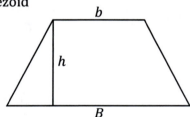

A trapezoid is a shape with exactly 1 pair of parallel sides called bases (B and b).

The area is $A = \frac{1}{2}h(B + b)$

Parallelogram

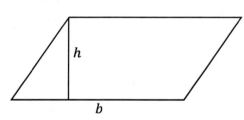

A parallelogram is a four sided figure with two pairs of parallel sides. Like any other four-sided figure, the sum of the interior angles is 360°.

The area is $A = bh$ and its diagonals bisect each other (divide each other in half)

Rhombus

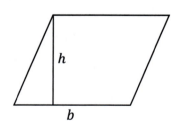

A rhombus is a parallelogram with four equal sides. The diagonals are perpendicular and bisect each other.

Area $= bh = \frac{1}{2}$ (product of diagonals)

Other Formulas

diameter of a circle $= 2r$ (twice the radius)

diagonal of a square $= s\sqrt{2}$

The term **congruent** means equal. We refer to geometric components such as angles and lines as being congruent when they are equal size or length.

Chapter 6 – Linear Equations of One Variable   **65**

We take a closer look at triangles in Chapter 15 but below are the three types of triangles you may encounter on the test.

Types of Triangle	Property
Equilateral	All three sides congruent
Isosceles	Two congruent sides
Scalene	No sides congruent

Example 6.1:

The perimeter of a rectangular pool is 60 feet. If the width of the pool is 8 feet, what is the length of the pool?

The problem involves the perimeter of a rectangle so we first find the equation for perimeter in the list of formulas on the previous page.

$P = 2l + 2w$

Now substitute the information given in the problem: the perimeter is 60 feet and the width is 8 feet

$60 = 2l + 2(8)$

Lastly, solve for the missing variable.

$60 = 2l + 16$

$2l = 60 - 16$

$2l = 44$

$l = \dfrac{44}{2} = 22$ feet

Therefore, the length of the pool is 22 feet.

Example 6.2:

Find the area of a circle whose diameter is 15 inches.

First find the formula for the area of a circle.

$A = \pi r^2$

We see that the formula requires the radius of the circle but we were only given the diameter. Since the diameter is twice the radius, we can find the radius from the diameter.

$d = 2r$

$15 = 2r$

$r = \dfrac{15}{2}$

$r = 7.5$ inches

Now use the radius to find the Area.

$A = \pi(7.5)^2 = 56.25\pi$ square inches

Therefore the area of the circle is 56.25π square inches.

66 Prep for Success: Florida's PERT Math Study Guide

** Note **

If the answer involves a fraction, simply use the fractional form of the radius in the calculation of the area.

$$A = \pi \left(\frac{15}{2}\right)^2 = \frac{225}{4}\pi \text{ square inches}$$

Algebraic Equations

Algebraic statements can be used to represent lengthy written phrases. Word problems can appear intimidating especially when you have to extract the information and then formulate an equation that can be solved. Here we focus on converting simple written phrases into algebraic expressions.

Some key words to look for and their meanings:

Key Words	Meaning
Difference	Subtract
Sum	Add
Product	Multiply
Less than	Subtract from
More than	Add to
Is	Equals

Example 6.3:

The difference between three times a number n and 6 less than twice the number is 26. Which of the following equations can be used to solve for the number n?

A. $(6n - 2) - n = 26$ B. $3n - (6 - 2n) = 26$

C. $3n - (2n - 6) = 26$ D. $3n - 6 - 2n = 26$

First identify the key words in the phrase. Sometimes it is easiest to break the entire phrase into smaller, more manageable phrases.

- three times a number n
- twice the number
- 6 less than twice the number
- is 26

Convert the phrases into algebraic expressions

three times a number n	$3n$
twice the number	$2n$
6 less than twice the number	$2n - 6$
is 26	$= 26$

Combine the algebraic expressions into one equation.

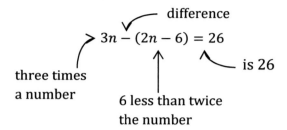

Therefore the answer is C.

Formulate and Solve First Degree Equations

An **equation** sets two mathematical expressions equal to each other to form an **equality**. The **solution** of an equation is one that when substituted for the variable, makes both sides of the equation the same and therefore makes the equality true. A **first degree equation** is one that has only one variable. To solve a first degree equation, simply move all the terms containing variables to one side of the equation and all the constants over to the other side of the equation before solving for the variable. We do this by using inverse operations as demonstrated below. Remember that to maintain equality, whatever is done to the left-hand side of the equation must also be done to the right-hand side.

To solve a first degree equation:
1. Eliminate any parentheses by distributing
2. Combine any like terms on each side
3. Move all the terms containing variables to the left and everything else to the right
4. Divide by the coefficient of the variable if necessary

To check your answer you can substitute the answer into the original equation to make sure it is valid.

Example 6.4:
If $2(6r - 5) = -2(r + 9)$ then $r =$

Before beginning to solve the equation, distribute to remove the parentheses.

$2(6r - 5) = -2(r + 9)$
$12r - 10 = -2r - 18$

Next, move all terms with variables to the left-hand side by adding $2r$ to each side. This is an inverse operation because $2r$ was negative.

$12r - 10 = -2r - 18$
$+2r \qquad\quad + 2r$

$14r - 10 = -18$

Now move all constants to the right-hand side by adding 10 to both sides.

$$14r - 10 = -18$$
$$\underline{+10 \quad\; + 10}$$
$$14r = -8$$

Finally, divide both sides by 14 to leave the variable r by itself.

$$\frac{14r}{14} = -\frac{8}{14}$$

$$r = -\frac{8}{14} = -\frac{4}{7}$$

Example 6.5:

Solve for x: $-2(2x - 5) + 2 = 4 - 6x$

Distribute to remove parentheses.
$$-4x + 10 + 2 = 4 - 6x$$

Combine like terms
$$-4x + 12 = 4 - 6x$$

Move variables to the left and constants to the right
$$-4x + 6x = -12 + 4$$
$$2x = -8$$

Divide by the coefficient
$$x = \frac{-8}{2} = -4$$

Example 6.6:

The sum of one-third of a number n and 5 less than the number is 31. What is the number?

Identify key phrases and convert to algebraic expressions.

one-third of a number n	$\frac{1}{3}n$
5 less than the number	$n - 5$

Combine the algebraic expressions into one equation

$$\frac{1}{3}n + (n - 5) = 31$$

Combine like terms then use inverse operations to solve for n.

$$\frac{1}{3}n + n - 5 = 31$$

$$\frac{4}{3}n - 5 = 31$$

$$\frac{4}{3}n = 36$$

$$\left(\frac{3}{4}\right) \cdot \frac{4}{3}n = 36 \cdot \left(\frac{3}{4}\right)$$

$$n = \frac{108}{4} = 27$$

Solving Equations involving Absolute Values

On the PERT, you may be required to solve equations involving absolute values. Since the result of an absolute value can never be negative, if the question sets an absolute value equal to a negative number there is no solution. If an absolute value is set equal to a positive number, there are two possible solutions. Since we do not know whether the expression *inside* the absolute value evaluates to a negative or a positive before we took the absolute value, we must consider both.

Example 6.7:

$|-5x + 4| = -3$

> This equation has no solution because it states that the absolute value of the expression $-5x + 4$ is -3. This is impossible because the absolute value of any number/expression always evaluates to a positive number.

Example 6.8:

$|2x - 5| = 19$

> This equation states that the absolute value of the expression $2x - 5$ is 19 but was the expression originally negative or positive? Since we do not know for certain, we must consider both options. The easiest way is to set the expression equal to both -19 and 19.

$$2x - 5 = 19 \qquad\qquad 2x - 5 = -19$$
$$2x = 24 \qquad\qquad 2x = -14$$
$$x = \frac{24}{2} \qquad\qquad x = -\frac{14}{2}$$
$$x = 12 \qquad\qquad x = -7$$

> The fact that one of the solutions is negative often confuses students. They assume that the variable cannot have a negative solution since we are dealing with absolute values. What we found was the value of x that would make the expression $2x - 5$ evaluate to -19 or 19 BEFORE taking the absolute value and so the value of x can be negative or positive.

Relevant vs. Irrelevant Information

Sometimes you may be given extra information than is needed to solve the problem. Extra information is deliberately included to test your ability to distinguish between the information that is relevant to solving the problem and the information that is not. This may be especially true in questions involving tables, charts or diagrams. Read the question carefully, formulate a plan of attack then extract the information necessary to execute the plan.

70 Prep for Success: Florida's PERT Math Study Guide

Example 6.9:

Using the table below, determine the amount of time, in seconds, taken by a blue ray to reach the wall's surface.

Color of Ray	Distance from wall (meters)	Speed of beam (meters/minute)	Time of travel (minutes)
Red	15	2.37	6.33
Blue	27	5.12	5.27
Yellow	9	1.83	4.92
Green	13	2.19	5.94

You may at first glance be tempted to use the rate formula with the given distances and speeds to solve for time, but note that we are already given time in minutes. Simply read off the time for the blue ray then convert the time to seconds as the question requested.

Time taken by blue ray to reach wall $= 5.27 \ mins = 5.27 \times 60 = 316.2 \ seconds$

Other Word Problems

Other arithmetic word problems are based on all the concepts you have just learned. It is important to read the question carefully and then apply the appropriate formulas or techniques. If the question appears too difficult, it is sometimes helpful to extract the data given and note what variable you were asked to find. With just the numbers from the problem it is often easier to figure out the type of question you are dealing with.

Example 6.10:

Sam has repaid $\frac{2}{5}$ of his loan from the bank. If he has repaid a total of $32,850$ then how much does he still have to repay?

We do not know the total amount of the loan so we call this x.

$\frac{2}{5}$ of the total loan is equal to $32,850$ so we turn this into equation form.

$$\frac{2}{5}x = 32850$$

The easiest way to solve for x is to multiply both sides by the reciprocal of $\frac{2}{5}$.

$$\left(\frac{5}{2}\right)\frac{2}{5}x = 32850\left(\frac{5}{2}\right)$$

$$x = \frac{32850}{1} \times \frac{5}{2} = \$82,125$$

Now that we know the total amount of the loan, subtract the amount repaid to find how much is left to be paid.

$82,125 - \$32,850 = \$49,275$

Therefore Sam still owes the bank $49,275$.

Chapter 6 – Linear Equations of One Variable   **71**

Example 6.11:
Dolly bought a dress, a belt and a pair of shoes for $108 before tax. If the dress cost 25% more than the shoes and the belt was $22 less than the shoes, what did she pay for the belt?

A. $10   B. $18   C. $22   D. $40

Since the cost of the dress and the cost of the belt are both based on the cost of the shoes, we use the cost of the shoes as the variable and develop expressions for the dress and belt.
Let the cost of shoes $= n$
25% more indicates a percentage increase. As discussed in Chapter 4, this can be written as $100\% + 25\% = 125\% = 1.25$
To find the cost of the dress, we take 1.25 of the cost of the shoes.
$Cost\ of\ dress = 1.25n$
The belt was $22 less than the shoes so we subtract 22 from n to find the cost of the belt.
$Cost\ of\ belt = n - 22$
We know that she paid $108 in total.
$Total\ cost = cost\ of\ dress + cost\ of\ belt + cost\ of\ shoes$
$1.25n + (n - 22) + n = 108$
Now solve for n
$3.25n - 22 = 108$
$3.25n = 130$
$$n = \frac{130}{3.25} = \frac{13000}{325} = 40$$

Recall that n is the cost of the shoes but we were asked for the cost of the belt.
Cost of belt $= n - 22 = 40 - 22 = \$18$

Example 6.12:
For a picnic, a company bought 15 pizzas at $5 each, 40 hot dogs at $1.50 each and 55 hamburgers. If the company spent $267 on food for the picnic, how much did each hamburger cost?

Since we are trying to find the price of a hamburger, we call this x.
15 pizzas at $5 = 15 \times \$5 = \$75$
40 hot dogs at $1.50 = 40 \times \$1.50 = \$60$
55 hamburgers at $\$x = 55 \times x = \$55x$

Add the cost of each type of food together, set this equal to the total spent and solve for x.
$75 + 60 + 55x = 267$
$135 + 55x = 267$
$55x = 267 - 135$
$55x = 132$
$$x = \frac{132}{55} = \$2.40$$

Therefore each hamburger cost $2.40

72 Prep for Success: Florida's PERT Math Study Guide

Example 6.13:

Charlie and Ethan both commute to and from work each day. Today, Charlie leaves the office at 5:34pm and heads east on the freeway, driving at 48 mph. Ethan leaves the office at 5:49pm and takes the freeway in the same direction, driving at 56 mph. At what time will Ethan overtake Charlie on the freeway?

A. 7:04pm   B.   7:19pm   C.   7:26pm   D.   7:34pm

Before Ethan can overtake Charlie, he has to catch up with him on the freeway. At this point they would have driven the same distance even though Ethan would have been driving for a shorter amount of time. We can use the fact that the distance is equal to formulate an equation that can be solved.

Since Charlie left at 5:34pm and Ethan left at 5:49pm, Ethan left 15 mins after Charlie i.e. his total driving time when he catches up will be 15 minutes less than Charlie's. Since the rate is in miles per hour we need time in hours so convert 15 minutes.

$$\frac{60 \; mins}{1 \; hr} = \frac{15 \; mins}{x \; hrs}$$

$$60x = 15$$

$$x = \frac{15}{60} = 0.25hr$$

So if Charlie has been driving for t hours, Ethan will have been driving for $t - 0.25$ hours i.e. quarter of an hour less than Charlie.

Let's create a table relating rate, time and distance with the calculated distance in the last column.

$rate \times time = distance$

	Rate (miles per hr)	Time (hr)	Distance (miles)
Charlie	48 mph	t	$48t$
Ethan	56 mph	$t - 0.25$	$56(t - 0.25)$

Since their distance will be equal:

$48t = 56(t - 0.25)$

$48t = 56t - 14$

$-8t = -14$

$$t = \frac{-14}{-8} = 1.75 \; hours = 1 \; hr \; 45 \; mins \text{ since 0.75 is } \frac{3}{4} \text{ of an hour}$$

Note that t is the amount of time that Charlie has been driving. Therefore we either add t to Charlie's departure time or we add $t - 0.25 = 1.50 \; hours = 1 \; hr \; 30 \; mins$ to Ethan's.

Charlie $\Rightarrow 5{:}34pm + 1hr \; 45 \; mins = 7{:}19pm$

Ethan $\Rightarrow 5{:}49pm + 1hr \; 30 \; mins = 7{:}19pm$

The answer is B.

Chapter 6 Exercises

1. What is the area of a rectangular box whose width is 5 inches, length is 8 inches and height is 3 inches?

   A. 44 sq inches   B. 79 sq inches   C. 120 sq inches   D. 158 sq inches

2. Sue's brother is four times as old as Sue. In 6 years, he will be three times as old as she is. How old is Sue now?

   A. 6 yrs   B. 12 yrs   C. 18 yrs   D. 24 yrs

3. The sum of 3 consecutive even integers is 162. Find the middle integer.

   A. 48   B. 51   C. 52   D. 54

4. Amelia has \$7.10 in nickels, dimes and quarters. If she has twice as many quarters as nickels and 6 more dimes than nickels, how many dimes does she have?

   A. 10   B. 14   C. 16   D. 18

5. What is the length of the base of an isosceles triangle if the length of the congruent side is 11 cm and the perimeter is 28 cm?

6. Choose the equation that is equivalent to the statement:
   "The difference between a number n and seven more than three times the number is 18."

   A. $n - 3n + 7 = 18$ C. $n(3n - 7) = 18$

   B. $3n + 7 = 18$ D. $n - (3n + 7) = 18$

7. If $5(2x - 9) + 7 = -3(x + 4)$ then $x =$

   A. $7\frac{1}{7}$ B. 2   C. $\frac{6}{13}$ D. 5

8. Solve the following equation for x:
   $$\frac{3}{4}x + \frac{1}{2} = \frac{2}{3}x + 2$$

   A. -30 B. -18 C. -9 D. 18

9. What is the width of a rectangle if its perimeter is 60 feet and its length is 3 less than twice its width?

   A. 11   B. 19   C. 120   D. 144

10. Solution A contains 45% acid while solution B contains 69% acid. How many liters of solution B should be mixed with solution A in order to create 60 liters of a mixture that is 50% acid?

    A. 12.5 liters   B. 22.7 liters   C. 37.3 liters   D. 47.5 liters

Solutions

1. **Answer: [D]**   A rectangular box is comprised of 3 pairs of rectangles: front and back, left and right, top and bottom.

   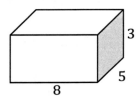

   Area of top and bottom = $2(l \times w) = 2(8 \times 5) = 2(40) = 80$
   Area of left and right = $2(w \times h) = 2(5 \times 3) = 2(15) = 30$
   Area of front and back = $2(l \times h) = 2(8 \times 3) = 2(24) = 48$
   Total area $= 80 + 30 + 48 = 158$ *square inches*

2. **Answer: [B]**   The method that we have found to be easiest for students to implement involves constructing a table defining their ages now and in the future. To find their age 6 years from now, we add 6 to their current ages.

	Now	Future
Sue	x	$x + 6$
Brother	4 times Sue = $4x$	$4x + 6$

   Next formulate an equation that can be used to solve for x. Note that in these types of problems, we base the equation on the second statement because it gives more information than the first. Set Sue's brother's age 6yrs from now equal to 3 times Sue's age 6yrs from now.

   Brother's future age $= 3 \times$ *Sue future age*
   $4x + 6 = 3(x + 6)$
   $4x + 6 = 3x + 18$
   $4x - 3x = 18 - 6$
   $x = 12$ *years old*
   Therefore, Sue is 12 yrs old now.

3. **Answer: [D]**   To go from one even number to the consecutive even number, we add 2.
   Let first integer $= x$
   Second integer $= x + 2$
   Third integer $= x + 4$
   $(x) + (x + 2) + (x + 4) = 162$
   $3x + 6 = 162$
   $3x = 156$
   $x = \dfrac{156}{3} = 52$
   Middle integer $= x + 2 = 52 + 2 = 54$

4. **Answer: [C]**   Since the number of dimes and quarters are based on the number of nickels:

Let the # of nickels $= x$

# of dimes $= x + 6$

# of quarters $= 2x$

Since the constant is in dollars, use the value of a nickel, dime & quarter to formulate an equation

$0.05x + 0.10(x + 6) + 0.25(2x) = 7.10$

Multiply the entire equation by 100 to remove the decimals

$5x + 10(x + 6) + 25(2x) = 710$

$5x + 10x + 60 + 50x = 710$

$65x + 60 = 710$

$65x = 650$

$x = \dfrac{650}{65} = 10 \ nickels$

# of dimes $= x + 6 = 10 + 6 = 16 \ dimes$

5. **Answer: [6]**   An isosceles triangle has 2 congruent (equal) sides.

Perimeter = length of base + 2(length of congruent side)

$28 = b + 2(11)$

$28 = b + 22$

$b = 28 - 22 = 6 \ cm$

6. **Answer: [D]**   The word "difference" indicates subtraction, the result of which is 18. The key words "more than" indicate addition while "times" indicates multiplication.

Number $= n$

7 more than 3 times the number $= 3n + 7$

So we subtract these two and set equal to 18. We put $3n + 7$ in parentheses because it is treated as one number. Therefore:

$n - (3n + 7) = 18$

7. **Answer: [B]** $5(2x - 9) + 7 = -3(x + 4)$

$10x - 45 + 7 = -3x - 12$

$10x - 38 = -3x - 12$

$10x + 3x = -12 + 38$

$13x = 26$

$x = \dfrac{26}{13} = 2$

8. **Answer: [D]**   Move all the terms involving variables to the left and all the constants to the right.

$$\frac{3}{4}x - \frac{2}{3}x = 2 - \frac{1}{2}$$

Combine the fractions on each side of the equation using LCDs.

$$\frac{(3)3}{(3)4}x - \frac{2(4)}{3(4)}x = \frac{2(2)}{1(2)} - \frac{1}{2}$$

$$\frac{9}{12}x - \frac{8}{12}x = \frac{4}{2} - \frac{1}{2}$$

$$\frac{1}{12}x = \frac{3}{2}$$

$$x = \frac{3}{2}\left(\frac{12}{1}\right) = \frac{36}{2} = 18$$

9. **Answer: [A]** Formulate a linear equation using the formula for perimeter.

Let width $= x$

Length $= 2x - 3$

Perimeter $= 2l + 2w$

$$= 2(2x - 3) + 2(x) = 60$$

Solve the equation

$2(2x - 3) + 2(x) = 60$

$4x - 6 + 2x = 60$

$6x - 6 = 60$

$6x = 66$

Width $x = 11$

10. **Answer: [A]** Since we are trying to find the number of liters of solution B:

Let the number of liters of solution B $= x$

There is a total of 60 liters so:

The number of liters of solution A $= 60 - x$

The amount of acid (called the concentration) in each solution is given by:

Concentration $= percentage\ acid \times number\ of\ liters$

Set up a table defining concentration of acid in solution A, solution B and the mixture.

	Percentage Acid	Number of Liters	Concentration
Solution A	45%	$60 - x$	$0.45(60 - x)$
Solution B	69%	x	$0.69x$
Mixture	50%	60	$0.50(60)$

The concentration of acid in A plus the concentration of acid in B equals the concentration of acid in the mixture.

$0.45(60 - x) + 0.69x = 0.50(60)$

To avoid working with decimals, we multiply the entire equation by 100.

$45(60 - x) + 69x = 50(60)$

$2700 - 45x + 69x = 3000$

$24x = 300$

$$x = \frac{300}{24} = \frac{50}{4} = 12.5\ liters\ of\ solution\ B$$

Chapter 7

Linear Equations of Two Variables

In the last chapter, we reviewed the formulation of linear equations in one variable. Here we take a look at linear equations in two variables x and y on the coordinate (two-dimensional) plane.

Linear Equations

A linear equation of x and y is used to define how one variable (y) changes in response to changes in the other variable (x). A linear equation is often called a first order equation because no variable in the equation is raised to a power greater than 1. When it comes to linear equations of x and y, there can be no variables under radicals (i.e. variables raised to a fractional power), no products of variables and no variables in the denominator (that cannot be removed using algebra without creating a product of variables). Here are some examples of linear and non-linear equations.

Linear	Non-Linear	
$5x - 2y = 9$	$y = x^2 - 5$	*(Power greater than 1)*
$y = 4x + 3$	$y = 2^x$	*(Variable in exponent)*
$x = 15y$	$xy = 4$	*(Product of two variables)*

Many of the equations created in the last chapter were linear equations in one variable. In this chapter, we examine linear equations involving x and y and their graphs on the coordinate plane.

The Coordinate Plane

Before we go into a description of functions and their properties, we define the regions of the coordinate plane. The coordinate plane consists of a two dimensional space where values of x lie on the horizontal line called the $x - axis$ and values of y lie on the vertical line called the $y - axis$. An x-coordinate can take on values anywhere between negative infinity ($-\infty$) and positive infinity ($+\infty$). The same goes for the y-coordinate.

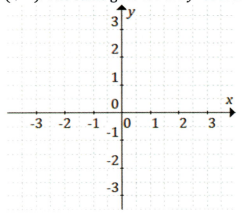

Take a look at the coordinate plane to the left and note where the values of x and y are positive versus negative.

Points on the coordinate plane are represented by an ordered pair (x, y). To plot a point on the coordinate plan, trace a vertical line through the x-coordinate and a horizontal line through the y-coordinate and mark the intersection of these lines with a point.

Example 7.1:
Plot the point $(-5, 3)$ on the coordinate plane

The x-coordinate is -5 and the y-coordinate is 3. The point will lie where the lines $x = -5$ and $y = 3$ intersect.

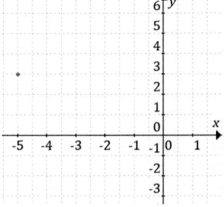

Equation of a Line
The equation of a line takes the form $y = mx + b$ where m is the slope and b is the y-intercept. This is called the **slope–intercept form**. To find the slope given two points (x_1, y_1) and (x_2, y_2) use the formula:
$$m = \frac{y_2 - y_1}{x_2 - x_1}$$
The y-intercept is the point where the graph crosses the y-axis and can be found by setting $x = 0$ in the equation of the line.

Given two points on a line, we can find the equation of the line by first finding the slope and then using the **point–slope formula**:
$$y - y_1 = m(x - x_1)$$ where m is the slope and (x_1, y_1) is any one of the points on the line.

Once we find the equation of the line, we often rearrange it into **standard form**: $ax + by = c$

Example 7.2:
Write the equation of the line passing through the points $(-3, 6)$ and $(1, 14)$.

Label the points
$(x_1, y_1) \quad (x_2, y_2)$
$(-3, 6) \quad (1, 14)$

Find the slope
$$m = \frac{14 - 6}{1 - (-3)} = \frac{8}{4} = 2$$
Use the point–slope equation

$$y - y_1 = m(x - x_1)$$
$$y - 6 = 2(x - (-3))$$
$$y - 6 = 2(x + 3)$$
$$y - 6 = 2x + 6$$
$$y = 2x + 12$$

Example 7.3:
Given $m = \frac{3}{2}$ and a point $(10, 4)$ on the line, find the equation of the line.
$$y - y_1 = m(x - x_1)$$
$$y - 4 = \frac{3}{2}(x - 10)$$
$$y - 4 = \frac{3}{2}x - \frac{3}{2}(10)$$
$$y - 4 = \frac{3}{2}x - 15$$
$$y = \frac{3}{2}x - 11$$

Intercepts

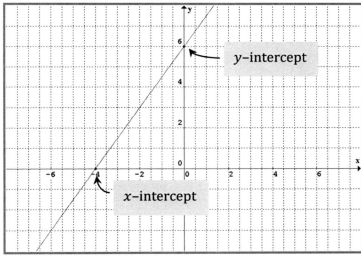

The x-intercept of a graph is the point where the graph crosses the x-axis. The y-intercept of a graph is the point where the graph crosses the y-axis. Note that a line will only have one x-intercept and one y-intercept and that the coefficients of x and y help determine the value of the slope and the intercepts.

To find the x-intercept, substitute 0 for y and solve for x.
To find the y-intercept, substitute 0 for x and solve for y.

Example 7.4:
Find the y-intercept of the line $2x - 3y = 9$
    Substitute $x = 0$ and solve for y
    $2(0) - 3y = 9$
    $-3y = 9$
    $y = \dfrac{9}{-3} = -3$
    Therefore the y-intercept occurs at $(0, -3)$

Example 7.5:

Find the x-intercept of the line $2y = 4x + 18$

Substitute $y = 0$ and solve for x

$2(0) = 4x + 18$

$0 = 4x + 18$

$-4x = 18$

$x = \dfrac{18}{-4} = -\dfrac{9}{2}$

Therefore the x − intercept occurs at $\left(-\dfrac{9}{2}, 0\right)$

Graphs of Linear Equations

One of the easiest ways to graph a linear equation is by first finding the x and y intercepts. Once the intercepts have been found, plot and connect the two points to graph the equation.

Example 7.6:

Graph the line $2x - y = -1$

Find the y-intercept by substituting $x = 0$

$2(0) - y = -1$

$-y = -1$

$y = 1$

$(0, 1)$

Find the x-intercept by substituting $y = 0$

$2x - (0) = -1$

$2x = -1$

$x = -\dfrac{1}{2}$

$\left(-\dfrac{1}{2}, 0\right)$

Plot and connect the intercepts to form the graph of the equation above.

Another method of graphing the equation is by using the y-intercept and the slope.
The slope of the graph determines the slant or steepness of the line. The slope therefore is the ratio of the distance moved vertically and the distance moved horizontally.

$$\text{slope} = \dfrac{\text{vertical distance moved}}{\text{horizontal distance moved}} = \dfrac{\text{rise}}{\text{run}}$$

A line that has a positive slope will rise from left to right while a line with a negative slope will fall from left to right.

Positive slope ↗   Negative slope ↘

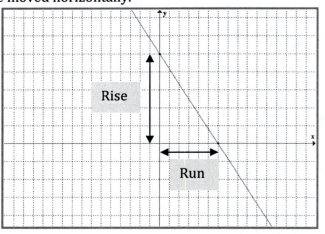

The distances are directional and so rise will be negative if you move downwards and positive if you move upwards. Likewise, run will be negative if you move to the left and positive if you move to the right.

To graph using this method, solve the equation for y so that the equation is arranged in the form:
$y = mx + b$ where m is the slope and b is the y-intercept.

Plot the y-intercept and use the slope to find another point on the line then connect the points to form the graph.

Example 7.7:
Graph the equation: $2y - 3x = 8$
    First solve the equation for y
    $2y - 3x = 8$
    $2y = 3x + 8$
    $y = \frac{3}{2}x + 4$

    From this equation, we see that the graph intersects the y-axis at the point $(0, 4)$ and that the slope is $m = \frac{3}{2}$.

    On the coordinate plane, plot the point representing the y-intercept.

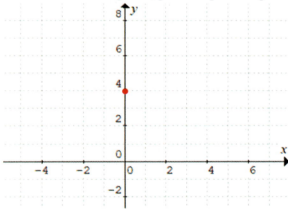

    Beginning at the plotted point, use the slope to arrive at another point on the line.

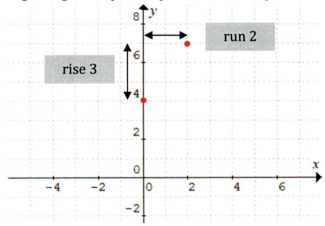

Notice the slope is positive so both rise and run are in the positive direction. Connect the two points to draw the graph of the line.

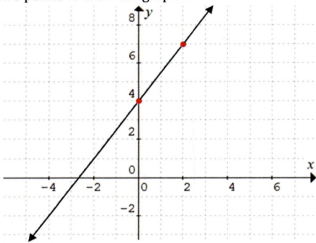

Example 7.8:

Graph the line: $-3x - 4y = 16$

First solve the equation for y

$-3x - 4y = 16$

$-4y = 3x + 16$

$y = -\frac{3}{4}x - 4$

From this equation, we see that the graph intersects the y–axis at the point $(0, -4)$ and that the slope is $m = -\frac{3}{4}$

On the coordinate plane, plot the point representing the y–intercept. Then, beginning at the plotted point, use the slope to arrive at another point on the line and connect the two points to draw the graph of the line.

Since the slope is negative, assign the negative sign to either rise or run but not to both. The resulting graph will be same so it does not matter which you assign the negative sign.

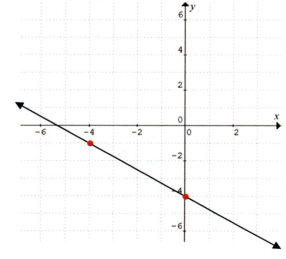

$$slope = \frac{rise}{run} = \frac{3}{-4}$$

Notice that since we assigned the negative sign to run, we moved to the left (towards the negative direction).

Alternative method of graphing

You can always create a table of values by choosing random values of x and calculating the corresponding values of y using the equation. Once the table of values is created, plot and connect the points to form the graph of the line. This method is more involved than simply using the x and y intercepts and should only be used as a last resort if you forget how to find the intercepts.

Example 7.9:
Graph the line $y = 2x - 3$

Create a table of values, plot the points and connect to form the graph.

x	y
-2	-7
-1	-5
0	-3
1	-1
2	1

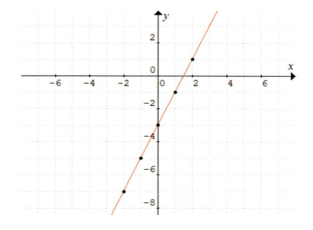

Graphs of Horizontal and Vertical Lines

A horizontal line has slope $m = 0$ and takes the form $y = k$ where k is any constant. A graph of this form is parallel to the x-axis and passes through k on the y-axis.

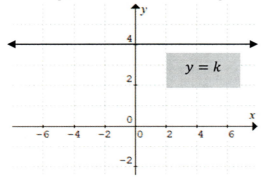

Example 7.10:
Graph the line: $y = -3$

This line will be parallel to the x-axis and pass through the y-axis at the point $(0, -3)$.

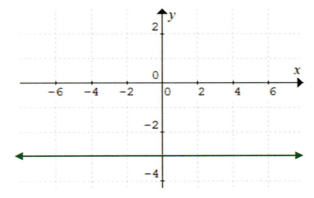

A vertical line has slope $m = $ undefined and takes the form $x = k$ where k is any constant. A graph of this form is parallel to the y-axis and passes through k on the x-axis.

Example 7.11:
Graph the line: $x = 5$

This line will be parallel to the y-axis and pass through the x-axis at the point $(5, 0)$.

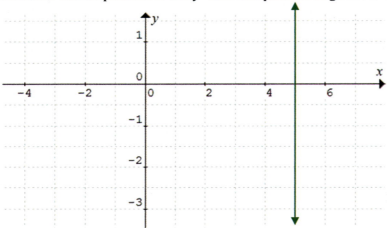

Example 7.12:

Graph the line parallel to the x-axis and passing through the point $(6, -2)$

A line parallel to the x-axis will take the form $y = k$. The value of k will be the y-coordinate of the given point. Therefore the equation of the line is $y = -2$.

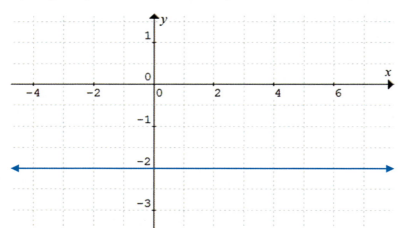

Parallel Lines

Parallel lines have the same slope. Given a linear equation and asked to find the equation of a line parallel to the first and passing through the point (x, y), write the first equation in slope–intercept form so that the slope is easily identified. This will also be the slope of the parallel line. Using the slope and the given point, find the equation of the new line by substituting into the point–slope formula.

Chapter 7 – Linear Equations of Two Variables **85**

Example 7.13:

Find the equation of a line that is parallel to the line $4x + 2y = 11$ and passing through the point $(5, -1)$.

$$4x + 2y = 11$$

$$2y = -4x + 11$$

$$y = \frac{-4}{2}x + \frac{11}{2} = -2x + \frac{11}{2}$$

The lines are parallel so the slope of the new line is also –2.
Using $m = -2$ and the point $(5, -1)$:
$$y - y_1 = m(x - x_1)$$
$$y - (-1) = -2(x - 5)$$
$$y + 1 = -2x + 10$$
$$y = -2x + 9$$

Perpendicular Lines

The process is the same as above except that the slope is the negative reciprocal of the given slope.

Example 7.14:

Find the equation of a line that is perpendicular to the line $x + 2y = 6$ and passing through the point $(1, 4)$.

$$x + 2y = 6$$
$$2y = -x + 6$$
$$y = \frac{-1}{2}x + \frac{6}{2}$$

$$y = -\frac{1}{2}x + 3$$

Since the slope of this line is $-\frac{1}{2}$, the slope of the perpendicular line will be the negative reciprocal 2.
Using $m = 2$ and the point $(1, 4)$:
$$y - y_1 = m(x - x_1)$$
$$y - 4 = 2(x - 1)$$
$$y - 4 = 2x - 2$$
$$y = 2x + 2$$

Distance and Midpoint

Given any two points (x_1, y_1) and (x_2, y_2), the **distance** between them can be found using the distance formula:
$$d = \sqrt{(x_2 - x_1)^2 + (y_2 - y_1)^2}$$

This formula is particularly useful for finding the length of a line in the coordinate plane.

The **midpoint** between two points on a line can be found using the formula:

$$M = \left(\frac{x_1 + x_2}{2}, \frac{y_1 + y_2}{2}\right)$$

Example 7.15:
Find the length of the line with endpoints $(5, 9)$ and $(2, 7)$.

$$\begin{aligned} d &= \sqrt{(2-5)^2 + (7-9)^2} \\ &= \sqrt{(-3)^2 + (-2)^2} \\ &= \sqrt{9 + 4} = \sqrt{13} \end{aligned}$$

Determine Equation of Line Given Graph

Given the graph of a linear equation, we can determine the equation of the line using a couple methods.
- Choose any two points on the line, find the slope and then the equation using the point-slope formula $y - y_1 = m(x - x_1)$.
- Read off the y-intercept b, and then use any other point on the line to determine the slope m. Choose the x-intercept if it is a whole number. Substitute both m and b in the slope-intercept form $y = mx + b$.

Example 7.16:
Which of the following is the equation of the graph to the right?

A. $10y = 3x - 50$ C. $-3y = 10x + 5$

B. $10x = 3y - 15$ D. $3y = 10x - 15$

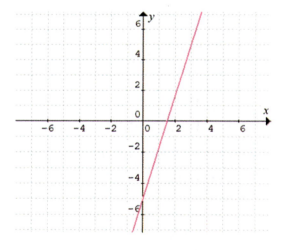

In this case, the y-intercept is a whole number but the x-intercept is not. We use the y-intercept and choose another point that is easy to work with.

Starting at the y-intercept $(0, -5)$ and ending at the point $(3, 5)$:

$$m = \frac{rise}{run} = \frac{10}{3}$$

$$y = mx + b$$

$$y = \frac{10}{3}x - 5$$

$$3\left(y = \frac{10}{3}x - 5\right)$$

$$3y = 10x - 15$$

Therefore, the answer is D.

Determine Equation of Line Given Table of Values

Matching a table of coordinates to the linear equation is the same procedure as in the previous section. Identify the x and y intercepts if given (x–intercept will have a y–coordinate of 0 and the y–intercept will have an x–coordinate of 0).

Example 7.17:

x	-2	-1	0	1	2
y	-5	0.5	4	8.5	13

Which of the following is the equation of the line represented by the table of coordinates given above?

A. $y + x = -7$

B. $3x - 2y = 6$

C. $2y - 9x = 8$

D. $y - 3x = 1$

Some students attempt to use points and substitute. If their chosen point is not a solution to the equation, they eliminate that equation. The problem is that this method is time consuming and can be frustrating if your chosen point works in more than one equation (for example, the first point $(-2, -5)$ is actually a solution to all four equations!).

Choose any two points.

(x_1, y_1) (x_2, y_2)
$(0, 4)$ $(2, 13)$

Find the slope

$$m = \frac{13 - 4}{2 - 0} = \frac{9}{2}$$

Note that we were given the y–intercept $(0, 4)$ so we know that $b = 4$

$y = mx + b$

$y = \frac{9}{2}x + 4$

$2\left(y = \frac{9}{2}x + 4\right)$

$2y = 9x + 8$

$2y - 9x = 8$

Therefore, the answer is C.

Applications

Here we look at word problems that can be modeled using linear equations. In such applications, the initial or starting value of the dependent variable becomes the y–intercept. The ratio of the dependent variable to the independent variable (usually indicated by key words such as "per" as in cost per year) becomes the slope. The independent variable x is usually a quantitative variable that can be measured on a numeric scale such as time, height, weight or test scores.

Example 7.18:

The monthly cost to operate a large printer is \$3, 200 while the printing cost per book is \$1.40. What is the total cost during a month where 53 books are printed?

The initial or base cost of operation is \$3, 200 so this is the y–intercept. From this and the wording of the problem, we can tell that monthly cost is plotted on the y–axis and the number of books printed is plotted on the x–axis. The cost per book (the change in y per the change in x) is the slope of the line.

Therefore the problem can be modeled by the linear equation:

$$y = 1.4x + 3200 \quad or \quad f(x) = 1.4x + 3200$$

We are asked to find the cost when 53 books are printed i.e. the value of y when $x = 53$

$$y = f(53) = 1.4(53) + 3200$$
$$= 74.20 + 3200$$
$$= \$3,274.20$$

Chapter 7 – Linear Equations of Two Variables   **89**

Chapter 7 Exercises

1. Find the equation of the line connecting the points $(-2, 0)$ and $(2, 3)$.

2. Find the equation of the line perpendicular to the line $3x + 2y = 6$ and passing through the point $(9, -4)$.

3. Find the equation of the line parallel to the line $y = 3 - 2x$ passing through the point $(5, -7)$.

4. Find the midpoint of the segment of a line with endpoints $(7, -11)$ and $(3, -5)$.

5. Find the length of the line with endpoints $(4, 1)$ and $(7, 5)$.

6. Find the perimeter of the triangle formed by the $x - axis$, the $y - axis$ and the points $(-5, 0)$ and $(0, 12)$.

7. Sketch the graph of $3x + 2y = 18$.

8. The weekly charge for renting a car includes a standard rental charge of \$27.00 per day plus a charge of \$0.22 per mile. How much does it cost to rent a car for a week if it is to be driven 52 miles?
    A. \$36.04   B. \$38.44   C. \$146.44   D. \$200.44

9. Which of the following does not lie on the line $7x + 3y = 4$?
    A. $(-2, 6)$ B. $(-1, 3)$ C. $(1, -1)$ D. $(4, -8)$

10. A water and sewage company charges a base fee of \$13.12 per month as well as \$0.43 per every 50 gallons used. How much water did the Thompson family use last month if their total bill was \$19.57?
    A. 15   B. 50   C. 225   D. 750

11. Which of the following most accurately describes the graph of the line $3x - y = 4$?
    A. The y–intercept lies below the x–axis and the line rises as the value of x increases
    B. The y–intercept lies above the x–axis and the line rises as the value of x increases
    C. The y–intercept lies below the x–axis and the line falls as the value of x increases
    D. The y–intercept lies above the x–axis and the line falls as the value of x increases

Solutions

1. Use the given points to find the slope
    $$m = \frac{y_2 - y_1}{x_2 - x_1} = \frac{3 - 0}{2 - (-2)} = \frac{3}{4}$$

Use the slope and one of the points in the point–slope formula

$y - y_1 = m(x - x_1)$

$y - 0 = \dfrac{3}{4}(x - (-2))$

$y = \dfrac{3}{4}x + \dfrac{3}{2}$

$4y = 3x + 6$

2. The slopes of perpendicular lines are the negative reciprocal of each other so we solve for y in the given line to find the slope.

$3x + 2y = 6$

$2y = -3x + 6$

$y = -\dfrac{3}{2}x + 3$

Slope $m_1 = -\dfrac{3}{2}$

Take the negative reciprocal of the slope

$m_2 = \dfrac{2}{3}$

Use the slope and the given point $(9, -4)$ in the point–slope formula

$y - y_1 = m(x - x_1)$

$y - (-4) = \dfrac{2}{3}(x - 9)$

$y + 4 = \dfrac{2}{3}x - 6$

$y = \dfrac{2}{3}x - 10$

$3y = 2x - 30$

The equation of the perpendicular line in standard form is $2x - 3y = 30$

3. The slope of the parallel line will be the same as the original line so first find the slope of that line.

$y = 3 - 2x$

$y = -2x + 3$

Therefore, the slope $m_2 = m_1 = -2$

Use the slope and the given point $(5, -7)$ in the point–slope formula

$y - y_1 = m(x - x_1)$

$y - (-7) = -2(x - 5)$

$y + 7 = -2x + 10$

$y = -2x + 3$

The equation of the parallel line in standard form is $2x + y = 3$

Chapter 7 – Linear Equations of Two Variables   91

4. Apply the midpoint formula
$$M = \left(\frac{x_1 + x_2}{2}, \frac{y_1 + y_2}{2}\right) = \left(\frac{7 + 3}{2}, \frac{-11 + (-5)}{2}\right) = \left(\frac{10}{2}, -\frac{16}{2}\right) = (5, -8)$$

5. To find the length of the line, apply the distance formula
$$d = \sqrt{(x_2 - x_1)^2 + (y_2 - y_1)^2} = \sqrt{(7-4)^2 + (5-1)^2} = \sqrt{3^2 + 4^2} = \sqrt{9 + 16} = \sqrt{25} = 5$$

6. First draw a sketch of the problem. Add and connect the points $(-5, 0)$ and $(0, 12)$ on a coordinate plane, then form the triangle with the x and y axes.

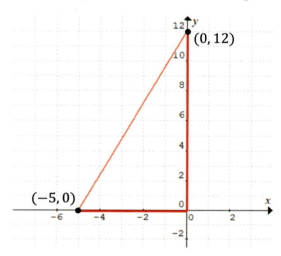

We need the lengths of the 3 sides to find the perimeter of the triangle. We can tell that the height is 12 units and the base is 5 units. Use the distance formula to find the length of the third side.
$$d = \sqrt{(0 - (-5))^2 + (12 - 0)^2} = \sqrt{5^2 + 12^2} = \sqrt{25 + 144} = \sqrt{169} = 13$$

The perimeter of a triangle is the sum of the lengths of its sides.
Perimeter $= 5 + 12 + 13 = 30 \; units$

7. We will sketch the graph using the x and y intercepts. Use any method you choose.

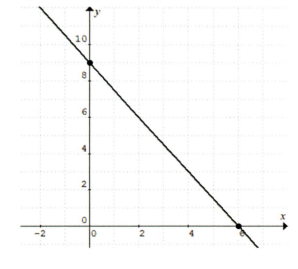

y-intercept

$3(0) + 2y = 18$

$2y = 18$

$y = \dfrac{18}{2} = 9$

$(0, 9)$

x-intercept

$3x + 2(0) = 18$

$3x = 18$

$x = \dfrac{18}{3} = 6$

$(6, 0)$

Plot the points and connect.

92 Prep for Success: Florida's PERT Math Study Guide

8. **Answer: [D]** The standard or base rental charge will be the initial charge or the y–intercept of the equation. The cost per mile is the ratio of the rate of change of cost (y) to the rate of change of miles (x). This is therefore the slope of the equation.

Note that the question references the charge per week but the base charge was given per day. So first find the base weekly charge.

$base\ weekly\ charge = daily\ charge \times 7\ days$
$$= \$27 \times 7 = \$189.00$$

Let number of miles driven $= x$

The formula for total weekly charge can therefore be modeled by:

$y = f(x) = 0.22x + 189$

To find the cost of driving 52 miles, substitute for $x = 52$

$y = f(52) = 0.22(52) + 189 = 11.44 + 189 = \$200.44$

9. **Answer: [B]** For this question we simply plug and chug the answers into the equation to determine which makes the equation false.

$(-2, 6)$	$(-1, 3)$
$7(-2) + 3(6) = 4$	$7(-1) + 3(3) = 4$
$-14 + 18 = 4$	$-7 + 9 = 4$
$4 = 4 \quad true$	$2 = 4 \quad false$

Therefore $(-1, 3)$ does not lie on the line $7x + 3y = 4$

10. **Answer: [D]** The amount paid depends on the amount of water used so the dependent variable y is the amount paid and the independent variable x is the amount of water used. In this case, we are given the total bill so $y = 19.57$ and we are asked to find the number of gallons of water. The problem can be modeled by the equation $y = 0.43x + 13.12$ where x is the number of 50 gallon units used.

When $y = \$19.57$

$19.57 = 0.43x + 13.12$

$0.43x = 19.57 - 13.12$

$0.43x = 6.45$

$x = \dfrac{6.45}{0.43} = \dfrac{645}{43} = 15$

Recall that x is the number of 50 gallon units used.

Therefore:

Number of gallons $= 15 \times 50 = 750\ gallons$

11. **Answer: [A]** Solving the equation for y:

$3x - y = 4$

$-y = -3x + 4$

$y = 3x - 4$

The y–intercept is -4 and the slope if 3. Since the y–intercept is negative, it lies below the x–axis. Since the slope is positive, y increases as x increases and the line rises. Therefore, the answer is A.

Chapter 8 — Inequalities

The last few chapters covered statements of equality. Here we take a look at inequalities in which the expressions are not equal. We also cover how inequalities can be applied to real world situations.

Linear Inequalities

A linear inequality is similar to a linear equation except that the solution to an inequality involves an interval of numbers rather than a single solution. The symbols used in equalities are:

>	Greater than
≥	Greater than or equal to
<	Less than
≤	Less than or equal to
○ ( )	Not including
● [ ]	Including

Solving Linear Inequalities

The procedure of solving inequalities is exactly the same as solving linear equations except for one rule: If you multiply or divide by a negative number, reverse the direction of the inequality sign.

Solutions to an inequality can be represented in one of three ways:

1. **Interval Notation**
   Example: $(-\infty, 2) \cup (2, \infty)$

2. **Set Notation**
   Example: $\{x \mid x \geq 7\}$

3. **Line Graph Notation**
   Example:

Example 8.1:

$3x + 10 \leq 5x - 2$

Move all terms with variables to the left-hand side of the equation and all constants to the right-hand side.

$3x \leq 5x - 2 - 10$

$3x \leq 5x - 12$

$3x - 5x \leq -12$

$-2x \leq -12$

Divide both sides by -2. Reverse the direction of the inequality sign because of division by a negative number.

$x \geq \dfrac{-12}{-2}$

$x \geq 6$

Interval Notation: $[6, \infty)$
Set Notation: $\{x | x \geq 6\}$
Line Graph Notation

Example 8.2:

$\dfrac{3}{5}x + 7 > \dfrac{1}{5}x - 3$

$\dfrac{3}{5}x > \dfrac{1}{5}x - 3 - 7$

$\dfrac{3}{5}x > \dfrac{1}{5}x - 10$

$\dfrac{3}{5}x - \dfrac{1}{5}x > -10$

$\dfrac{2}{5}x > -10$

$\left(\dfrac{5}{2}\right) \cdot \dfrac{2}{5}x > -10 \cdot \left(\dfrac{5}{2}\right)$

$x > -\dfrac{50}{2}$

$x > -25$

Interval Notation: $(-25, \infty)$
Set Notation: $\{x | x > -25\}$
Line Graph Notation

Example 8.3:

Solve the inequality: $2 \leq 3x + 5 < 17$

To solve a relationship involving two inequality signs, solve for the variable in the middle using inverse operations, remembering to perform the same operations on the ends of the inequality.

$2 \leq 3x + 5 < 17$

Subtract 5

$$\begin{aligned} 2 &\leq 3x + 5 < 17 \\ -5 & -5 \phantom{<} -5 \end{aligned}$$

$-3 \leq 3x < 12$

Divide by 3

$\dfrac{-3}{3} \leq \dfrac{3x}{3} < \dfrac{12}{3}$

$-1 \leq x < 4$

This means that values of x between -1 and 4, but not including 4, satisfy the inequality.
Writing this solution using set, interval and line notations:

Set Notation: $\{x|-1 \leq x < 4\}$

Interval Notation: $[-1, 4)$

Line Notation:

Solving Linear Inequalities Graphically

In order to solve a problem involving inequalities graphically, we first learn how to graph an inequality.

Graphs of Linear Inequalities

To graph a linear inequality, follow the same steps to graphing a linear equation.
The graph of the line, known as the **boundary line**, divides the coordinate plane into two **half-planes**. A half-plane is a region containing all the points on one side of a straight line.

If the inequality sign is $\leq$ or $\geq$ then the graph will be a closed half-plane which includes the points on the boundary line. Use a solid line for the boundary line.
If the inequality sign is $<$ or $>$ then the graph will be an open half-plane which does not include the points on the boundary line. Use a dashed line for the boundary line.

Once done graphing the line, choose a test point on the graph and substitute into the inequality. If the test point makes the inequality true, shade that area otherwise shade the area on the opposite side of the line. The origin is often a perfect test point unless the boundary line passes through it.

Example 8.4:

Graph the inequality: $y - 3x > 5$

First graph the line $y - 3x = 5$. The graph will be dashed because of the inequality sign >.

$y - 3x = 5$

$y = 3x + 5$

y-intercept $= 5$ and slope $= \frac{3}{1}$

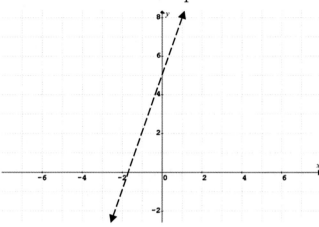

Using the origin as a test point:

$y > 3x + 5$

$(0) > 3(0) + 5$

$0 > 5 \ \ false$

Since the inequality is false at the origin, shade the opposite region that does not contain the origin.

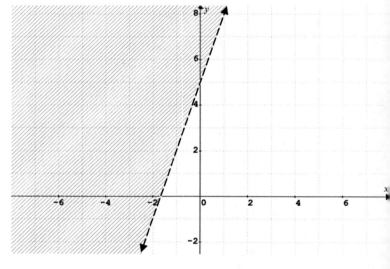

Example 8.5:

Graph the inequality: $y \leq -3x$

First graph the line $y = -3x$. The graph will be a solid line because of the inequality sign $\leq$

y-intercept $= 0$ and slope $= \frac{-3}{1}$

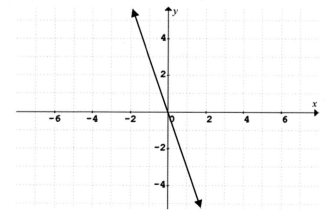

Notice that the graph passes through the origin so choose another test point. We chose $(1, -1)$.

$y \leq -3x$
$(-1) \leq -3(1)$
$-1 \leq -3$

Since the inequality is false at the chosen point, shade the opposite region that does not contain the point $(1, -1)$.

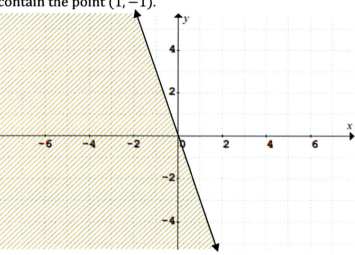

Example 8.6:
Which of the following points is not a part of the solution to the inequality: $2y - 3x \geq -6$?
A. $(-1, 1)$ B. $(-2, 5)$ C. $(2, 3)$ D. $(3, -2)$

We could solve this using algebra by substituting each point into the inequality. The point that makes the inequality false does not belong to the solution. We will however demonstrate how to solve graphically. Solve for y to determine which half-plane to shade.

$2y - 3x \geq -6$
$2y \geq 3x - 6$
$y \geq \frac{3}{2}x - 3$

The inequality sign is $\geq$ so we will shade above the boundary line.

y-intercept
$2y - 3(0) = -6$
$2y = -6$
$\frac{2y}{2} = \frac{-6}{2}$
$y = -3$

x-intercept
$2(0) - 3x = -6$
$-3x = -6$
$\frac{-3x}{-3} = \frac{-6}{-3}$
$x = 2$

Plot and connect the points. Use a solid boundary line and shade above the line. See the graph to the right.

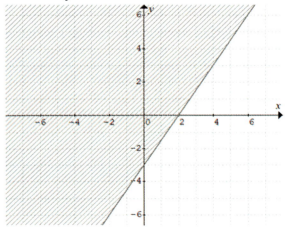

Now just plot the given points to determine which lies outside of the shaded region.

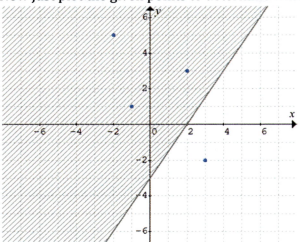

We see that $(3, -2)$ lies outside the shaded region and is therefore not part of the solution.

Example 8.7:

$3y + x < 0$

$y > 2x + 4$

Which of the following graphs represents the solution to the inequalities above?

A.

C.

B.

D.

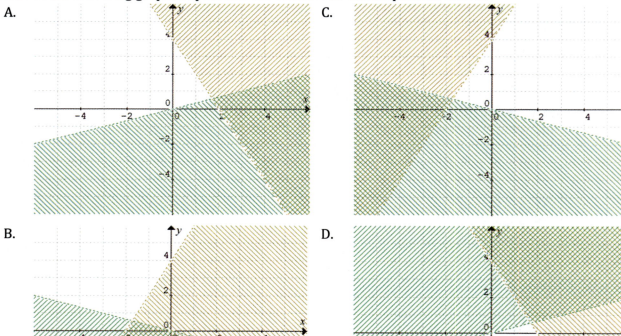

The solution will be the shaded region common to and satisfying both inequalities. Graph the two inequalities on the same coordinate plane.

$3y + x < 0$
$3y < -x$
$y < -\frac{1}{3}x$
The boundary line will be dashed and we shade the half-plane below the line.

$y > 2x + 4$
The boundary line will be dashed and we shade the half-plane above the line.

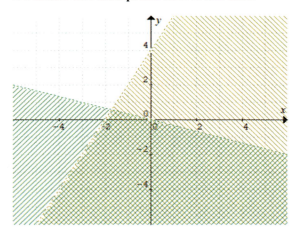

The area shaded both brown and green represents the solution common to both inequalities so the answer is B.

Absolute values and the less than inequality

When the absolute value of an expression is set less than a number, place the expression between the negative and positive of the number then remove the absolute value symbol. The inequality signs point left. Solve for the variable, remembering that whatever operations are done on the middle must also be performed on both the left and right-hand sides of the inequality.

Example 8.8:
$|4x + 2| - 7 < 3$

First add 7 to both sides of the inequality to isolate the absolute value.
$|4x + 2| < 10$

Now remove the absolute value sign and set the expression between -10 and 10 with the original inequality signs separating the 3 terms.
$-10 < 4x + 2 < 10$

Solve for x in the middle using inverse operations, remembering to perform the same operations on the ends of the inequality.

Subtract 2
$-10 < 4x + 2 < 10$
$-2 \qquad -2 \quad -2$
$-12 < 4x < 8$

Divide by 4
$\frac{-12}{4} < \frac{4x}{4} < \frac{8}{4}$

$-3 < x < 2$

Set Notation: $\{x | -3 < x < 2\}$
Interval Notation: $(-3, 2)$
Line Notation:

Absolute values and the greater than inequality

When the absolute value is set greater than a number, the solution involves two separate inequalities. For the first solution, set the expression greater than the positive of the number and for the second solution, set the expression less than the negative of the number.

Example 8.9:
$|5x - 3| + 2 \geq 11$

First subtract 2 from both sides of the inequality to isolate the absolute value.
$|5x - 3| \geq 9$
Now separate into two separate inequalities and solve as follows:

$5x - 3 \geq 9 \qquad\qquad 5x - 3 \leq -9$
$5x \geq 12 \qquad\qquad 5x \leq -6$
$x \geq \dfrac{12}{5} \qquad\qquad x \leq -\dfrac{6}{5}$

Set Notation:
$\left\{x \,\middle|\, x \leq -\dfrac{6}{5} \text{ or } x \geq \dfrac{12}{5}\right\}$

Interval Notation (the symbol ∪ means union or combination):
$\left(-\infty, -\dfrac{6}{5}\right] \cup \left[\dfrac{12}{5}, \infty\right)$

Line Notation:

Word Problems Involving Inequalities

Solving word problems involving inequalities is much like solving word problems involving linear equations. Read the problem carefully and use the key words to determine which of the symbols in the table on page 93 to apply.

Example 8.10:
A factory that manufactures mp3 players sells them to retail stores for $45 apiece. If the factory pays $18,400 per month in operation costs and $13 to produce each mp3 player, at least how many players do they need to produce and sell each month to see a profit?

A. 573   B. 574   C. 575   D. 576

The factory needs to sell enough players to supersede both the operation costs and the cost to manufacture the players. In other words, their *revenue* must be **greater than** their *cost*.

Let the number of players produced and sold $= x$

$Revenue = selling\ price \times number\ of\ players\ sold$

$\qquad = 45x$

$Cost = operation + production$

$\qquad = operation + (cost\ per\ player \times number\ of\ players\ produced)$

$\qquad = 18400 + 13x$

Formulate the inequality and solve for x

Revenue > Cost

$$45x > 18400 + 13x$$
$$\underline{-13x \qquad\qquad -13x}$$
$$32x > 18400$$
$$x > \frac{18400}{32}$$
$$x > 575$$

Since the factory needs to sell more than 575 players, they need to sell a minimum of 576 players to see a profit and the answer is D.

Chapter 8 Exercises

1. Which of the following represents the solution to the inequality $-2 \leq 3x + 4 < 19$?

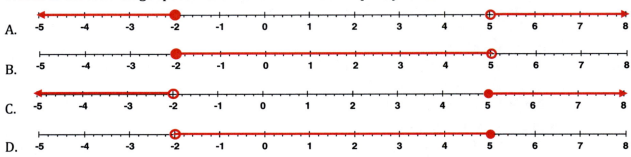

2. Graph the inequaltiy $x \leq \frac{1}{4}y - 2$

3. Graph the solution of the following inequalities:
   $2y < x + 2$
   $y + 2x < 6$

4. If $8 \leq |x| \leq 8$ then $x =$

5. Solve the inequality: $-|2x + 7| \geq -3$

6. Solve the inequality: $|3x - 1| + 5 > 9$

7. Felicia can purchase season tickets to see her favorite basketball team for $320.00 or she can purchase single game admission for $21.00. How many games would she need to attend to justify purchasing season tickets?
   A. 14   B. 15   C. 16   D. 17

8. Customers of Budget N Save Car Rental can rent a car with unlimited mileage for $82.00 per day or they can pay $45.00 per day with the added cost of $0.70 per mile driven. What is the farthest whole number of miles a customer can drive using the second option and pay no more than the cost of the first option?
   A. 50   B. 52   C. 54   D. 56

9. You are offered a sales job that either pays a flat rate salary of $65, 000 per year or a base salary of $22, 000 plus 12% commission on sales. What would you need to earn in sales in order to justify working on commission rather than accepting the flat rate salary?

10. Which of the following inequalities matches the following line graph solution?

   A. $3x - 2 > 4$ B. $x + 3 > 1$ C. $5 - 4x \geq 13$ D. $1 - 2x \leq 5$

Chapter 8 – Inequalities   103

Solutions

1. **Answer: [B]**   Solve for the variable in the middle using inverse operations.
   $$-2 \leq 3x + 4 < 19$$
   $$-4 \quad\quad -4 \quad -4$$

   $$-6 \leq 3x < 15$$
   $$-\frac{6}{3} \leq x < \frac{15}{3}$$
   $$-2 \leq x < 5$$

   Therefore the solution includes all the values between $[-2, 5)$ which includes -2 (closed circle) but not 5 (open circle). Therefore, the answer is B.

2. Graph the line: $x = \frac{1}{4}y - 2$

   $y - intercept$ $x - intercept$
   $0 = \frac{1}{4}y - 2$ $x = \frac{1}{4}(0) - 2$
   $\frac{1}{4}y = 2$ $x = 0 - 2$
   $y = 2(4) = 8$ $x = -2$
   $(0, 8)$ $(-2, 0)$

   Connect and plot the points to form the solid boundary line. Solve the inequality for y to determine which half-plane to shade.

   $x \leq \frac{1}{4}y - 2$
   $-\frac{1}{4}y \leq -x - 2$
   $y \geq 4x + 8$

   Therefore, we shade above the boundary line.
   We could have also used the origin as a test point
   $0 \leq \frac{1}{4}(0) - 2$
   $0 \leq -2$
   This statement is false so shade the other half-plane.

   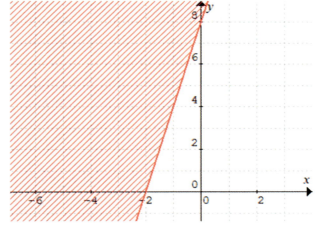

3. Both lines will be dashed (use this to eliminate answers with solid lines on the test).

   Graph the line $2y = x + 2$ Graph the line $y + 2x = 6$
   $y - intercept$ $x - intercept$ $y - intercept$ $x - intercept$
   $2y = 0 + 2$ $2(0) = x + 2$ $y + 2(0) = 6$ $0 + 2x = 6$
   $y = \frac{2}{2} = 1$ $x = -2$ $y = 6$ $x = 3$
   $(0, 1)$ $(-2, 0)$ $(0, 6)$ $(3, 0)$

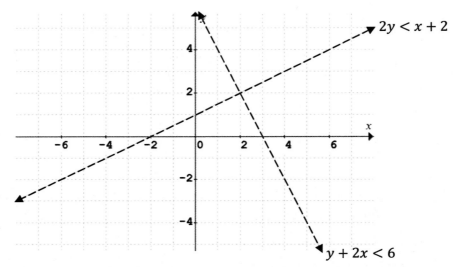

Solve both inequalities for y to determine which areas to shade.

$2y < x + 2$ $\qquad$ $y + 2x < 6$

$y < \dfrac{1}{2}x + 2$ $\qquad$ $y < -2x + 6$

Therefore, we shade the area under both lines. The solution is the shaded area common to both lines.

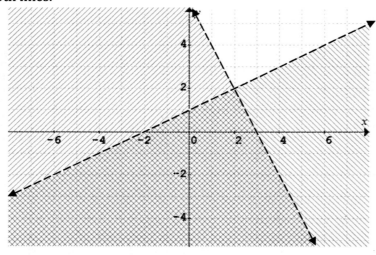

4. The only way a number could be both less than or equal to 8 and greater than or equal to 8 is if the number is equal to 8. Therefore, $x = 8$.

5. Divide by –1 and reverse the direction of the inequality sign.
$|2x + 7| \leq 3$
Remove the absolute value sign and set the expression between -3 and 3 with the less than or equal to inequality sign separating the 3 terms.
$-3 \leq 2x + 7 \leq 3$
$-7 \qquad -7 \ -7$

$-10 \leq 2x \leq -4$

$$\frac{-10}{2} \le x \le \frac{-4}{2}$$

$$-5 \le x \le -2$$

$$[-5, -2]$$

6. Remove the absolute value sign by separating the problem into two separate inequalities.

    $|3x - 1| + 5 > 9$

    $|3x - 1| > 4$

$3x - 1 > 4$	$3x - 1 < -4$
$3x > 5$	$3x < -3$
$x > \dfrac{5}{3}$	$x < -1$

    $(-\infty, -1) \cup \left(\dfrac{5}{3}, \infty\right)$

7. **Answer: [C]**   Let the number of games $= n$

    $Cost\ of\ n\ games = 21 \times n = 21n$

    To justify the cost of season tickets, she would need to be attending more single games than $320 would buy. Therefore:

    $Cost\ of\ individual\ tickets\ >\ Cost\ of\ season\ tickets$

    $21n > 320$

    $n > \dfrac{320}{21}$

    $n > 15\dfrac{5}{21}$

    Therefore, if she plans on attending at least 16 or more games, she should purchase season tickets.

8. **Answer: [B]**   We need to find the number of miles driven that would make option 2 less than or equal to option 1.

    Let the number miles $= m$

	Option 1	Option 2
Total Cost	82	$45 + 0.70m$

    Formulate an inequality and solve for m

    $45 + 0.70m \le 82$

    $0.70m \le 37$

    $m \le \dfrac{37}{0.70}$

    $m \le \dfrac{370}{7}$

    $m \le 52\dfrac{6}{7}$

    The customer can therefore drive 52 miles or less.

9. The base salary plus commission would need to exceed (be greater than) the flat rate annual salary.

Let total amount made from sales $= x$

Commission-based Annual Salary > Flat Rate Annual Salary

$22,000 + (12\% \text{ of } x) > 65,000$

Formulate an inequality and solve for x

$22000 + 0.12x > 65000$

$0.12x > 65000 - 22000$

$0.12x > 43000$

$x > \dfrac{43000}{0.12}$

$x > \dfrac{4300000}{12}$

$x > \$358,333.33$

You would need to sell more than $\$358,333.33$ in order to earn more on the commission–based salary rather than the flat rate salary.

10. **Answer: [D]**   Immediately eliminate answer choices A and B because the line graph contains a closed circle indicating that the inequality sign will either be $\le$ or $\ge$.

Now just solve the remaining two choices algebraically. Remember to reverse the inequality sign when dividing or multiplying by a negative.

$$5 - 4x \ge 13 \qquad\qquad\qquad 1 - 2x \le 5$$
$$-5 \qquad\quad -5 \qquad\qquad -1 \qquad -1$$

$$-4x \ge 8 \qquad\qquad\qquad\quad -2x \le 4$$
$$\dfrac{-4x}{-4} \le \dfrac{8}{-4} \qquad\qquad\qquad \dfrac{-2x}{-2} \ge \dfrac{4}{-2}$$
$$x \le -2 \qquad\qquad\qquad\qquad x \ge -2$$

The solution represents the interval $x \ge -2$ so the answer is D.

Chapter 9 – Terms, Factors & Expressions **107**

Chapter 9

Terms, Factors & Expressions

Before the next chapter on polynomials, we review some other necessary algebraic concepts such as the laws of exponents and basic factoring.

Factors

Coefficients, Variables, Terms & Factors

A **term** can be described as a number or the product of numbers and variables. A number by itself is often called a **constant**. A **variable**, expressed as a letter, represents a number that can assume a range of values. When a term consists of both numbers and variables, the numbers are called the **coefficients** of the variable. A **factor** is any number (or variable) that divides into a term without leaving a remainder. Some examples of terms, coefficients, variables and factors are given in the table below.

Term	Coefficient	Variable(s)	Some Possible Factors
$15ab$	15	a and b	$3, 5, 15, a, b, 3a$
$7xy$	7	x and y	$7, x, y, 7x$
42	This term is constant		$2, 6, 7$
$3m^2n$	3	m and n	$3, m, m^2, 3n$

Note that m is a factor of m^2 because $m^2 = m \times m$

An **expression** is the combination of one or more terms. We name an expression by the number of terms it contains.

> Monomial – one term
> Binomial – two terms
> Trinomial – three terms
> Quadrinomial – four terms

Polynomial is the general name given to an expression with two or more terms.

Expressions and Equations

Solving Literal Equations

Part of the exam involves the ability to rearrange formulas and equations to isolate some given variable. The trick to getting the desired variable by itself is to eliminate all the other variables around it by performing their opposite operation. See the examples on the next page.

108   Prep for Success: Florida's PERT Math Study Guide

Example 9.1:

Solve $F = \dfrac{GmM}{r^2}$ for M

The right side is divided by r^2 so eliminate by multiplying both sides b r^2

$$Fr^2 = GmM$$

Next divide both sides by G and m to isolate M.

$$\frac{Fr^2}{Gm} = \frac{GmM}{Gm}$$

$$M = \frac{Fr^2}{Gm}$$

Example 9.2:

Solve $w + 2(y + x) = 5z$ for y

Subtract w from both sides
$$2(y + x) = 5z - w$$

Divide both sides by 2
$$y + x = \frac{5z - w}{2}$$

Subtract x from both sides
$$y = \frac{5z - w}{2} - x$$

Evaluating Expressions

Given a function or expression, you may be asked to evaluate the function at a provided value. Simply substitute the variable(s) with the given value(s) and simplify.

Example 9.3:

Given $4x^2 + 6x - 2$, find the value of the expression at $x = -1$

$$4(-1)^2 + 6(-1) - 2$$
$$= 4(1) - 6 - 2$$
$$= 4 - 6 - 2 = -4$$

Given a function $f(x)$ you may be asked to find the value of the function at a specific x value. Simply plug in the value of x and solve.

Example 9.4:

Given $f(x) = 2x^3 - 5x + 3$, find $f(2)$

Replace x with the value 2
$$f(2) = 2(2)^3 - 5(2) + 3$$

$$= 2(8) - 10 + 3$$
$$= 16 - 10 + 3$$
$$= 9$$

Example 9.5:

What is the value of the expression $\dfrac{-27a^2b^3}{\sqrt{c}}$ when $a = 2, b = -1$ and $c = 9$?

Substitute the values of the variables into the expression.

$$\frac{-27a^2b^3}{\sqrt{c}} = \frac{-27(2)^2(-1)^3}{\sqrt{9}}$$

$$= \frac{-27(4)(-1)}{3}$$

$$= \frac{108}{3} = 36$$

Order of Operations with Algebraic Terms

We have already covered the order in which arithmetic operations should be performed on numbers but let us look at a few examples using algebraic terms.

Example 9.6:

Simplify: $11m + (5m - 3 + 2m) - 2(4 - 3m)$

If the expression within the parentheses can be simplified, do so first.
$$= 11m + (7m - 3) - 2(4 - 3m)$$
Distribute to remove the parentheses
$$= 11m + 7m - 3 - 8 + 6m$$
Combine like terms
$$= 11m + 7m + 6m - 3 - 8$$
$$= 24m - 12$$

Example 9.7:

Simplify: $15x^2 - 7(2 - 3x) - 2(2x)^2 + 6$

None of the expressions within the parentheses can be simplified so we handle the exponents first.
$$= 15x^2 - 7(2 - 3x) - 2(4x^2) + 6$$
Next, distribute to remove the parentheses
$$= 15x^2 - 14 + 21x - 8x^2 + 6$$
Finally, combine like terms
$$= 15x^2 - 8x^2 + 21x - 14 + 6$$
$$= 7x^2 + 21x - 8$$

Exponential Expressions

A variable has two components: the **base** and the **exponent**. In previous chapters, we mainly dealt with variables of exponent 1. Here we look at variables containing exponents greater than 1 and the laws of exponents that govern them.

Base $\Longrightarrow x^a \Longleftarrow$ Exponent

The laws of exponents are used to combine and simplify variables and are as follows:

1. $(x^a)^b = x^{ab}$
2. $(xy)^a = x^a y^a$
3. $\left(\dfrac{x}{y}\right)^a = \dfrac{x^a}{y^a}$
4. $x^a x^b = x^{a+b}$
5. $\dfrac{x^a}{x^b} = x^{a-b}$
6. $x^{-a} = \dfrac{1}{x^a}$ and $\dfrac{1}{x^{-a}} = x^a$
7. $x^{1/a} = \sqrt[a]{x}$

Note that $x^a + x^b \neq x^{a+b}$

The laws above can be used to reduce the exponential expressions to their simplest forms.

Example 9.8:

$$\frac{(3xy^2)^3 (x^2 y^{-1})^2}{x^2 y^5} =$$

Remember, any variable appearing not to have an exponent, actually has an exponent of 1.

$$\frac{(3x^1 y^2)^3 (x^2 y^{-1})^2}{x^2 y^5}$$

Using law #1

$$\frac{3^3 x^3 y^6 \cdot x^4 y^{-2}}{x^2 y^5}$$

Using law #4 on the numerator

$$\frac{3^3 \cdot x^3 \cdot x^4 \cdot y^6 \cdot y^{-2}}{x^2 y^5} = \frac{27 x^{3+4} \cdot y^{6+(-2)}}{x^2 y^5} = \frac{27 x^7 y^4}{x^2 y^5}$$

Using law #5

$$27 x^{7-2} y^{4-5}$$

$$= 27 x^5 y^{-1}$$

Using law #6

$$\frac{27 x^5}{y}$$

Example 9.9:

$$\left(\frac{y}{x}\right)^4 \left(\frac{zy^2}{x}\right)^{-3} =$$

Using law #6

$$\left(\frac{y}{x}\right)^4 \left(\frac{x}{zy^2}\right)^3$$

Using law #3 and law #1

$$\frac{y^4}{x^4} \cdot \frac{x^3}{z^3 y^6}$$

Using law #5

$$\frac{y^{4-6} \cdot x^{3-4}}{z^3} = \frac{y^{-2} \cdot x^{-1}}{z^3}$$

Using law #6

$$\frac{1}{xy^2 z^3}$$

Solving Exponential Equations

To solve an exponential function, try to make the bases the same on either side of the equation, then set the exponents equal to each other and solve.

Example 9.10:

Solve: $9^{3x} = 27^{x-2}$

The bases 9 and 27 are both powers of 3 so convert the bases to base 3 first.

$$9^{3x} = 27^{x-2}$$
$$(3^2)^{3x} = (3^3)^{x-2}$$

Using law #1, simplify the exponents.

$$3^{6x} = 3^{3x-6}$$

Now that the bases are the same, set the exponents equal to each other and solve.

$$6x = 3x - 6$$
$$3x = -6$$
$$x = -\frac{6}{3} = -2$$

Basic Factoring

On the test, factoring is usually restricted to no more than four terms. In order to factor a polynomial, first determine whether there is a factor that is common to all the terms. Divide each term by this highest common factor. Let us first look at factoring simple binomials.

112   Prep for Success: Florida's PERT Math Study Guide

Example 9.11:

Factor: $36x^2y^3 - 30xy^2$

> List the prime factors of each number and then multiply the factors common to each number to determine the greatest common factor (GCF) of the numbers.
>
> $36 = \mathbf{2} \times 2 \times \mathbf{3} \times 3$
> $30 = \mathbf{2} \times \mathbf{3} \times 5$
>
> There are two common prime factors (in bold): 2 and 3. Therefore, the greatest common factor of the numbers is:
>
> $2 \times 3 = 6$
>
> If a variable is common to all terms in the expression, use the occurrence with the lowest exponent. Therefore in the given expression, x and y^2 are both factors and so the greatest common factor of the expression is $6xy^2$.
>
> Divide each term by $6xy^2$ to determine the other factor of the expression. Review law #5 of exponents on page 110 of this chapter for help with dividing exponential expressions if necessary.
>
> $$\frac{36x^2y^3}{6xy^2} = 6xy$$
>
> $$\frac{-30xy^2}{6xy^2} = -5$$
>
> The factored expression is therefore: $6xy^2(6xy - 5)$

Example 9.12:

Factor: $48ab^2c^4 + 4a^2b^3 - 24b^2c^3$

> $48 = \mathbf{2} \times \mathbf{2} \times 2 \times 2 \times 3$
> $4 = \mathbf{2} \times \mathbf{2}$
> $24 = \mathbf{2} \times \mathbf{2} \times 2 \times 3$
>
> There are two common prime factors: 2 and 2. Therefore the greatest common factor of the numbers is $2 \times 2 = 4$
>
> Only the variable b is common to all terms and the lowest exponent of b is 2.
>
> The greatest common factor of the expression is therefore $4b^2$.
>
> $$\frac{48ab^2c^4}{4b^2} = 12ac^4$$
>
> $$\frac{4a^2b^3}{4b^2} = a^2b$$
>
> $$-\frac{24b^2c^3}{4b^2} = -6c^3$$
>
> The factored expression is therefore: $4b^2(12ac^4 + a^2b - 6c^3)$

Chapter 9 – Terms, Factors & Expressions   **113**

Chapter 9 Exercises

1. Find the value of the expression $12x^2y^4 - 3xy^5 + 9x^3y^2$ when $x = 2$ and $y = -1$

2. If $h(x) = \frac{3}{2}x^2 - 5x + 2$, find $h\left(\frac{1}{3}\right)$

3. A ball dropped from a building follows the equation $h(t) = 2t^2 - 17t + 50$ where t is the time in seconds and h is the height in feet. What is the height after 3 seconds?

4. Solve for x in the equation: $y = b + 3x - 6$

5. $5n + 2n(5 - 7)^3 - 8n \div (4 - 2) =$
   A. $17n$ B. $-13n - 2$ C. $-15n$ D. $19n - 2$

6. Factor: $25a^2b^3c + 15ab^2c^5$

7. Which of the following is not equal to the expression $x^{3/2}$?
   A. $\sqrt{x^3}$ B. $\dfrac{x^2}{\sqrt{x}}$ C. $\sqrt[3]{x^2}$ D. $x^{1/2} \cdot x^{1/2} \cdot x^{1/2}$

8. Factor: $12x^7y^4z^2 - 18x^4yz^3 - 24x^5y^2z^6$

9. Simplify: $(a^2bc^{-3})^{-2}(b^4c^6)^{\frac{1}{2}}$

10. Simplify: $\dfrac{(m^3n)^{-2}(2mn^{-1})^3}{(m^{-4}n^5)^{-1}}$

Solutions

1. Substitute the values of x and y and simplify
   $12x^2y^4 - 3xy^5 + 9x^3y^2$
   $= 12(2)^2(-1)^4 - 3(2)(-1)^5 + 9(2)^3(-1)^2$
   $= (12)(4)(1) - 3(2)(-1) + 9(8)(1)$
   $= 48 + 6 + 72 = 126$

2. Substitute the value of x and simplify
   $$h\left(\frac{1}{3}\right) = \frac{3}{2}\left(\frac{1}{3}\right)^2 - 5\left(\frac{1}{3}\right) + 2$$
   $$= \frac{3}{2}\left(\frac{1}{9}\right) - \frac{5}{3} + \frac{2}{1}$$

$$= \frac{1}{6} - \frac{5}{3} + \frac{2}{1}$$

$$= \frac{1}{6} - \frac{10}{6} + \frac{12}{6}$$

$$= \frac{3}{6} = \frac{1}{2}$$

3. Since we are asked to find the height after 3 seconds, we substitute $t = 3$ into the equation for $h(t)$.

$$h(t) = 2t^2 - 17t + 50$$
$$h(3) = 2(3)^2 - 17(3) + 50$$
$$= 18 - 51 + 50 = 17$$

4. Use inverse operations to rearrange the equation

$$y = b + 3x - 6$$

$$3x = y - b + 6$$

$$x = \frac{y - b + 6}{3}$$

5. **Answer: [C]**   Simplify the expressions within the parentheses
$$= 5n + 2n(-2)^3 - 8n \div 2$$
Simplifying exponents
$$= 5n + 2n(-8) - 8n \div 2$$
Simplify multiplication and division left to right
$$= 5n - 16n - 8n \div 2$$
$$= 5n - 16n - 4n$$
Simplify addition and subtraction from left to right
$$= -11n - 4n$$
$$= -15n$$

6. Determine whether there is a factor common to both terms
$$25 = 5 \times 5$$
$$15 = 3 \times 5$$
The greatest common factor of 25 and 15 is 5. If a variable is common to all terms in the expression, use the occurrence with the lowest exponent. Therefore in the given expression, a, b^2 and c are all factors and so:
$$GCF = 5ab^2c$$
Divide each term by the GCF
$$25a^2b^3c + 15ab^2c^5 = 5ab^2c(5ab + 3c^4)$$

7. **Answer: [C]**   Using law #1 of exponents:
$$x^{3/2} = (x^3)^{1/2} = \left(x^{1/2}\right)^3$$

Chapter 9 – Terms, Factors & Expressions   **115**

We know from law #7 that a fractional exponent can be written as a root so the above can be written as:

$$x^{3/2} = \sqrt{x^3} = (\sqrt{x})^3$$

Therefore choice A is equivalent.

$\left(x^{1/2}\right)^3$ can also be written as $\left(x^{1/2}\right)\left(x^{1/2}\right)\left(x^{1/2}\right)$

Therefore choice D is equivalent.

Choice B can be simplified using law #5 and law #7

$$\frac{x^2}{\sqrt{x}} = \frac{x^2}{x^{1/2}} = x^{2-1/2} = x^{3/2}$$

Therefore choice B is equivalent.

Choice C is therefore the one that is not equivalent.

$$\sqrt[3]{x^2} = (x^2)^{1/3} = x^{2/3} \neq x^{3/2}$$

8. Find the greatest common factor between all three terms

$12 = 2 \times 2 \times 3$

$18 = 2 \times 3 \times 3$

$24 = 2 \times 2 \times 2 \times 3$

The greatest common factor of 12, 18 and 24 is $2 \times 3 = 6$. If a variable is common to all terms in the expression, use the occurrence with the lowest exponent. Therefore in the given expression, x^4, y and z^2 are all factors and so:

$$\text{GCF} = 6x^4yz^2$$

Divide each term by the GCF

$$12x^7y^4z^2 - 18x^4yz^3 - 24x^5y^2z^6 = 6x^4yz^2(2x^3y^3 - 3z - 4xyz^4)$$

9. Using law #1 of exponents

$$a^{2(-2)}b^{1(-2)}c^{-3(-2)}b^{4\left(\frac{1}{2}\right)}c^{6\left(\frac{1}{2}\right)}$$

$$= a^{-4}b^{-2}c^6b^2c^3$$

Using law #4 of exponents

$$= a^{-4}b^{-2+2}c^{6+3}$$

$$= a^{-4}c^9$$

Using law #6 of exponents

$$= \frac{c^9}{a^4}$$

10. Using law #1 of exponents

$$= \frac{m^{-3(2)} \cdot n^{1(-2)} \cdot 2^{1(3)} \cdot m^{1(3)} \cdot n^{-1(3)}}{m^{-4(-1)}n^{5(-1)}}$$

$$= \frac{8m^{-6}n^{-2}m^3n^{-3}}{m^4n^{-5}}$$

Using law #4 of exponents

$$= \frac{8m^{-6+3}n^{-2+(-3)}}{m^4 n^{-5}}$$

$$= \frac{8m^{-3}n^{-5}}{m^4 n^{-5}}$$

Using law #5 of exponents

$$= 8m^{-3-4}n^{-5-(-5)}$$

$$= 8m^{-7}$$

Using law #6 of exponents

$$= \frac{8}{m^7}$$

Chapter 10 – Polynomials **117**

Chapter 10 — Polynomials

Recall that polynomial is the general name given to an expression with two or more terms. This chapter reviews arithmetic operations on polynomials.

Adding and Subtracting Polynomials

Standard Form

A polynomial is considered to be written in standard form if all its terms are written in order of descending degree (power) from left to right.

$y = a_n x^n + a_{n-1} x^{n-1} + a_{n-2} x^{n-2} + \cdots + a_3 x^3 + a_2 x^2 + a_1 x + a_0$ where the coefficients a_i are constants.

After applying arithmetic operations, we normally write the resulting polynomial in standard form.

Like terms

We can only combine **like terms** in polynomials. Like terms have the same variables with the same exponents.

$4xy^2$ and $-5xy^2$ are like terms

$15x^2 y$ and $9xy^2$ are NOT like terms

Pay special attention and only add/subtract like terms.

Example 10.1:

Simplify: $(-4m^2 + 13mn + n^2) - (-9m^2 + 2m - 5n^2)$

Distribute the negative sign between the two expressions.

$= -4m^2 + 13mn + n^2 + 9m^2 - 2m + 5n^2$

Gather like terms together and combine.

$= -4m^2 + 9m^2 + n^2 + 5n^2 + 13mn - 2m$

$= 5m^2 + 6n^2 + 13mn - 2m$

Example 10.2:

Simplify: $2(7x^3 - 4x^2 + x - 3) + (5x^3 - 7x)$

Distribute the 2 and drop the second set of parentheses because we are performing addition.

$= 14x^3 - 8x^2 + 2x - 6 + 5x^3 - 7x$

Gather like terms and combine.

$= 14x^3 + 5x^3 - 8x^2 + 2x - 7x - 6 = 19x^3 - 8x^2 - 5x - 6$

Factoring Polynomials

Recall that a factor is a number that divides into another number perfectly leaving no remainder. The same concept applies to polynomials.

The expression $x^2 - 3x - 28$ can be written as $(x + 4)(x - 7)$ so we say that $x + 4$ and $x - 7$ are factors of the expression.

Factoring Quadratic Expressions

There is a special type of trinomial with the highest exponent being 2. It is called a **quadratic expression** and typically takes the form $ax^2 + bx + c$ where a, b & c are constants.

On the test you may be given a quadratic expression and asked to find its linear factors. Not all quadratic expressions are factorable but we focus on the ones that are in this section. If you have a method of factoring that works for you, feel free to skip this section.

There are two scenarios of which you need to be aware:

When $a = 1$

When $a = 1$ factoring is simple. Simply find two numbers that when multiplied gives c but when added gives b. Remember to pay attention to the sign of c. It is often helpful to list all the possible combinations of negative and positive.

Example 10.3:
Factor: $x^2 + 5x - 6$

There are four ways to obtain -6.
-1×6
-6×1
-2×3
-3×2

Determine which of these pairs of factors add to yield 5.
$-1 + 6 = 5$
$-6 + 1 = -5$
$-2 + 3 = 1$
$-3 + 2 = -1$

Therefore, the factors to be used are -1 and 6 and the quadratic expression factors into: $(x - 1)(x + 6)$
We say that $(x - 1)$ and $(x + 6)$ are both linear factors of the quadratic expression.

When $a \neq 1$

Factor the first term ax^2 and the last term c. Choose the combination of factors that yield the middle term as demonstrated in the example on the next page.

Example 10.4:

Factor: $2m^2 - 13m + 15$

$2m^2 = 2m \times m$

There are four ways to obtain +15:

1×15
-3×-5
3×5
-1×-15

Test to find the combination that would yield $-13m$.

$(2m + 1)(m + 15)$ Multiply the ends of the arrows then add the results

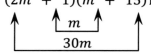

$30m + m \neq -13m$

Switch the position of one of the pairs of factors and try again.

$(2m + 15)(m + 1)$ Multiply the ends of the arrows then add the results

$15m + 2m \neq -13m$

Select another pair of factors and try again.

$(2m - 3)(m - 5)$ Multiply the ends of the arrows then add the results

$-3m - 10m = -13m$

This pair of factors works!

$2m^2 - 13m + 15 = (2m - 3)(m - 5)$

This method works well when there are only a few possible combinations to test. It becomes really time consuming when the first and last terms have many possible factors. The following alternative method is faster than trial and error and involves factoring by grouping.

1. Multiply a and c.
2. List the possible pairs of factors for the result of step 1.
3. Choose the pair that adds to give b.
4. Rewrite the expression, substituting the pair of factors for b.
5. Factor the expression by grouping.

For demonstration purposes, we are going to factor the same expression as in the above example.

Example 10.5:

Factor: $2m^2 - 13m + 15$

Step 1:

$15 \times 2 = 30$

Step 2:

There are eight ways to obtain +30:

1×30 3×10

-1×-30 -3×-10

2×15 5×6

-2×-15 -5×-6

Step 3:

Only one pair will add to yield $-13m$: $-3m$ and $-10m$

Step 4:

Rewrite $-13m$ as $-10m - 3m$ so the new expression is:

$2m^2 - 10m - 3m + 15$

The order of replacement does not matter so you could also write:

$2m^2 - 3m - 10m + 15$

Step 5:

Group the first two terms and the last two terms together. Factor the greatest common factor of each group so that the remaining factor of the first group matches the remaining factor of the second group. Combine the greatest common factor of each group and write the matching factor once.

$2m^2 - 10m - 3m + 15$

$2m(m - 5) - 3(m - 5)$

Combine the GCFs to yield $2m - 3$ and write the other factor $(m - 5)$ once. Therefore, the factored expression is $(2m - 3)(m - 5)$.

Example 10.6:

Factor: $4x^2 + 11x - 3$

Step 1:

$4 \times -3 = -12$

Steps 2 & 3:

Write the factors of -12 and choose the pair that adds to 11

1×-12 2×-6

$\mathbf{-1 \times 12}$ 3×-4

-2×6 -3×4

Step 4:

Rewrite $11x$ as $-1x + 12x$ so the new expression is:

$4x^2 - 1x + 12x - 3$

Step 5:

Factor by grouping

$4x^2 - 1x + 12x - 3$

$x(4x - 1) + 3(4x - 1)$

$(x + 3)(4x - 1)$

Therefore, the factored expression is $(x + 3)(4x - 1)$.

Chapter 10 – Polynomials   **121**

Difference and Sum of Cubes

A difference of cubes is identified by a binomial of the form $a^3 - b^3$. There is an established rule for factoring a difference of cubes and so it is only necessary to identify a and b and substitute into the formula.

$$a^3 - b^3 = (a - b)(a^2 + ab + b^2)$$

There is also a rule for the sum of cubes.

$$a^3 + b^3 = (a + b)(a^2 - ab + b^2)$$

Example 10.7:

Factor: $27x^3 - y^3$

Rewrite the expression so that a and b can be clearly determined.

$$27x^3 - y^3$$
$$3^3 \cdot x^3 - y^3$$
$$(3x)^3 - (y)^3$$

We can clearly see that $a = 3x$ and $b = y$.

Substitute into the formula for difference of cubes.

$$27x^3 - y^3 = (3x - y)[(3x)^2 + (3x)(y) + (y)^2]$$
$$= (3x - y)(9x^2 + 3xy + y^2)$$

Example 10.8:

Factor: $m^3 + \frac{1}{8}n^3$

Again rewrite the expression so that a and b can be clearly determined.

$$m^3 + \frac{1}{8}n^3$$

$$(m)^3 + \left(\frac{1}{2}\right)^3 \cdot n^3$$

$$(m)^3 + \left(\frac{1}{2}n\right)^3$$

Therefore we see that $a = m$ and $b = \frac{1}{2}n$.

Substitute into the formula for the sum of cubes.

$$m^3 + \frac{1}{8}n^3 = \left(m + \frac{1}{2}n\right)\left((m)^2 - (m)\left(\frac{1}{2}n\right) + \left(\frac{1}{2}n\right)^2\right)$$
$$= \left(m + \frac{1}{2}n\right)\left(m^2 - \frac{1}{2}mn + \frac{1}{4}n^2\right)$$

Factoring 4ᵗʰ Degree Polynomials

The PERT will not feature any polynomials that are difficult to factor because the exam tests knowledge of concepts rather than focusing on lengthy computations. Let's look at an example of factoring a simple 4th degree polynomial.

Example 10.9:

Factor $x^4 - 5x^2 + 4$

We can use the same method we used to factor quadratics. Write the factors of 4 and choose the pair that adds to -5.

1×4 -1×-4

2×2 -2×-2

Rewrite the equation

$x^4 - x^2 - 4x^2 + 4$

Factor by grouping

$x^2(x^2 - 1) - 4(x^2 - 1)$

$(x^2 - 4)(x^2 - 1)$

Now notice that we ended up with two sets of **difference of squares** which can be factored further using the difference of squares formula: $a^2 - b^2 = (a - b)(a + b)$

Therefore $x^4 - 5x^2 + 4 = (x - 2)(x + 2)(x - 1)(x + 1)$

Multiplying and Dividing Polynomials

When multiplying or dividing polynomials, there is no need to gather like terms. You will however use the laws of exponents and some factoring.

FOIL

The concept of FOIL is used when multiplying two binomials. The acronym stands for:

First

Outer

Inner

Last

This gives the order in which the terms of the two binomials should be multiplied. The example below demonstrates the concept of FOIL.

Example 10.10:

$(x - 4)(2x + 5) =$

Multiply the First term of each binomial

$(x - 4)(2x + 5)$

$(x)(2x) = 2x^2$

Next multiply the Outer terms

$(x - 4)(2x + 5)$

$(x)(+5) = 5x$

Multiply the Inner terms
$(x - 4)(2x + 5)$

$(-4)(2x) = -8x$

Finally, multiply the Last terms
$(x - 4)(2x + 5)$

$(-4)(+5) = -20$
$(x - 4)(2x + 5) = 2x^2 + 5x - 8x - 20$
Combine like terms
$(x - 4)(2x + 5) = 2x^2 - 3x - 20$

Multiplying Polynomials

To ensure that no terms are overlooked, multiply the number of terms in the first expression by the number of terms in the second expression to determine the number of terms you should have after multiplying but before combining like terms.

Example 10.11:
Simplify: $(x - 3)(x^2 + 4x - 5)$

There are 2 terms in the first expression and 3 terms in the second, so there should be $2 \times 3 = 6$ terms after multiplying.

Multiply the first term of the first expression by all 3 terms of the second expression then multiply the second term of the first expression by all 3 terms of the second expression.

$(x - 3)(x^2 + 4x - 5)$

$x(x^2 + 4x - 5) - 3(x^2 + 4x - 5)$
Remember the law of exponents: $a^x \cdot a^y = a^{x+y}$
$x^3 + 4x^2 - 5x - 3x^2 - 12x + 15$
Check that there are 6 terms before combining like terms
$x^3 + x^2 - 17x + 15$

Example 10.12:
Simplify: $2(x^2 - 4x + 3)(-5x^2 + 3x + 1)$

There are 3 terms in the first expression and 3 terms in the second, so there should be $3 \times 3 = 9$ terms after multiplying.

Follow the arrows showing the order in which the terms should be multiplied.

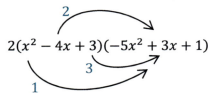

124   Prep for Success: Florida's PERT Math Study Guide

$$2[x^2(-5x^2 + 3x + 1) - 4x(-5x^2 + 3x + 1) + 3(-5x^2 + 3x + 1)]$$
$$2[-5x^4 + 3x^3 + x^2 + 20x^3 - 12x^2 - 4x - 15x^2 + 9x + 3]$$

Check that there are 9 terms within the brackets before combining like terms
$$2[-5x^4 + 3x^3 + 20x^3 + x^2 - 12x^2 - 15x^2 - 4x + 9x + 3]$$
$$2[-5x^4 + 23x^3 - 26x^2 + 5x + 3]$$

$$-10x^4 + 46x^3 - 52x^2 + 10x + 6$$

Dividing Polynomials

The PERT mainly focuses on dividing polynomials by monomials and binomials. To divide by a monomial, place each term of the polynomial over the monomial and simplify using the laws of exponents.

Example 10.13:

Divide $6m^2n^4 - 18m^7n^3 + 27m^3n^8$ by $-3m^2n^3$

Set up the problem like a fraction
$$\frac{6m^2n^4 - 18m^7n^3 + 27m^3n^8}{-3m^2n^3}$$

Place each term of the numerator over the denominator
$$\frac{6m^2n^4}{-3m^2n^3} - \frac{18m^7n^3}{-3m^2n^3} + \frac{27m^3n^8}{-3m^2n^3}$$

Remember that when dividing variables with exponents subtract the exponent in the denominator from the exponent in the numerator to find the new exponent.
$$-2m^{2-2}n^{4-3} + 6m^{7-2}n^{3-3} - 9m^{3-2}n^{8-3} = -2n + 6m^5 - 9mn^5$$

To divide a polynomial by a binomial, we either use factoring or long division.

Division by Factoring

You should always first attempt to factor both the numerator and denominator and then simplify by eliminating factors common to both the numerator and the denominator.

Example 10.14:

Divide: $\dfrac{x^2 + 7x + 10}{x^2 - 25}$

Factor both the numerator and the denominator (difference of squares)
$$\frac{(x + 2)(x + 5)}{(x - 5)(x + 5)}$$

Eliminate factors common to both the numerator and the denominator
$$\frac{x + 2}{x - 5}$$

Chapter 10 – Polynomials **125**

Division by Long Division

Before we demonstrate long division, it might be worth going over the names of the pieces involved.

$$\text{Divisor } \sqrt{\underline{\overset{\text{Quotient}}{\text{Dividend}}}}$$

$$\vdots$$

$$\text{Remainder}$$

Example 10.15:

Divide: $\dfrac{x^2 + 7x + 10}{x^2 - 25}$

Even though both the numerator and denominator can be factored, we do long division to demonstrate the process.

Step 1: Set up the long division problem with the numerator as the dividend and the denominator as the divisor.

$$x^2 - 25\,\sqrt{x^2 + 7x + 10}$$

Step 2: Divide the first term of the dividend by the first term of the divisor to obtain the first term of the quotient.

$$\overset{1}{x^2 - 25\,\sqrt{x^2 + 7x + 10}}$$

Step 3: Multiply the first term of the quotient by the entire divisor then place the result under the dividend and subtract.

$$
\begin{array}{r}
1 \\
x^2 - 25\,\overline{\smash{)}\,x^2 + 7x + 10} \\
\underline{-(x^2 \qquad - 25)} \\
7x + 35
\end{array}
$$

Step 4: The result of subtraction becomes the new dividend and we repeat Steps 2 and 3. The process ends when the degree of the divisor (highest exponent) is greater than the degree of the new dividend which in this case becomes the remainder.

$7x$ cannot be divided by x^2 so the process ends here.

Step 5: Write the final answer as follows:

$$\text{Quotient} + \frac{\text{Remainder}}{\text{Divisor}}$$

$$1 + \frac{7x + 35}{x^2 - 25}$$

Factor and simplify if possible

$$1 + \frac{7(x + 5)}{(x - 5)(x + 5)} = 1 + \frac{7}{x - 5}$$

126 Prep for Success: Florida's PERT Math Study Guide

Note

> The result of long division is algebraically the same as the result if we had factored and simplified except that it may be written differently.

To demonstrate this:

$$1 + \frac{7}{x-5}$$

$$\frac{1}{1} + \frac{7}{x-5}$$

Find the common denominator

$$\frac{(x-5)1}{(x-5)1} + \frac{7}{x-5}$$

$$\frac{x-5+7}{x-5}$$

$$\frac{x+2}{x-5} \quad \text{which is the same result as in Example 10.14}$$

Expanding Polynomials

Earlier in this chapter, the concept of FOIL and multiplying simple polynomials was introduced. Here we cover the expansion of factors containing exponents of 3 or greater.

Example 10.16:

Expand: $(x+2)^3$

$$(x+2)^3 = (x+2)(x+2)(x+2)$$

Multiply the first two factors first and simplify.

$$\underbrace{(x+2)(x+2)}(x+2)$$

$$(x^2 + 2x + 2x + 4)(x+2)$$

$$(x^2 + 4x + 4)(x+2)$$

Now multiply with the last factor.

$$(x^2 + 4x + 4)(x+2)$$

$$x^2(x+2) + 4x(x+2) + 4(x+2)$$
$$x^3 + 2x^2 + 4x^2 + 8x + 4x + 8$$
$$x^3 + 6x^2 + 12x + 8$$

Chapter 10 – Polynomials **127**

Chapter 10 Exercises

1. $(8x^2y - 2xy^2 + x^2) + (-5x^2y + 3xy^2 - y^2) =$

2. $(4m^3 - 5m^2 - 3) - 2(m^3 + 7m + 9) =$

3. Simplify: $(4x - 1)(3x + 7)$
   - A. $12x^2 - 7$
   - B. $12x^2 + 31 - 7$
   - C. $7x - 6$
   - D. $12x^2 + 25x - 7$

4. Factor: $2x^2 + 18x + 40$

5. Factor: $8x^3 - y^3z^6$

6. If $(x + 3)$ is a factor of the expression $x^2 + nx + 24$, what is the value of n?

7. If $x^2 + 9y^2 = -4$ and $2xy = 5$, then $(x + 3y)^2 =$?

8. $(2x - 5\sqrt{3})(2x + 5\sqrt{3}) =$

9. Divide $8x^2 - 18x - 5$ by $2x^2 - 9x + 10$

10. Divide: $\dfrac{16x^7y^4 - 12x^6y^2}{4x^4y^3}$

Solutions

1. Since we are dealing with addition and there is nothing to distribute outside the parentheses, simply remove them. Then group like terms and simplify.
   $8x^2y - 5x^2y - 2xy^2 + 3xy^2 + x^2 - y^2$
   $= 3x^2y + xy^2 + x^2 - y^2$

2. First distribute to remove the parentheses
   $4m^3 - 5m^2 - 3 - 2m^3 - 14m - 18$
   Next group like terms and simplify
   $= 4m^3 - 2m^3 - 5m^2 - 14m - 3 - 18$
   $= 2m^3 - 5m^2 - 14m - 21$

3. **Answer: [D]**   Using FOIL:
   $(4x - 1)(3x + 7)$
   $= (4x)(3x) + (4x)(7) + (-1)(3x) + (-1)(7)$
   $= 12x^2 + 28x - 3x - 7$
   $= 12x^2 + 25x - 7$

4. Since all the coefficients are even, we factor by 2
    $2(x^2 + 9x + 20)$
    Next factor the quadratic expression
    $= 2(x + 4)(x + 5)$

5. Since all terms can be written as a power of 3 this is a difference of cubes
    $(2x)^3 - (yz^2)^3 = (2x - yz^2)((2x)^2 + (2x)(yz^2) + (yz^2)^2)$
    $$= (2x - yz^2)(4x^2 + 2xyz^2 + y^2z^4)$$

6. Remember that the product of the last term in each factor gives the last term of the quadratic expression
    $3 \times 8 = 24$ therefore the other factor is $(x + 8)$
    $(x + 3)(x + 8) = x^2 + 8x + 3x + 24 = x^2 + 11x + 24$
    Comparing this to the given expression we see that $n = 11$

7. The left side of the first equation is not a difference of squares so we cannot use that to factor. Instead we expand the last expression.
    $(x + 3y)^2 = (x + 3y)(x + 3y) = x^2 + 6xy + 9y^2$
    Now we can see that this expression contains $x^2 + 9y^2$ and $2xy$ so we substitute for their values
    Since $x^2 + 9y^2 = -4$ and $2xy = 5$
    $x^2 + 6xy + 9y^2$
    $\quad = x^2 + 9y^2 + 3(2xy)$
    $\quad = -4 + 3(5)$
    $\quad = -4 + 15 = 11$

8. Recognizing that this is a difference of squares
    $\left(2x - 5\sqrt{3}\right)\left(2x + 5\sqrt{3}\right)$
    $\quad = (2x)^2 - (5\sqrt{3})^2$
    $\quad = 4x^2 - 25(3) = 4x^2 - 75$

9. Factor the numerator and the denominator to determine whether they have any factors in common.
    $\dfrac{8x^2 - 18x - 5}{2x^2 - 9x + 10}$
    $= \dfrac{(4x + 1)(2x - 5)}{(x - 2)(2x - 5)} = \dfrac{4x + 1}{x - 2}$

10. Since the divisor is a single term, we can divide each term in the numerator individually.
    $\dfrac{16x^7y^4}{4x^4y^3} - \dfrac{12x^6y^2}{4x^4y^3}$
    $= 4x^3y - \dfrac{3x^2}{y}$

Chapter 11 – Radicals & Exponents   **129**

Chapter 11

Radicals & Exponents

Complex Roots and Exponents

Complex Exponents

Exponents were already introduced in Chapter 9 but we now demonstrate use of the laws of exponents in more complex expressions.

As a reminder, here are the laws of exponents again:

1. $(x^a)^b = x^{ab}$
2. $(xy)^a = x^a y^a$
3. $\left(\dfrac{x}{y}\right)^a = \dfrac{x^a}{y^a}$
4. $x^a x^b = x^{a+b}$
5. $\dfrac{x^a}{x^b} = x^{a-b}$
6. $x^{-a} = \dfrac{1}{x^a}$ and $\dfrac{1}{x^{-a}} = x^a$
7. $x^{1/a} = \sqrt[a]{x}$

Note that $x^a + x^b \neq x^{a+b}$

Now let's look at a few examples of the type of calculations required in this section.

Example 11.1:

Simplify: $3^{-5} \cdot 27^{2/3} \cdot \dfrac{1}{9}^{-1/2}$

> First, attempt to convert all the bases to the same number so that the terms can be combined using the laws of exponents.
>
> $3^{-5} \cdot (3^3)^{2/3} \cdot \left(\dfrac{1}{3^2}\right)^{-1/2}$
>
> Using law #6
> $3^{-5} \cdot (3^3)^{2/3} \cdot (3^{-2})^{-1/2}$
>
> Next use law #1 to simplify the exponents of each term if possible.
> $3^{-5} \cdot 3^{3 \times \frac{2}{3}} \cdot 3^{-2 \times -\frac{1}{2}}$
> $3^{-5} \cdot 3^2 \cdot 3^1$
>
> Use law #4 to combine all three terms.

$$3^{-5+2+1} = 3^{-2}$$

Finally, use law #6 to simplify the answer.

$$\frac{1}{3^2} = \frac{1}{9}$$

Example 11.2:

Simplify: $5^{3/2} + 5^{1/2}$

Since there is no law for the addition or subtraction of exponential expressions, we employ factoring techniques. Check the term with the smaller exponent to determine if it is a factor of the other term (it usually is).

Factor using $5^{1/2}$ as the first factor, dividing each term b $5^{1/2}$ to find the other factor.

$$\frac{5^{3/2}}{5^{1/2}} = 5^{3/2-1/2} = 5^{2/2} = 5^1 = 5$$

$$\frac{5^{1/2}}{5^{1/2}} = 5^{1/2-1/2} = 5^0 = 1 \quad \text{Since anything to the power 0 equals 1}$$

Therefore:

$$5^{3/2} + 5^{1/2} = 5^{1/2}(5 + 1)$$
$$= 5^{1/2}(6) = 6\sqrt{5} \quad \text{Using law #7}$$

Example 11.3:

Simplify: $\left(\dfrac{16}{36}\right)^{-3/2}$

Separate the fractional exponent as follows:

$$\left(\left(\frac{16}{36}\right)^{1/2}\right)^{-3}$$

$$\left(\sqrt{\frac{16}{36}}\right)^{-3}$$

$$\left(\frac{\sqrt{16}}{\sqrt{36}}\right)^{-3} \qquad \text{Take the square root of the numerator and denominator}$$

$$\left(\frac{4}{6}\right)^{-3} \qquad \text{Reduce fraction}$$

$$\left(\frac{2}{3}\right)^{-3}$$

Using law #6

$$\left(\frac{3}{2}\right)^3 = \frac{3^3}{2^3} = \frac{27}{8}$$

Chapter 11 – Radicals & Exponents **131**

You can choose to flip the positions of the exponents around in the first step but it is far easier to take the square root and then the cube of the result than do it the other way around. This is because a cube gives a larger number to square root as opposed to the square root giving a smaller number to cube. Remember that some institutions do not allow the use of calculator and it is important to make calculations easier where possible.

Now let's apply the above technique to algebraic expressions.

Example 11.4:
Simplify the following using only positive exponents:

$$\frac{\left(x^{-\frac{1}{2}}y^3\right)^{-2}}{x^4y^{-5}}$$

A. $\dfrac{x^3}{y^{11}}$ B. $\dfrac{1}{x^3y}$ C. $\dfrac{y}{x^5}$ D. $x^{\frac{3}{2}}y^6$

Apply law #1 of exponents to the numerator

$$\frac{x^{\left(-\frac{1}{2}\right)(-2)}y^{(3)(-2)}}{x^4y^{-5}} = \frac{x^{\frac{2}{2}}y^{-6}}{x^4y^{-5}} = \frac{xy^{-6}}{x^4y^{-5}}$$

Next apply law #5 of exponents

$$\frac{xy^{-6}}{x^4y^{-5}} = x^{1-4}y^{-6-(-5)} = x^{-3}y^{-1}$$

Since we were asked to write the answer using only positive exponents, apply law #6

$$x^{-3}y^{-1} = \frac{1}{x^3y}$$

Example 11.5:
Simplify: $\dfrac{x^{\frac{7}{2}} - x^{\frac{11}{2}}}{x^3}$

We factor to remove the fractional exponent in the numerator first

$$\frac{x^{7/2}}{x^{1/2}} = x^{7/2 - 1/2} = x^{6/2} = x^3 \qquad \frac{x^{11/2}}{x^{1/2}} = x^{11/2 - 1/2} = x^{10/2} = x^5$$

Therefore:

$$\frac{x^{\frac{7}{2}} - x^{\frac{11}{2}}}{x^3} = \frac{x^{\frac{1}{2}}(x^3 - x^5)}{x^3} = \frac{\sqrt{x}(x^3 - x^5)}{x^3}$$

The expression within the parentheses can be factored further.

$$\frac{\sqrt{x}(x^3 - x^5)}{x^3} = \frac{(\sqrt{x})(x^3)(1 - x^2)}{x^3}$$

Eliminate x^3 because it is common to both the numerator and the denominator.

$$\frac{(\sqrt{x})(x^3)(1 - x^2)}{x^3} = \sqrt{x}(1 - x^2)$$

Complex Roots

Before now, we have mostly tackled square roots. Here we introduce simplification of higher roots. The same rules and procedures apply especially that radicals cannot be added or subtracted unless they have the same radicand (number under the root sign).

Example 11.6:

Simplify: $\sqrt[4]{81} \cdot 2\sqrt[4]{32}$

The fourth root indicates that in order to simplify, we should find factors that appear in groups of 4.

$32 = 16 \times 2$ $81 = 27 \times 3$
$16 = 8 \times 2$ $27 = 9 \times 3$
$8 = 4 \times 2$ $9 = 3 \times 3$
$4 = 2 \times 2$ $3 = 1 \times 3$
$2 = 1 \times 2$ Therefore: $81 = 3 \times 3 \times 3 \times 3$
Therefore: $32 = 2 \times 2 \times 2 \times 2 \times 2$

Therefore:

$\sqrt[4]{81} \cdot 2\sqrt[4]{32}$

$= \sqrt[4]{3 \times 3 \times 3 \times 3} \cdot 2\sqrt[4]{2 \times 2 \times 2 \times 2 \times 2}$

$= (3) \cdot 2 \cdot \left(2\sqrt[4]{2}\right)$

$= 3 \times 2 \times 2\sqrt[4]{2}$

$= 12\sqrt[4]{2}$

Example 11.7:

$4\sqrt[3]{54} - \sqrt[3]{16} =$

The cube root indicates that in order to simplify, we should find factors that appear in groups of 3.

$16 = 8 \times 2$ $54 = 27 \times 2$
$8 = 4 \times 2$ $27 = 9 \times 3$
$4 = 2 \times 2$ $9 = 3 \times 3$
$2 = 1 \times 2$ $3 = 1 \times 3$

Therefore: Therefore:
$16 = 2 \times 2 \times 2 \times 2$ $54 = 2 \times 3 \times 3 \times 3$

Therefore:

$4\sqrt[3]{54} - \sqrt[3]{16}$

$= 4\sqrt[3]{2 \times 3 \times 3 \times 3} - \sqrt[3]{2 \times 2 \times 2 \times 2}$

$= 4\left(3\sqrt[3]{2}\right) - \left(2\sqrt[3]{2}\right)$

$= 12\sqrt[3]{2} - 2\sqrt[3]{2}$

$= 10\sqrt[3]{2}$

Complex Roots with Variables

To take the root of a variable with an exponent, divide the exponent of the variable by the root. The quotient becomes the exponent of the variable outside of the radical and the remainder becomes the exponent of the variable under the radical.

Example 11.8:

Simplify: $4a^2b\sqrt{32a^5b^7c^4}$

$\sqrt{32} = \sqrt{16 \times 2} = \sqrt{16}\sqrt{2} = 4\sqrt{2}$

$\sqrt{a^5} = a^2\sqrt{a}$ since $5 \div 2 = 2R1$

$\sqrt{b^7} = b^3\sqrt{b}$ since $7 \div 2 = 3R1$

$\sqrt{c^4} = c^2$ since $4 \div 2 = 2R0$

$4a^2b\sqrt{32a^5b^7c^4} = 4a^2b\left(4a^2b^3c^2\sqrt{2ab}\right) = 16a^4b^4c^2\sqrt{2ab}$

Example 11.9:

$4a^2\sqrt[3]{12a^3b^5c} \cdot ab\sqrt[3]{6a^2c^3}$

Since we are taking the cube root in both terms, we multiply the expressions outside and under the radicals to combine into one expression.

$(4a^2)(ab)\sqrt[3]{12a^3b^5c \cdot 6a^2c^3}$

When multiplying exponential terms with the same base, add the exponents.

$4a^3b\sqrt[3]{72a^5b^5c^4}$

$\sqrt[3]{72} = \sqrt[3]{2 \times 2 \times 2 \times 3 \times 3} = 2\sqrt[3]{9}$

$\sqrt[3]{a^5} = a\sqrt[3]{a^2}$ since $5 \div 3 = 1R2$

$\sqrt[3]{b^5} = b\sqrt[3]{b^2}$ since $5 \div 3 = 1R2$

$\sqrt[3]{c^4} = c\sqrt[3]{c}$ since $4 \div 3 = 1R1$

So:

$4a^3b\sqrt[3]{72a^5b^5c^4}$

$= 4a^3b\left(2abc\sqrt[3]{9a^2b^2c}\right)$

$= 8a^4b^2c\sqrt[3]{9a^2b^2c}$

Rationalizing the denominator

When given a rational (fractional) expression with a radical in the denominator, we can use a technique called **rationalizing the denominator** to remove the radical in the denominator and write the rational in its simplest form. To rationalize a denominator of the form $a \pm b\sqrt{c}$, multiply both the numerator and denominator by the **conjugate** of the denominator then expand both the numerator and the denominator and reduce the expression as far as possible. The conjugate of the denominator is the same expression but using the opposite sign between the terms. If $a = 0$ and the denominator is of the form $b\sqrt{c}$, multiply by $\sqrt{c}$. If $a = 0$ and the root is larger than a square root, choose a rationalizing factor that would make a perfect root in the denominator.

For example, with a denominator of $\sqrt[4]{x^3}$ we choose to multiply by $\sqrt[4]{x}$ so that the denominator becomes the fourth root of a power divisible by 4.

$$\sqrt[4]{x^3} \cdot \sqrt[4]{x} = \sqrt[4]{x^3 \cdot x} = \sqrt[4]{x^4} = x$$

The PERT is normally limited to rationalizing square and cube roots.

Example11.10:

$$\frac{3}{2 + \sqrt{5}} =$$

The conjugate of the denominator is $2 - \sqrt{5}$

$$= \frac{3(2 - \sqrt{5})}{(2 + \sqrt{5})(2 - \sqrt{5})}$$

Expand and simplify

$$= \frac{3(2) - (3)\sqrt{5}}{2(2) - 2(\sqrt{5}) + 2(\sqrt{5}) - \sqrt{5}\sqrt{5}}$$

$$= \frac{6 - 3\sqrt{5}}{4 - 5}$$

$$= \frac{6 - 3\sqrt{5}}{-1}$$

$$= \frac{6}{-1} - \frac{3}{-1}\sqrt{5}$$

$$= -6 + 3\sqrt{5}$$

Example 11.11:

$$\frac{18 - \sqrt[3]{12}}{\sqrt[3]{4}} =$$

To rationalize the denominator we need to create a perfect cube under the cube root.
$$\sqrt[3]{4} = \sqrt[3]{2 \times 2}$$

Therefore, we only need to multiply by $\sqrt[3]{2}$. Make sure to multiply by both the numerator and the denominator.

$$\frac{18 - \sqrt[3]{12}}{\sqrt[3]{4}} = \frac{(18 - \sqrt[3]{12})\sqrt[3]{2}}{(\sqrt[3]{4})\sqrt[3]{2}}$$

$$= \frac{18\sqrt[3]{2} - \sqrt[3]{24}}{\sqrt[3]{8}}$$

$$= \frac{18\sqrt[3]{2} - \sqrt[3]{8 \times 3}}{2}$$

$$= \frac{18\sqrt[3]{2} - 2\sqrt[3]{3}}{2}$$

$$= \frac{2(9\sqrt[3]{2} - \sqrt[3]{3})}{2}$$

$$= 9\sqrt[3]{2} - \sqrt[3]{3}$$

Example 11.12:

Rationalize the denominator: $\sqrt{\dfrac{32a^5b^2}{c^7}}$

There is nothing in common between the numerator and the denominator so we separate into two radicals.

$$= \frac{\sqrt{32a^5b^2}}{\sqrt{c^7}}$$

Since we are taking the square root, we simply need to make the power under the radical in the denominator an even power. So we multiply by $\sqrt{c}$.

$$= \frac{\sqrt{32a^5b^2}}{\sqrt{c^7}} \cdot \frac{\sqrt{c}}{\sqrt{c}} = \frac{\sqrt{32a^5b^2c}}{\sqrt{c^8}}$$

Now take the square root using the technique learned in the section on complex roots with variables.

$$\sqrt{32} = \sqrt{16 \times 2} = 4\sqrt{2}$$

$$\sqrt{a^5} = a^2\sqrt{a} \qquad because\ 5 \div 2 = 2R1$$

$$\sqrt{b^2} = b \qquad because\ 2 \div 2 = 1R0$$

$$\sqrt{c} \qquad cannot\ be\ simplified$$

$$\sqrt{c^8} = c^4 \qquad because\ 8 \div 2 = 4R0$$

Therefore:

$$\frac{\sqrt{32a^5b^2c}}{\sqrt{c^8}} = \frac{4a^2b\sqrt{2ac}}{c^4}$$

Note that even though the expression does not look any simpler, it is important to know how to eliminate radicals from the denominator for upper level college courses.

Solving Equations involving Radicals

To solve an equation involving a square root or cube root, we need to eliminate the radical by using the inverse operation (square or cube). First isolate the root before squaring or cubing both sides. If the equation involves more than one root, you may need to repeat the inverse operation more than once. Always test the solution(s) in the original statement to ensure that they are valid. Any values that result in a negative under an even root are discarded as invalid because we can only take the even root of a positive number.

Example 11.13:

Solve the equation: $x = \sqrt{x + 13} + 7$

Move 7 to the other side of the equation in order to isolate the radical

$$\sqrt{x + 13} = x - 7$$

Square both sides to eliminate the radical. FOIL the right side.

$$\left(\sqrt{x + 13}\right)^2 = (x - 7)^2$$

$$x + 13 = x^2 - 14x + 49$$

Move all terms to one side then factor

$$x^2 - 14x - x + 49 - 13 = 0$$

$$x^2 - 15x + 36 = 0$$

$$(x - 3)(x - 12) = 0$$

$$x - 3 = 0 \qquad x - 12 = 0$$

$$x = 3 \qquad\qquad x = 12$$

Testing the two values:

When $x = 3$	When $x = 12$
$3 = \sqrt{3 + 13} + 7$	$12 = \sqrt{12 + 13} + 7$
$3 = \sqrt{16} + 7$	$12 = \sqrt{25} + 7$
$3 = 4 + 7$	$12 = 5 + 7$
$3 = 11 \quad false$	$12 = 12 \quad true$

Only the solution $x = 12$ works when you test in the original equation.

Example 11.14:

$$\sqrt[3]{12 - x} = -2$$

Take the cube of both sides

$$(\sqrt[3]{12 - x})^3 = (-2)^3$$

$$12 - x = -8$$

$$-x = -20$$

$$x = 20$$

Testing the solution

When $x = 20$

$$\sqrt[3]{12 - 20} = -2$$

$$\sqrt[3]{-8} = -2$$

$$-2 = -2 \quad true$$

Therefore, the solution $x = 20$ is valid.

Chapter 11 – Radicals & Exponents **137**

Chapter 11 Exercises

1. Simplify: $\dfrac{4^{\frac{3}{2}} \cdot 16^{-1}}{2^{-5}}$

2. Simplify: $27^{\frac{1}{2}} - \dfrac{1}{3^{-\frac{1}{2}}}$

3. $\dfrac{\sqrt{20} - \sqrt{80}}{2\sqrt{5}} =$

4. Simplify: $7\sqrt[3]{24} - \dfrac{15}{\sqrt[3]{9}}$

5. Simplify: $8x^2y^3\sqrt{45x^5y^3} + 3x\sqrt{20x^7y^9}$

6. Rationalize the denominator: $\dfrac{3\sqrt{x}}{\sqrt{x} + 2}$

7. Simplify: $\dfrac{5z\sqrt{32x^4y^9z^7}}{2xy\sqrt{4y^4z^3}}$

8. Simplify: $5\sqrt[3]{-27} - 3\sqrt[3]{48}$

9. Solve the equation: $\sqrt{3x + 1} = 3 + \sqrt{x - 4}$

10. Rationalize the denominator: $\dfrac{8x^2y}{\sqrt[3]{x^2} \cdot \sqrt{y}}$

Solutions

1. Change all terms to base 2

$$\frac{(2^2)^{\frac{3}{2}} \cdot (2^4)^{-1}}{2^{-5}}$$

Simplify exponents in the numerator using law #1 of exponents

$$= \frac{2^3 \cdot 2^{-4}}{2^{-5}}$$

Combine the terms in the numerator using law #4 of exponents

$$= \frac{2^{3+(-4)}}{2^{-5}} = \frac{2^{-1}}{2^{-5}}$$

Combine the numerator and the denominator using law #5 of exponents then simplify

$$= 2^{-1-(-5)} = 2^4 = 16$$

138 Prep for Success: Florida's PERT Math Study Guide

2. Use law #6 of exponents on the 2nd term

$$27^{\frac{1}{2}} - \frac{1}{3^{-\frac{1}{2}}} = 27^{\frac{1}{2}} - 3^{\frac{1}{2}}$$

Change all terms to base 3

$$= (3^3)^{\frac{1}{2}} - 3^{\frac{1}{2}} = 3^{\frac{3}{2}} - 3^{\frac{1}{2}}$$

Factor as in Examples 11.2 & 11.5 and then simplify

$$= 3^{\frac{1}{2}}(3^1 - 1) = 3^{\frac{1}{2}} \cdot 2 = 2\sqrt{3}$$

3. We could go directly into rationalizing the denominator but notice that 5 is a factor of both 20 and 80 so division of radicals would simplify the expression.

$$\frac{\sqrt{20} - \sqrt{80}}{2\sqrt{5}} = \frac{\sqrt{20}}{2\sqrt{5}} - \frac{\sqrt{80}}{2\sqrt{5}}$$

Combine the radicals in the numerator and the denominator of each rational expression

$$= \frac{1}{2}\sqrt{\frac{20}{5}} - \frac{1}{2}\sqrt{\frac{80}{5}} = \frac{\sqrt{4}}{2} - \frac{\sqrt{16}}{2} = \frac{2}{2} - \frac{4}{2} = 1 - 2 = -1$$

4. Before we attempt to subtract, we need to rationalize the denominator of the second term.

$$7\sqrt[3]{8 \times 3} - \frac{15}{\sqrt[3]{3 \times 3}}$$

We need to multiply the denominator of the second term by $\sqrt[3]{3}$ to form a perfect cube

$$7\sqrt[3]{8 \times 3} - \frac{15\sqrt[3]{3}}{\sqrt[3]{3 \times 3}\sqrt[3]{3}} = 7\sqrt[3]{8 \times 3} - \frac{15\sqrt[3]{3}}{3} = 7(2\sqrt[3]{3}) - 5\sqrt[3]{3} = 14\sqrt[3]{3} - 5\sqrt[3]{3} = 9\sqrt[3]{3}$$

5. Simplify the radicals first

$$8x^2y^3\sqrt{45x^5y^3} + 3x\sqrt{20x^7y^9}$$

$$= 8x^2y^3\sqrt{9 \times 5x^5y^3} + 3x\sqrt{4 \times 5x^7y^9}$$

$$= 8x^2y^3\left(3x^2y\sqrt{5xy}\right) + 3x\left(2x^3y^4\sqrt{5xy}\right)$$

$$= 24x^4y^4\sqrt{5xy} + 6x^4y^4\sqrt{5xy}$$

Now we can add the two terms because the expressions under the radicals are the same.

$$= 30x^4y^4\sqrt{5xy}$$

6. To rationalize the denominator, multiply by the conjugate of the denominator $\sqrt{x} + 2$ which is $\sqrt{x} - 2$

$$\frac{3\sqrt{x}}{\sqrt{x} + 2} \cdot \frac{\left(\sqrt{x} - 2\right)}{\left(\sqrt{x} - 2\right)}$$

Distribute the numerator, use difference of squares in the denominator then simplify

$$\frac{(3\sqrt{x})(\sqrt{x}) - (3\sqrt{x})(2)}{(\sqrt{x})^2 - (2)^2} = \frac{3x - 6\sqrt{x}}{x - 4}$$

7. Since there are variables common to both, combine the radicals in the numerator and denominator.

$$\frac{5z}{2xy}\sqrt{\frac{32x^4y^9z^7}{4y^4z^3}}$$

Use law #5 of exponents to simplify the rational under the radical

$$\frac{5z}{2xy}\sqrt{8x^4y^{9-4}z^{7-3}} = \frac{5z}{2xy}\sqrt{8x^4y^5z^4}$$

Now simplify the expression under the radical

$$\frac{5z}{2xy}\left(2x^2y^2z^2\sqrt{2y}\right)$$

$$\frac{10x^2y^2z^3}{2xy}\sqrt{2y} = 5xyz^3\sqrt{2y}$$

8. Reduce the radicals and simplify
$$= 5\sqrt[3]{-3 \times -3 \times -3} - 3\sqrt[3]{2 \times 2 \times 2 \times 2 \times 3}$$
$$= 5(-3) - 3\left(2\sqrt[3]{6}\right)$$
$$= -15 - 6\sqrt[3]{6}$$

9. Since there are radicals on each side rather than just one radical, we skip the step of trying to isolate the radical and go directly to squaring both sides. FOIL the right side.
$$\left(\sqrt{3x+1}\right)^2 = \left(3 + \sqrt{x-4}\right)^2$$
$$3x + 1 = 9 + 3\sqrt{x-4} + 3\sqrt{x-4} + \left(\sqrt{x-4}\right)\left(\sqrt{x-4}\right)$$
$$3x + 1 = 9 + 6\sqrt{x-4} + x - 4$$
Move the term containing the radical to the one side of the equation and everything else to the other side
$$6\sqrt{x-4} = 3x + 1 - 9 - x + 4$$
$$6\sqrt{x-4} = 2x - 4$$
Now again square both sides to eliminate the square root. FOIL the right side
$$\left(6\sqrt{x-4}\right)^2 = (2x-4)^2$$
$$36(x-4) = 4x^2 - 8x - 8x + 16$$
Move everything to one side, combine like terms then factor and solve
$$36x - 144 = 4x^2 - 16x + 16$$
$$4x^2 - 52x + 160 = 0$$
$$4(x^2 - 13x + 40) = 0$$
$$4(x-5)(x-8) = 0$$
$$x = 5 \qquad x = 8$$

Check both answers

When $x = 5$

$$\left(\sqrt{3(5) + 1}\right)^2 = \left(3 + \sqrt{5 - 4}\right)^2$$
$$\left(\sqrt{16}\right)^2 = \left(3 + \sqrt{1}\right)^2$$
$$16 = 4^2$$
True

When $x = 8$

$$\left(\sqrt{3(8) + 1}\right)^2 = \left(3 + \sqrt{8 - 4}\right)^2$$
$$\left(\sqrt{25}\right)^2 = \left(3 + \sqrt{4}\right)^2$$
$$25 = (3 + 2)^2$$
$$25 = 5^2$$
True

Therefore both solutions are valid.

10. Since there are two different roots in the denominator, we make each a perfect root.

To make $\sqrt[3]{x^2}$ a perfect cube root, we need to multiply by $\sqrt[3]{x}$

To make $\sqrt{y}$ a perfect square root, we need to multiply by $\sqrt{y}$

$$\frac{8x^2y}{\sqrt[3]{x^2} \cdot \sqrt{y}} \cdot \frac{\sqrt[3]{x}\sqrt{y}}{\sqrt[3]{x}\sqrt{y}}$$

$$= \frac{8x^2y\left(\sqrt[3]{x}\sqrt{y}\right)}{\sqrt[3]{x^3} \cdot \sqrt{y^2}}$$

$$= \frac{8x^2y\left(\sqrt[3]{x}\sqrt{y}\right)}{xy}$$

Simplify by eliminating variables common to the numerator and denominator or by using law #5 of exponents.

$$= 8x\left(\sqrt[3]{x}\sqrt{y}\right)$$

Chapter 12 – Quadratics   **141**

Chapter 12

Quadratics

Complex Numbers

Before we delve into solving quadratic equations, we must introduce the concept of imaginary and real numbers. Real numbers include all the numbers you should already be familiar with such as: integers, whole numbers, rational (decimals, fractions) etc.

For simplicity, in earlier courses students are taught that the square root of a negative number does not exist. This is because no real number multiplied by itself would yield a negative result.

Remember:
$(+)(+) = (+)$
$(-)(-) = (+)$
So $\sqrt{16} = 4$ because $\sqrt{16} = \sqrt{4 \times 4}$ BUT $\sqrt{-16} \neq -4$ because $-4 \times -4 \neq -16$

At this level, the concept of **imaginary numbers** is introduced. An imaginary number represents the square root of a negative number by using the relationship $i^2 = -1$.

Example 12.1:
$\sqrt{-400}$

$\quad\quad \sqrt{-400} = \sqrt{-1 \times 400}$
$\quad\quad$ Replace -1 with i^2 and simplify
$\quad\quad = \sqrt{400i^2}$
$\quad\quad = \sqrt{20i \times 20i}$
$\quad\quad = 20i$
$\quad\quad 20i$ is the imaginary number used to represent $\sqrt{-400}$

A **complex number** is one that combines a real number and an imaginary number in the form $a \pm bi$. In an expression such as $5 + 3i$ we say that 5 is the real part and $3i$ is the imaginary part.

Algebraic Operations involving Complex Numbers

To add complex numbers, add their corresponding real parts and imaginary parts.

Example 12.2:
$(2 + 3i) + (-4 + 7i) =$
$\quad\quad$ Combine the real parts and the imaginary parts.

$$= (2 - 4) + (3i + 7i) = -2 + 10i$$

To subtract complex numbers, subtract their corresponding real parts and imaginary parts.

Example 12.3:
$(5 - 3i) - (2 - 9i) =$

> Combine the real parts and the imaginary parts.
> $$= 5 - 2 - 3i - (-9i) = 3 - 3i + 9i$$
> $$= 3 + 6i$$

To multiply complex numbers, use FOIL if necessary and then combine like terms. In the case of a perfect square, you can use the rule:
$(a + b)^2 = a^2 + 2ab + b^2$

Remember to simplify by replacing i^2 by –1 if possible.

Example 12.4:
$(2 - 3i)(5 - i) =$

> FOIL
> $$(2)(5) + (2)(-i) + (-3i)(5) + (-3i)(-i)$$
> $$= 10 - 2i - 15i + 3i^2$$
> $$= 10 - 17i + 3(-1)$$
> $$= 10 - 17i - 3 = 7 - 17i$$

To divide complex numbers, write the problem as a fraction if necessary then multiply the numerator and the denominator by the complex conjugate of the denominator.

Complex Conjugates
Complex conjugates differ only in that the sign of the imaginary parts are opposite.
Therefore $3 + 2i$ and $3 - 2i$ are complex conjugates while $5 - 7i$ and $7 + 5i$ are not.

Example 12.5:
$$\frac{4i}{1 + 3i} =$$

> Multiply the numerator and the denominator by $1 - 3i$ (the conjugate of the denominator).
> $$\frac{4i}{1 + 3i} \cdot \frac{(1 - 3i)}{(1 - 3i)}$$
> $$\frac{4i - 12i^2}{1 - 3i + 3i - 9i^2} = \frac{4i - 12i^2}{1 - 9i^2}$$
> Replace i^2 by –1
> $$= \frac{4i - 12(-1)}{1 - 9(-1)}$$

Simplify and write in standard form $a \pm bi$
$$\frac{12+4i}{1+9} = \frac{12}{10} + \frac{4}{10}i = \frac{6}{5} + \frac{2}{5}i$$

Powers of i

If the power of i is odd, first factor an i so that the remaining power is even. Next divide the even power by 2 and rewrite the expression in the form $(i^2)^n$ where n is the result of the division. Finally, replace i^2 by -1 and simplify.

Example 12.6:
Evaluate i^{1005}

If the exponent is odd, factor an i so that the remaining exponent is even.
$(i^{1004})i$
Since we know that $i^2 = -1$, divide the even power by 2 and replace i^2 by -1 as shown.
$(i^2)^{502}i$
$(-1)^{502}i$
-1 raised to an even power is positive 1
$(1)i = i$

Therefore $i^{1005} = i$

Quadratic Equations

Here we take the process of factoring a quadratic a step further and introduce solving quadratic equations after factoring. The graph of a typical quadratic equation is shown to the right. A quadratic equation is a quadratic expression with an equal sign.
$y = f(x) = ax^2 + bx + c$
The points where the graph intersects the x-axis are known as the **x-intercepts** or **zeros** of the equation. When we set a quadratic expression equal to 0 and solve for x, we are finding the zeros.

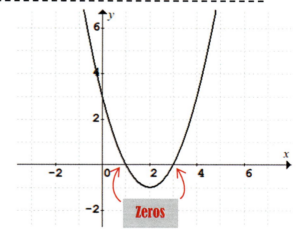

Solving Quadratic Equations by Factoring

When asked to find the zeros or solve the quadratic equation, we need to find the x-intercepts. Use inverse operations to ensure that all the terms are on one side so that the equation is equal to zero. Attempt to factor the equation, set the factors equal to zero and solve for the variables. Sometimes the quadratic can be factored perfectly; other times it will be necessary to use the quadratic formula as demonstrated in the next section.

144 Prep for Success: Florida's PERT Math Study Guide

Example 12.7:

If $6x^2 = 7x + 5$, then $x =$

First move all the terms to the left side of the equation so that the equation equals zero.

$6x^2 - 7x - 5 = 0$

Next factor the left-hand side of the equation. We leave this up to you to factor and make sure that you know how we arrived at our answer.

$(3x - 5)(2x + 1) = 0$

Lastly, set each linear factor equal to zero and solve for x.

$$3x - 5 = 0 \qquad\qquad 2x + 1 = 0$$
$$3x = 5 \qquad\qquad 2x = -1$$
$$x = \frac{5}{3} \qquad\qquad x = -\frac{1}{2}$$

Therefore $x = -\frac{1}{2}$ and $x = \frac{5}{3}$

Example 12.8:

If $m^2 - 6m = 0$ then $m =$

$$m^2 - 6m = 0$$
$$m(m - 6) = 0$$
$$m = 0 \qquad\qquad m - 6 = 0$$
$$m = 6$$

Therefore $m = 0$ and $m = 6$

Example 12.9:

If $2c^2 - 4c = 9(2 - c)$ then $c =$

$$2c^2 - 4c = 9(2 - c)$$
$$2c^2 - 4c = 18 - 9c$$
$$2c^2 - 4c - 18 + 9c = 0$$
$$2c^2 + 5c - 18 = 0$$
$$(2c + 9)(c - 2) = 0$$
$$2c + 9 = 0 \qquad\qquad c - 2 = 0$$
$$2c = -9 \qquad\qquad c = 2$$
$$c = -\frac{9}{2}$$

Therefore $c = 2$ and $c = -\frac{9}{2}$

The Quadratic Formula

To solve quadratic equations that cannot be factored, use the quadratic formula.

Given a quadratic equation $ax^2 + bx + c = 0$

$$x = \frac{-b \pm \sqrt{b^2 - 4ac}}{2a}$$

Note that the quadratic formula can also be used to solve quadratic equations that can be factored. Obtain the values of a, b and c from the quadratic equation and then substitute into the formula to find the value(s) of x.

Example 12.10:
Find the solution(s) of the following: $3x^2 = 5x + 10$

Put the equation in the form $ax^2 + bx + c = 0$ and then identify the values of a, b and c.
$3x^2 - 5x - 10 = 0$
$a = 3, b = -5, c = -10$

Substitute the values of a, b and c into the formula.
$$x = \frac{-(-5) \pm \sqrt{(-5)^2 - 4(3)(-10)}}{2(3)}$$
$$x = \frac{5 \pm \sqrt{25 + 120}}{6}$$
$$x = \frac{5 \pm \sqrt{145}}{6}$$

$\sqrt{145}$ cannot be simplified.

Therefore the values of x are:
$$x = \frac{5 + \sqrt{145}}{6}, \quad x = \frac{5 - \sqrt{145}}{6}$$

Imaginary numbers often appear when we solve for x using the quadratic formula. As stated before, solutions to a quadratic equation represent the x-intercepts. If a quadratic equation has only imaginary solutions (often called imaginary roots), then there are no x-intercepts and the graph will be similar to the one below where the graph does not touch the x-axis.

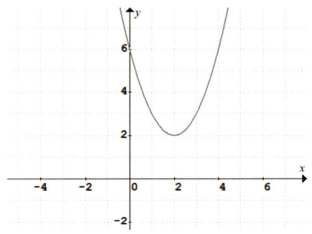

Pay attention to if the question asks for the real roots of the equation or just asks for the solutions. The solutions to a quadratic equation can include imaginary or real roots. When the problem asks for the real roots of the equation, any imaginary roots become invalid solutions.

Example 12.11:

Find the solution(s) of the following: $6x^2 - 2x + 1 = 0$

$a = 6, b = -2, c = 1$

$$x = \frac{-(-2) \pm \sqrt{(-2)^2 - 4(6)(1)}}{2(6)}$$

$$x = \frac{2 \pm \sqrt{4 - 24}}{12}$$

$$x = \frac{2 \pm \sqrt{-20}}{12}$$

Replace -1 with i^2 under the radical.

$$x = \frac{2 \pm \sqrt{20i^2}}{12}$$

Simplify.

$$x = \frac{2 \pm \sqrt{5 \times 4i^2}}{12}$$

$$x = \frac{2 \pm 2i\sqrt{5}}{12}$$

Reduce by 2.

$$x = \frac{1 \pm i\sqrt{5}}{6}$$

Example 12.12:

Find the real roots of the following: $x^2 - 4x + 5 = 0$

$a = 1, b = -4, c = 5$

$$x = \frac{-(-4) \pm \sqrt{(-4)^2 - 4(1)(5)}}{2(1)}$$

$$x = \frac{4 \pm \sqrt{16 - 20}}{2}$$

$$x = \frac{4 \pm \sqrt{-4}}{2}$$

$$x = \frac{4 \pm \sqrt{4i^2}}{2}$$

$$x = \frac{4 \pm 2i}{2}$$

$$x = 2 \pm i$$

Since the roots are imaginary, there are no real roots.

Graphs of Quadratic Equations

To graph a quadratic equation, find the x and y intercepts and the vertex then connect the points. The same rules as before apply to solving for the x and y intercepts.

To find the vertex, first find the x-coordinate using $x = -\frac{b}{2a}$ then substitute the value of x into the equation to find the y-coordinate.

Example 12.13:

Graph the quadratic equation $y = x^2 + 2x - 8$

Comparing the equation to $ax^2 + bx + c$, we find that $a = 1, b = 2$ and $c = -8$

To find the y-intercept, set $x = 0$
$y = (0)^2 + 2(0) - 8$
$y = -8$
$(0, -8)$

To find the x-intercept, set $y = 0$
$x^2 + 2x - 8 = 0$
$(x - 2)(x + 4) = 0$
$x = 2$ and $x = -4$
$(2, 0)$ and $(-4, 0)$

To find the vertex
$$x = -\frac{b}{2a} = -\frac{2}{2(1)} = -1$$

Substitute to find y
$y = (-1)^2 + 2(-1) - 8 = 1 - 2 - 8 = -9$
$(-1, -9)$

Plot the points and connect

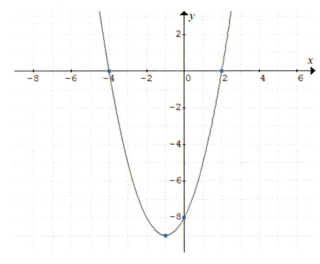

If $a < 0$ the vertex will be a **maximum** point and the graph will be shaped like an "n".

If $a > 0$ the vertex will be a **minimum** point and the graph will be shaped like a "u".

If the equation is written in the form $y = a(x - h)^2 + k$ then the vertex is given by the point (h, k).

Domain and Range

A **function** is a relationship that maps each value of x (independent variable) unto a single value of y (dependent variable). The set of values of x that define the function represent its **domain** and the corresponding values of y are called its **range**.

The notation $f(x)$ is used interchangeably with y.

For a relationship to be defined as a function, each value of x must have exactly one corresponding value of y. We demonstrate this concept with two diagrams.

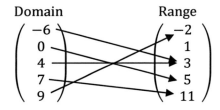
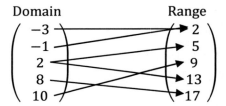

The relationship to the left is a function because only one arrow originates from each one of the values of the domain. Note that it does not matter if two or more x values lead to the same y value. We just cannot have each x mapping to more than one value of y. Hence the relationship to the right is NOT a function. A **one-to-one function** maps exactly one value of x unto one value of y. Therefore, even though the relationship to the left is a function, it is not a one-to-one function.

To determine whether a set of coordinates belong to a function, verify that no x-coordinate is repeated. To determine whether a set of coordinates belong to a one-to-one function, verify that no x or y coordinate is repeated.

Example 12.14:
Which of the following relationships does not represent a function?
A. $[(-1, 5), (2, 3), (4, 11), (3, 5)]$
C. $[(-3, 12), (0, 4), (6, 2), (7, -10)]$
B. $[(1, 1), (2, 2), (3, 2), (4, 4)]$
D. $[(-2, 1), (5, 9), (3, -4), (-2, 8)]$

    Remember that we can have 2 or more values of x mapping to the same values of y but not the other way around. So in the answer choices look for any repeating values of x which would indicate that the relationship is not a function.
    Choice D maps $x = -2$ unto two different values of y and is therefore not a function.

We can test to determine whether a graph is a function by using the **vertical line test**. If we draw a vertical line through any value of x and the line intercepts the graph at more than one point, the graph is NOT a function (because the value of x has more than one corresponding value of y). We can test to determine whether a graph is a one-to-one function by using the **horizontal line test**. If we draw a horizontal line through any value of y and the line intercepts the graph at more than one point, the graph is NOT a one-to-one function (because the value of y has more than one corresponding value of x).

Example 12.15:
Determine whether the graph below represents a function.

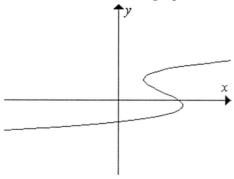

If we draw a vertical line through the curves in the middle we can see that the line intersects the graph at more than one point indicating that the graph does not represent a function.

When given a function and asked to find its domain, we start with the domain of all real numbers $(-\infty, \infty)$ and then exclude all the values for which the function is not defined. Polynomials and linear equations have a domain of all real numbers. Here are two of the main situations where the domain may not be all real numbers.

1. **Vertical Asymptotes**

A **vertical asymptote** is a vertical line that the graph of a function does not cross. They are found in rational (fractional) functions and correspond to the values of x that would make the denominator zero. Remember that we can never divide by 0 so any values of x that cause the denominator to evaluate to 0, must be excluded from the domain.

Example 12.16:
What is the domain of the following function?

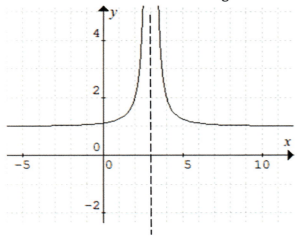

The dashed line indicates an asymptote at $x = 3$. We therefore exclude that value from the domain. The graph extends out to the left and right with no visible endpoint so the domain is:
$(-\infty, 3) \cup (3, \infty)$

2. Radicals

Functions that involve even radicals need to exclude all values of x that would cause the inside of the radical to evaluate to a negative number.

If we are dealing with an odd root, the domain is the set of all real numbers. When dealing with an even root, to find the domain set the inside of the root ≥ 0 and solve for x. If the above two cases are combined and the even root is in the denominator, set the inside of the root > 0 and solve for x.

Example 12.17:
Find the domain of the function $y = \sqrt[4]{5-x}$.

Since the root is even (4th root), we need to make sure that the domain does not include values that would make the expression under the root evaluate to a negative number. Set the expression greater than or equal to 0 and solve for the domain.
$5 - x \geq 0$
$-x \geq -5$
$x \leq 5$ We reverse the inequality sign because of multiplication by a negative.
Therefore the domain is $(-\infty, 5]$.

Example 12.18:
What is the range of the graph below?

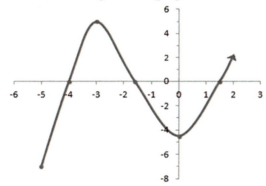

The lowest y-value on the graph is -7. The dot at the point $(-6, -7)$ indicates an end boundary so the graph does not go further downwards. The endpoint is a closed circle so -7 is included in the range. You may think that the highest point is at $(0, 5)$ but the arrow on the other end of the graph indicates that the graph will rise infinitely because there is no endpoint. Therefore the range is $[-7, \infty)$.

Chapter 12 Exercises

1. Solve for p: $p^2 - 4p = 5$

2. Find the zeros of the equation: $2x^3 + 18x = 0$

3. Find all the solutions of the equation: $x^2 - 6x + 59 = 0$

4. Graph the quadratic equation: $y = x^2 + 8x + 15$

5. What is the domain of the function $f(x) = \sqrt{x^2 - 25}$?
   A. $[-5, 5]$ B. $[5, \infty)$ C. $(-\infty, -5] \cup [5, \infty)$ D. All real numbers

6. What is the minimum value of y in the function $f(x) = 3x^2 + 12x + 8$?
   A. -4 B. -2 C. 0 D. 8

7. Which of the following does not represent the graph of a function?

   A.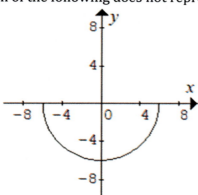

   C.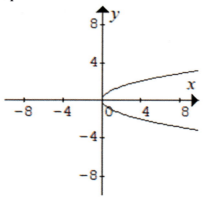

   B.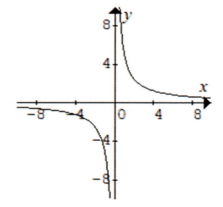

   D.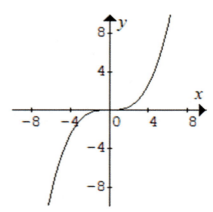

8. Which of the following is a root of the equation of $x^2 - 2x + 5 = 0$?
   A. $2i$ B. $1 - 2i$ C. $1 + \dfrac{i}{2}$ D. $1 - i\sqrt{5}$

152 Prep for Success: Florida's PERT Math Study Guide

9. Which of the following is a root of the equation $3x^2 + 2x = -5$?

    A.   1   B. $\dfrac{5}{3}$ C. $-\dfrac{1}{3} - \dfrac{i\sqrt{14}}{3}$ D. $-\dfrac{1}{3} + 3i$

10. What is the domain of the function below?

$$f(x) = \frac{x+3}{\sqrt[3]{x}}$$

    A. $\{x : x \geq 0\}$ B. $\{x : 0 < x \leq 3\}$ C. $\{x : x \neq 0\}$ D. All real numbers

Solutions

1. Move all terms to one side then attempt to factor the quadratic.
$$p^2 - 4p - 5 = 0$$
$$(p + 1)(p - 5) = 0$$
$$p = -1 \text{ and } p = 5$$

2. We can tell that the terms of the binomial have the factor $2x$ in common.
$$2x(x^2 - 9) = 0$$
Factor further using difference of squares
$$2x(x - 3)(x + 3) = 0$$
$$2x = 0 \qquad x - 3 = 0 \qquad x + 3 = 0$$
$$x = -3, 0, 3$$

3. We go directly into using the quadratic formula
$$a = 1 \qquad b = -6 \qquad c = 59$$

$$x = \frac{-b \pm \sqrt{b^2 - 4ac}}{2a}$$

$$x = \frac{-(-6) \pm \sqrt{(-6)^2 - 4(1)(59)}}{2(1)}$$

$$x = \frac{6 \pm \sqrt{36 - 236}}{2} = \frac{6 \pm \sqrt{-200}}{2}$$

Replace -1 with i^2 and simplify

$$x = \frac{6 \pm \sqrt{100i^2 \times 2}}{2} = \frac{6 \pm 10i\sqrt{2}}{2} = 3 \pm 5i\sqrt{2}$$

4. Find, plot and connect the y–intercept, x–intercept and vertex.
$$y = x^2 + 8x + 15$$
$$a = 1, \quad b = 8, \quad c = 15$$

$y - intercept$	$x - intercept$	$vertex$
$y = 0^2 + 8(0) + 15$	$0 = x^2 + 8x + 15$	$x = -\dfrac{b}{2a} = -\dfrac{8}{2} = -4$

$y = 15$ $(x+3)(x+5) = 0$ $y = (-4)^2 + 8(-4) + 15 = -1$
$(0, 15)$ $(-3, 0)$ and $(-5, 0)$ $(-4, -1)$

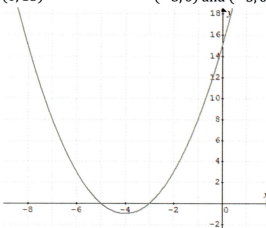

5. **Answer [C]:** Remembering that we cannot find the even root of a negative number without the use of complex numbers:
$x^2 - 25 \geq 0$
$(x-5)(x+5) \geq 0$
Solve for the critical values of x
$x = -5$ and $x = 5$
Place the critical values on a number line and choose test points in each of the 3 regions formed.

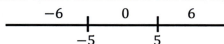

Substitute the test points into original inequality to see which satisfy the inequality.

When $x = -6$	When $x = 0$	When $x = 6$
$(-6)^2 - 25 \geq 0$	$(0)^2 - 25 \geq 0$	$(6)^2 - 25 \geq 0$
$36 - 25 \geq 0$	$0 - 25 \geq 0$	$36 - 25 \geq 0$
$11 \geq 0$	$-25 \geq 0$	$11 \geq 0$
True	False	True

Therefore, the domain of the function $f(x)$ is $(-\infty, -5] \cup [5, \infty)$.

6. **Answer [A]:** To find the x-coordinate of the minimum or maximum point of a quadratic equation, use the formula:
$$x = -\frac{b}{2a} = -\frac{12}{2(3)} = -\frac{12}{6} = -2$$
Substitute to find the value of y
$$y = f(-2) = 3(-2)^2 + 12(-2) + 8$$
$$= 12 - 24 + 8 = -4$$

7. **Answer [C]:** A function maps every value of x unto only one value of y. Use the vertical line test to determine whether each graph is the graph of a function. If the vertical line

intersects the graph at more than one point, the graph does not represent a function. Therefore, the correct answer is C.

8. **Answer [B]:** All the solutions are imaginary roots so we go directly to using the quadratic formula. From the given equation:

$$a = 1, \quad b = -2, \quad c = 5$$

$$x = \frac{-(-2) \pm \sqrt{(-2)^2 - 4(1)(5)}}{2(1)}$$

$$x = \frac{2 \pm \sqrt{4 - 20}}{2} = \frac{2 \pm \sqrt{-16}}{2}$$

Replace -1 with i^2 under the radical.

$$x = \frac{2 \pm \sqrt{16i^2}}{2} = \frac{2 \pm 4i}{2} = 1 \pm 2i$$

Therefore the answer is B.

9. **Answer [C]:** First move everything to the left side of the equation

$$3x^2 + 2x + 5 = 0$$

You can try to factor but we suggest using the quadratic equation outright because some of the answers include i indicating that there may be imaginary roots.

$$a = 3, \quad b = 2, \quad c = 5$$

$$x = \frac{-b \pm \sqrt{b^2 - 4ac}}{2a}$$

$$x = \frac{-2 \pm \sqrt{2^2 - 4(3)(5)}}{2(3)}$$

$$= \frac{-2 \pm \sqrt{4 - 60}}{6} = \frac{-2 \pm \sqrt{-56}}{6} = \frac{-2 \pm \sqrt{4i^2 \times 14}}{6}$$

$$= \frac{-2 \pm 2i\sqrt{14}}{6} = -\frac{2}{6} \pm \frac{2i\sqrt{14}}{6} = -\frac{1}{3} \pm \frac{i\sqrt{14}}{3}$$

Therefore the answer is C.

10. **Answer [C]:** We focus on the denominator because the numerator is linear and all real numbers are valid for linear functions.

All real numbers are valid for the cube root because it is an odd root. However, we exclude $x = 0$ because the denominator cannot equal 0 else the function will be undefined.
Therefore, the domain is all real numbers except 0 and the solution is C.

Chapter 13 – Rational Expressions **155**

Chapter 13

Rational Expressions

Adding and Subtracting Rational Expressions

To add or subtract rational expressions containing polynomials, first factor the numerator and denominator then simplify each rational by eliminating common factors if possible. Find the common denominator and convert each rational so that they all contain the common denominator. Combine and simplify the numerators.

Example 13.1:

$$\frac{x^2 + x - 20}{x^2 + 3x - 10} + \frac{3x}{2x - 2} =$$

Factor any polynomials that can be factored.

$$\frac{(x + 5)(x - 4)}{(x + 5)(x - 2)} + \frac{3x}{2(x - 1)}$$

Eliminate factors common to both numerator and denominator.

$$\frac{(x - 4)}{(x - 2)} + \frac{3x}{2(x - 1)}$$

Find the common denominator and convert fractions so they all have the same denominators.

LCD: $2(x - 1)(x - 2)$

$$\frac{2(x - 1)}{2(x - 1)} \cdot \frac{(x - 4)}{(x - 2)} + \frac{3x}{2(x - 1)} \cdot \frac{(x - 2)}{(x - 2)}$$

Simplify numerators by expanding and then combine fractions.

$$\frac{2(x^2 - 4x - x + 4)}{2(x - 1)(x - 2)} + \frac{3x^2 - 6x}{2(x - 1)(x - 2)}$$

$$\frac{2x^2 - 8x - 2x + 8 + 3x^2 - 6x}{2(x - 1)(x - 2)}$$

$$\frac{5x^2 - 16x + 8}{2(x - 1)(x - 2)}$$

Factor the numerator and simplify further if possible. In this case, the numerator cannot be factored and so the rational cannot be simplified further.

156   Prep for Success: Florida's PERT Math Study Guide

Example 13.2:
$$\frac{x}{x^2 - 3x + 2} - \frac{x + 1}{x^2 - 4} + \frac{3x}{x^2 + x - 2} =$$

Factor any polynomials that can be factored.
$$\frac{x}{(x - 1)(x - 2)} - \frac{x + 1}{(x - 2)(x + 2)} + \frac{3x}{(x - 1)(x + 2)}$$

None of the rational expressions can be reduced so find the lowest common denominator and convert fractions so they all have the same denominators.
LCD: $(x - 1)(x + 2)(x - 2)$

$$\frac{(x + 2)}{(x + 2)} \cdot \frac{x}{(x - 1)(x - 2)} - \frac{(x - 1)}{(x - 1)} \cdot \frac{x + 1}{(x - 2)(x + 2)} + \frac{(x - 2)}{(x - 2)} \cdot \frac{3x}{(x - 1)(x + 2)}$$

Simplify the numerators by expanding and combine fractions.
$$\frac{x^2 + 2x}{(x - 1)(x + 2)(x - 2)} - \frac{x^2 - 1}{(x - 1)(x + 2)(x - 2)} + \frac{3x^2 - 6x}{(x - 1)(x + 2)(x - 2)}$$

$$\frac{x^2 + 2x - (x^2 - 1) + 3x^2 - 6x}{(x - 1)(x + 2)(x - 2)}$$

$$\frac{3x^2 - 4x + 1}{(x - 1)(x + 2)(x - 2)}$$

Factor the numerator and simplify further if possible.
$$\frac{(3x - 1)(x - 1)}{(x - 1)(x + 2)(x - 2)}$$

$$\frac{(3x - 1)}{(x + 2)(x - 2)}$$

Multiplying and Dividing Rational Expressions

There is no need to find a common denominator when multiplying and dividing rational expressions involving polynomials. You will still need to factor the polynomials in order to simplify.

Multiplying Rational Expressions
When multiplying rational expressions:
1. Factor as much as possible
2. Multiply horizontally
3. Eliminate any factors common to both the numerator and denominator

Example 13.3:
$$\frac{2x^2 - 7x - 4}{x^2 - 6x + 8} \cdot \frac{x^2 + 4x + 3}{2x^2 + 7x + 3} =$$

Step 1: Factor
$$\frac{(2x+1)(x-4)}{(x-2)(x-4)} \cdot \frac{(x+1)(x+3)}{(2x+1)(x+3)}$$

Step 2: Multiply horizontally
$$\frac{(2x+1)(x-4)(x+1)(x+3)}{(x-2)(x-4)(2x+1)(x+3)}$$

Step 3: Eliminate common factors
$$\frac{(\cancel{2x+1})(\cancel{x-4})(x+1)(\cancel{x+3})}{(x-2)(\cancel{x-4})(\cancel{2x+1})(\cancel{x+3})}$$

$$\frac{(x+1)}{(x-2)}$$

Example 13.4:
$$\frac{x^2+5x+6}{x^2-9} \cdot \frac{2x^2-6x}{x^2+7x+10} =$$

Step 1: Factor
$$\frac{(x+2)(x+3)}{(x-3)(x+3)} \cdot \frac{2x(x-3)}{(x+2)(x+5)}$$

Step 2: Multiply horizontally
$$\frac{2x(x+2)(x+3)(x-3)}{(x-3)(x+3)(x+2)(x+5)}$$

Step 3: Eliminate common factors
$$\frac{2x(\cancel{x+2})(\cancel{x+3})(\cancel{x-3})}{(\cancel{x-3})(\cancel{x+3})(\cancel{x+2})(x+5)}$$

$$\frac{2x}{(x+5)}$$

Dividing Rational Expressions

The steps to dividing rational polynomials are the same as for multiplying except that the second fraction is inverted and the division operator is changed to multiplication.

1. Factor as much as possible
2. Take the reciprocal of the second rational and change from division to multiplication
3. Multiply horizontally
4. Eliminate any factors common to both the numerator and denominator

Example 13.5:
$$\frac{x^2+5x+4}{x^2-8x+7} \div \frac{x^2+9x+20}{x^2+4x-5} =$$

Step 1: Factor as much as possible

$$\frac{(x+1)(x+4)}{(x-1)(x-7)} \div \frac{(x+5)(x+4)}{(x-1)(x+5)}$$

Step 2: Invert the second rational and change from division to multiplication

$$\frac{(x+1)(x+4)}{(x-1)(x-7)} \cdot \frac{(x-1)(x+5)}{(x+5)(x+4)}$$

Step 3: Multiply horizontally

$$\frac{(x+1)(x+4)(x-1)(x+5)}{(x-1)(x-7)(x+5)(x+4)}$$

Step 4: Eliminate any factors common to both the numerator and denominator

$$\frac{(x+1)\cancel{(x+4)}\cancel{(x-1)}\cancel{(x+5)}}{\cancel{(x-1)}(x-7)\cancel{(x+5)}\cancel{(x+4)}}$$

$$\frac{(x+1)}{(x-7)}$$

Example 13.6:

$$\frac{x^2 - 5x - 6}{6x - x^2} \div \frac{3x^2 + 4x + 1}{2x} =$$

Step 1: Factor as much as possible

$$\frac{(x+1)(x-6)}{x(6-x)} \div \frac{(3x+1)(x+1)}{2x}$$

Step 2: Invert the second rational and change from division to multiplication

$$\frac{(x+1)(x-6)}{x(6-x)} \cdot \frac{2x}{(3x+1)(x+1)}$$

Step 3: Multiply horizontally

$$\frac{2x(x+1)(x-6)}{x(6-x)(3x+1)(x+1)}$$

Step 4: Eliminate any factors common to both the numerator and denominator

Notice that $6 - x$ can be reversed to become $x - 6$ by factoring -1

$$\frac{2x(x+1)(x-6)}{-x(x-6)(3x+1)(x+1)}$$

$$\frac{-2}{3x+1}$$

Solving Equations involving Rational Expressions

Cross multiplication is the most effective way to solve an equation involving rational expressions. If there is more than one rational expression on either side of the equation, first find the LCD and then combine the rational expressions. Always check your answers to ensure that they fall within the domain of the original equation. Sometimes solutions may cause the denominator to equal zero which is not valid.

Chapter 13 – Rational Expressions   **159**

Example 13.7:
$$\frac{4}{x+2} = \frac{7}{x-4}$$

Cross multiply. It does not matter what goes on which side of the equation formed.

$4(x-4) = 7(x+2)$

Distribute and solve for x.

$4x - 16 = 7x + 14$

$4x = 7x + 14 + 16$

$4x = 7x + 30$

$4x - 7x = 30$

$-3x = 30$

$x = \dfrac{30}{-3}$

$x = -10$

Check the solution:

$$\frac{4}{-10+2} = \frac{7}{-10-4}$$

$$\frac{4}{-8} = \frac{7}{-14}$$

$$-\frac{1}{2} = -\frac{1}{2}$$

The statement is true and so the solution is valid.

Example 13.8:
$$\frac{8}{x} + \frac{3}{x} = 4$$

The LCD of the left-hand side is x. The denominator of both rational expressions is already x so combine.

$$\frac{8+3}{x} = 4$$

$$\frac{11}{x} = \frac{4}{1}$$

Cross multiply and solve.

$4x = 11$

$x = \dfrac{11}{4}$

Check solution:

$$\frac{8}{x} + \frac{3}{x} = 4$$

$$\frac{8}{\frac{11}{4}} + \frac{3}{\frac{11}{4}} = 4$$

$$\left(\frac{8}{1}\right)\left(\frac{4}{11}\right) + \left(\frac{3}{1}\right)\left(\frac{4}{11}\right) = 4$$

$$\frac{32}{11} + \frac{12}{11} = 4$$

$$\frac{44}{11} = 4$$

$$4 = 4$$

The statement is true and so the solution is valid.

Example 13.9:
$$\frac{4}{2-m} + \frac{7}{m-2} = \frac{m}{5}$$

The LCD of the left-hand side of the equation is $m - 2$. Remember that $2 - m$ can be reversed by factoring -1.

$$\frac{4}{-(m-2)} + \frac{7}{m-2} = \frac{m}{5}$$

$$\frac{-4}{m-2} + \frac{7}{m-2} = \frac{m}{5}$$

$$\frac{-4+7}{m-2} = \frac{m}{5}$$

$$\frac{3}{m-2} = \frac{m}{5}$$

Cross multiply and solve for m.
$$m(m-2) = 3(5)$$
$$m^2 - 2m = 15$$
$$m^2 - 2m - 15 = 0$$
$$(m+3)(m-5) = 0$$
$$m = -3 \text{ and } m = 5$$

Check your answers and eliminate any that are not valid. We leave this step up to you but both solutions are valid.

Inequalities and Rational Expressions

The temptation when faced with an inequality involving a rational is to cross multiply and solve. Inequalities are different from equations however and using this method will result in missing or incomplete solutions.

To solve an inequality involving rational expressions:
1. Move all terms to the left so that the right-hand side becomes 0.
2. Find the LCD if necessary and combine all the terms on the left-hand side.
3. Find the critical values of the inequality by setting both the numerator and denominator equal to 0 and solving.
4. Use these critical values to choose test points.
5. Use the test points to determine the range of values that satisfy the original inequality.

Example 13.10:
$$\frac{2x+5}{x+1} \leq 3$$

Step 1:
$$\frac{2x+5}{x+1} - 3 \leq 0$$

Step 2:

LCD: $x+1$

$$\frac{2x+5}{x+1} - \frac{3}{1} \cdot \frac{(x+1)}{(x+1)} \leq 0$$

$$\frac{2x+5-3(x+1)}{x+1} \leq 0$$

$$\frac{2x+5-3x-3}{x+1} \leq 0$$

$$\frac{2-x}{x+1} \leq 0$$

Step 3:

$2-x=0 \qquad x+1=0$
$x=2 \qquad x=-1$

Step 4:

Place the critical values on a number line and choose test points in each of the 3 regions formed.

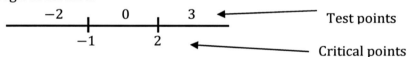

Test points

Critical points

Step 5:

Substitute the test points into original inequality to see which satisfy the inequality.

When $x=-2$	When $x=0$	When $x=3$
$\frac{2(-2)+5}{-2+1} \leq 3$	$\frac{2(0)+5}{0+1} \leq 3$	$\frac{2(3)+5}{3+1} \leq 3$
$\frac{-4+5}{-1} \leq 3$	$\frac{0+5}{1} \leq 3$	$\frac{6+5}{4} \leq 3$
$\frac{1}{-1} \leq 3$	$\frac{5}{1} \leq 3$	$\frac{11}{4} \leq 3$
$-1 \leq 3$	$5 \leq 3$	$2\frac{3}{4} \leq 3$
True	False	True

Therefore the inequality is true in two of the three regions.

Set Notation
$\{x | x \leq -1 \text{ or } x \geq 2\}$

Interval Notation
$(-\infty, -1] \cup [2, \infty)$

162   Prep for Success: Florida's PERT Math Study Guide

Chapter 13 Exercises

1. Simplify: $\dfrac{5x+4}{x^2+x-2} - \dfrac{3x-12}{x^2-5x+4}$

2. Simplify: $\dfrac{x^2+4x-21}{2x^2-17x+8} \cdot \dfrac{2x^2+9x-5}{x^2+2x-15}$

3. Simplify: $\dfrac{8+2x-x^2}{3x^2+4x-4} \div \dfrac{x^2-3x-4}{6x^3-x^2-2x}$

4. Simplify: $\dfrac{4x^2-9y^2}{2x^2-7xy+6y^2} \cdot \dfrac{2x^2-2xy-4y^2}{2x^2+5xy+3y^2}$

5. Simplify: $\dfrac{1}{x+3} + \dfrac{4}{x-2}$

6. Solve the equation: $\dfrac{3}{x} + \dfrac{5}{x+2} = 2$

7. Solve the inequality: $\dfrac{3+x}{1-x} \geq 2$

Solutions

1. Factor then simplify if possible

   $\dfrac{5x+4}{(x-1)(x+2)} - \dfrac{3(x-4)}{(x-1)(x-4)} = \dfrac{5x+4}{(x-1)(x+2)} - \dfrac{3}{(x-1)}$

   $\text{LCD} = (x-1)(x+2)$

   $\dfrac{5x+4}{(x-1)(x+2)} - \dfrac{3(x+2)}{(x-1)(x+2)}$

   $\dfrac{5x+4-3(x+2)}{(x-1)(x+2)}$

   $\dfrac{5x+4-3x-6}{(x-1)(x+2)} = \dfrac{2x-2}{(x-1)(x+2)} = \dfrac{2(x-1)}{(x-1)(x+2)} = \dfrac{2}{x+2}$

2. Factor as much as possible

   $\dfrac{(x+7)(x-3)}{(2x-1)(x-8)} \cdot \dfrac{(2x-1)(x+5)}{(x-3)(x+5)}$

   $\dfrac{(x+7)(x-3)(2x-1)(x+5)}{(2x-1)(x-8)(x-3)(x+5)} = \dfrac{x+7}{x-8}$

3. Factor as much as possible. Change division to multiplication and take the reciprocal of the second rational expression

$$\frac{-(x^2 - 2x - 8)}{3x^2 + 4x - 4} \div \frac{x^2 - 3x - 4}{x(6x^2 - x - 2)}$$

$$\frac{-(x + 2)(x - 4)}{(3x - 2)(x + 2)} \div \frac{(x + 1)(x - 4)}{x(3x - 2)(2x + 1)}$$

Change division to multiplication and flip the second rational expression
$$\frac{-(x + 2)(x - 4)}{(3x - 2)(x + 2)} \cdot \frac{x(3x - 2)(2x + 1)}{(x + 1)(x - 4)}$$

Multiply straight across then eliminate factors common to both the numerator and denominator.
$$\frac{-x(x + 2)(x - 4)(3x - 2)(2x + 1)}{(3x - 2)(x + 2)(x + 1)(x - 4)} = \frac{-x(2x + 1)}{(x + 1)}$$

4. Factor as much as possible
$$\frac{(2x + 3y)(2x - 3y)}{(2x - 3y)(x - 2y)} \cdot \frac{2(x + y)(x - 2y)}{(2x + 3y)(x + y)}$$

Eliminate any factors common to both the numerator and denominator.

$$\frac{2(2x + 3y)(2x - 3y)(x + y)(x - 2y)}{(2x - 3y)(x - 2y)(2x + 3y)(x + y)} = 2$$

5. Combine the rational expressions using the LCD
LCD $= (x + 3)(x - 2)$
$$\frac{(x - 2)1}{(x - 2)(x + 3)} + \frac{4(x + 3)}{(x - 2)(x + 3)}$$

$$\frac{x - 2 + 4(x + 3)}{(x - 2)(x + 3)}$$

$$\frac{x - 2 + 4x + 12}{(x - 2)(x + 3)}$$

$$\frac{5x + 10}{(x - 2)(x + 3)} = \frac{5(x + 2)}{(x - 2)(x + 3)}$$

6. Combine the fractions on the left
LCD: $x(x + 2)$
$$\frac{3(x + 2)}{x(x + 2)} + \frac{5(x)}{(x + 2)(x)} = 2$$

$$\frac{3x + 6 + 5x}{x(x + 2)} = 2$$

$$\frac{8x + 6}{x(x + 2)} = \frac{2}{1}$$
Cross multiply and solve

$2x(x+2) = 8x + 6$
$2x^2 + 4x - 8x - 6 = 0$
$2x^2 - 4x - 6 = 0$
$2(x^2 - 2x - 3) = 2(x-3)(x+1) = 0$
$x - 3 = 0 \qquad x + 1 = 0$
$x = 3 \qquad x = -1$
Both solutions are valid.

7. Move all terms to the left side of the equation.
$$\frac{3+x}{1-x} - \frac{2}{1} \geq 0$$
Find the LCD and combine the two rational expressions on the left
$$\frac{3+x}{1-x} - \frac{2(1-x)}{1-x} \geq 0$$
$$\frac{3+x-2(1-x)}{1-x} \geq 0$$
$$\frac{3x+1}{1-x} \geq 0$$

Solve for the critical values
$3x + 1 = 0 \qquad 1 - x = 0$
$x = -\frac{1}{3} \qquad x = 1$

Place the critical values on a number line and choose test points in each of the 3 regions formed.

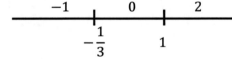

When $x = -1$ \qquad When $x = 0$ \qquad When $x = 2$

$\frac{3+(-1)}{1-(-1)} \geq 2 \qquad \frac{3+0}{1-0} \geq 2 \qquad \frac{3+2}{1-2} \geq 2$

$\frac{2}{2} \geq 2 \qquad \frac{3}{1} \geq 2 \qquad \frac{5}{-1} \geq 2$

$1 \geq 2$ false \qquad $3 \geq 2$ true \qquad $-5 \geq 2$ false

Therefore the solution is $\left[-\frac{1}{3}, 1\right)$

Chapter 14 – Systems of Equations   **165**

Chapter 14 | Systems of Equations

Solving Systems of Equations

Given two equations, a solution exists if the two equations intersect so that there is at least one common point.

Some Common key words:

Independent	One solution (lines intersect at one point)
Dependent	Infinitely many solutions (equations represent the same line)
Consistent	At least one solution
Inconsistent	No solutions (lines are parallel or do not intersect)

Solving by Inspection

If you can easily identify patterns in the coefficients of x and y, you can use inspection to predict the number of solutions that a system of linear equations will have. This can save you the trouble of solving in cases where there is no solution or infinitely many solutions.

The slope and the y–intercept of the lines are impacted by the values of the coefficients of x and y so we inspect the coefficients as follows:

One Solution

Systems that have one solution will have different slopes. If in the form $y = mx + b$, their slopes will be different. If in the form $ax + by = c$, then a_1 and b_1 are **NOT** multiples of a_2 and b_2

Examples:

$y = 5x + 9$
$y = -2x + 3$
$m_1 = 5 \ \& \ m_2 = -2$
Different slopes = 1 solution

$x - 5y = 11$
$3x + 2y = -9$
Different values of a and b and no visible multiples = 1 solution

No Solution

Lines that have no solution will be parallel and have the same slope. If in the form $y = mx + b$, their slopes will be the same but their y–intercepts will be different. If in the form $ax + by = c$, a_1 and b_1 are multiples of a_2 and b_2 by some factor but c_1 is not a multiple of c_2 by the same factor.

Examples:

$y = 3x + 11$
$y = 3x + 4$
Same slope,
Different y–intercepts = No solution

$-2x + 3y = 5$
$4x - 6y = 7$
$a_1 = -2$ & $b_1 = 3$ and we can obtain
a_2 & b_2 by multiplying by –2
$-2(-2x + 3y) = 4x - 6y$
Same slope = No solution
Note that $-2c_1 \neq c_2$ so y–intercepts
are different

Infinitely Many Solutions

These lines have the same slope and the same y–intercept so they are just different representations of the same line. Therefore when graphed, they lie on top of each other. If in the form $y = mx + b$, then $m_1 = m_2$ and $b_1 = b_2$. If in the form $ax + by = c$, then a_1, b_1 and c_1 are multiples of a_2, b_2 and c_2 by the same factor.

Example:

$-2x + y = 4$
$6x - 3y = -12$
$a_1 = -2, b_1 = 1$ & $c_1 = 4$ and we can obtain a_2, b_2 & c_2 by multiplying by –3
$-3(-2x + y = 4) \implies 6x - 3y = -12$
Both equations represent the same line and there are infinitely many solutions.

Solving Algebraically

There are two ways to solve a system of equations algebraically: **elimination** and **substitution**. If the result of substitution or elimination is a single point (x, y) then we say that the graphs are consistent and independent. If substitution or elimination yields a true statement that has no variables (e.g. 0 = 0), then we say that the graphs are consistent and dependent. If the result is a false statement (e.g. 0 = 5), then we say that the graphs are parallel and inconsistent.

Elimination

To solve a system of equations by elimination:
1. Multiply one or more of the equations by some factor so that the coefficients of the same variable in both equations match but with opposite signs.
2. Add the equations so that the chosen variable is eliminated.
3. Solve for the other variable.
4. Substitute the value of the solved variable into either of the original equations and solve for the value of the other variable.

Example 14.1:

Solve the system of equations:
$2x + 3y = 8$
$2x - y = 0$

Step 1:
>
> The coefficients of the variable x already match so multiply either equation by -1.
> $-(2x - y = 0)$
> $-2x + y = 0$

Step 2:
>
> Add the equations.
> $$2x + 3y = 8$$
> $$\underline{-2x + y = 0}$$
> $$4y = 8$$

Step 3:
>
> Solve for the remaining variable.
> $4y = 8$
> $y = \dfrac{8}{4} = 2$

Step 4:
>
> Substitute $y = 2$ into either original equation.
> $2x + 3(2) = 8$
> $2x + 6 = 8$
> $2x = 2$
> $x = 1$

There is one solution $(1, 2)$ to the system so the equations are consistent and independent.

Example 14.2:

Solve the system of equations:

$3x - 2y = 11$
$-9x + 6y = -33$

Step 1:
>
> Choose one of the variables to eliminate. We randomly chose y. Multiply the top equation by 3 so that the coefficients of y match. They are already opposite in sign.
> $3(3x - 2y = 11)$
> $9x - 6y = 33$

Step 2:
>
> Add the equations.
> $$9x - 6y = 33$$
> $$\underline{-9x + 6y = -33}$$
> $$0 = 0$$
>
> Since the result is a true statement but there is no variable, we conclude that the equations are consistent and dependent. You may also see the answer choice in coordinate format which requires that either of the original equations be solved for y.
> $3x - 2y = 11$
> $-2y = -3x + 11$
> $y = \dfrac{-3}{-2}x + \dfrac{11}{-2}$

168   Prep for Success: Florida's PERT Math Study Guide

$$y = \frac{3}{2}x - \frac{11}{2}$$

Solution is:

$$\left(x, \frac{3}{2}x - \frac{11}{2}\right)$$

Substitution

It is often easier to use the substitution method instead of elimination when one of the variables in the two equations has a coefficient of 1.

To solve a system of equations by substitution:
1. Solve either equation for x or for y.
2. Substitute for the chosen variable in the other equation.
3. Solve the new equation.
4. Substitute the value of the solved variable into either of the original equations and solve for the value of the other variable.

Example 14.3:

Solve the system of equations:

$2y = x - 3$

$3y + 2x = -8$

Step 1:

Solving the first equation for x would be easier because it has a coefficient of 1.

$2y = x - 3$

$x = 2y + 3$

Step 2:

Substitute for x in the second equation.

$3y + 2(2y + 3) = -8$

Step 3:

Solve the new equation for y.

$3y + 4y + 6 = -8$

$7y + 6 = -8$

$7y = -14$

$y = \frac{-14}{7}$

$y = -2$

Step 4:

Substitute for y in either of the two original equations. Using the first equation would be easier.

$2y = x - 3$

$2(-2) = x - 3$

$-4 = x - 3$

$x = -4 + 3 = -1$

There is one solution (−1, −2) to the system so the equations are consistent and independent.

Example 14.4:
Solve the system of equations:
$x + 4y = 8$
$3x + 12y = 3$

Step 1:
Solving the first equation for x would be easier because it has a coefficient of 1.
$x + 4y = 8$
$x = 8 - 4y$

Step 2:
Substitute for x in the second equation
$3(8 - 4y) + 12y = 3$

Solve for y
$24 - 12y + 12y = 3$
$24 = 3$

Since the result is a false statement and there is no variable, we conclude that the equations are inconsistent. You may see the answer choice as inconsistent or parallel.

Example 14.5:
Solve the system of equations:
$y - 2x = -5$
$y = -x^2 + 4x - 2$

The systems of equations most commonly found on the PERT involve linear equations but may occasionally involve parabolas.

Since the line has variable of coefficient equal to 1, we can use substitution.
$y - 2x = -5$
$y = 2x - 5$

Substitute into the quadratic equation.
$y = -x^2 + 4x - 2$
$2x - 5 = -x^2 + 4x - 2$
$x^2 - 2x - 3 = 0$
$(x + 1)(x - 3) = 0$
$x = -1 \quad x = 3$

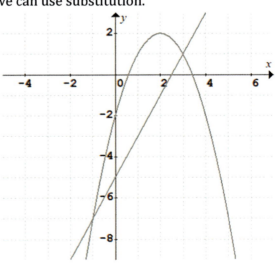

Now substitute the values of x into either of the two original equations to find the values of y.

When $x = -1$ 　　　　When $x = 3$
$y = 2x - 5$ 　　　　　$y = 2x - 5$
　$= 2(-1) - 5$ 　　　　$= 2(3) - 5$
　$= -2 - 5 = -7$ 　　　$= 6 - 5 = 1$
$(-1, -7)$ 　　　　　　$(3, 1)$

There are therefore two solutions as shown on the sketch of the graphs.

Solving Systems of Equations Graphically

If the graphs are not given, sketch the graphs using the techniques learned in Chapter 7. Then inspect the graphs to see if they intersect at one point, are parallel or are the same line.

Example 14.6:
$2x + 3y = -9$
$y + x = -2$
Find the solution to the pair of lines above graphically.

Solving both equations for y:

$2x + 3y = -9$ $\qquad$ $y + x = -2$
$3y = -2x - 9$ $\qquad$ $y = -x - 2$
$y = -\frac{2}{3}x - 3$

The lines have different slopes and y-intercepts so they will intersect at one point. Graph the lines on the same coordinate plane. If you are attempting to do this by hand, try to keep the scale as consistent as possible so the solution will be easier to read off your sketch.

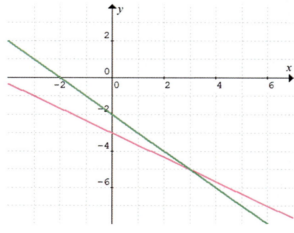

From our sketch we see that the graphs intersect at one point: $(3, -5)$.

** Note **

If the graphs are not given, most students prefer to solve using algebra to avoid the potential inaccuracy of graphing by hand. If however you are allowed the onscreen graphing calculator on these types of questions, it may actually be easier to solve by graphing.

Applications

Some word problems on the PERT involve formulating a system of equations before solving. In such problems, you are usually provided with a couple of statements involving two variables, then asked to find the value of one or both.

Example 14.7:
Susan deposited a total of $18,900 into two separate interest bearing savings accounts. If the first account earns 4% interest annually and the second account earns 6% annually, how much would she need to deposit into the account earning 6% in order to accrue a total of $970 in interest by the end of one year?

A. $2,850 $\qquad$ B. $8,200 $\qquad$ C. $10,700 $\qquad$ D. $16,050

Let the amount of money deposited in the first account $= x$
Let the amount of money deposited in the second account $= y$

Since the total deposited equals $18,900$, the first equation is given by:
$x + y = 18,900$

The second equation is based on the amount of interest earned.
The first account earns $4\% = 4\%$ of $x = 0.04x$
The second account earns $6\% = 6\%$ of $y = 0.06y$

Total interest earned:
$0.04x + 0.06y = 970$
To avoid working with decimals, multiply the entire equation by 100.
$100(0.04x + 0.06y = 970)$
$4x + 6y = 97,000$

Now we have two equations that we can solve via substitution because the first equation has variables with coefficients of 1. Note that we were asked to find the amount deposited into the account earning 6% so we need to find the value of y. So we substitute for x.
$x + y = 18,900$
$x = 18,900 - y$
Substitute into the second equation.
$4x + 6y = 97,000$
$4(18,900 - y) + 6y = 97,000$
$75,600 - 4y + 6y = 97,000$
$2y = 21,400$
$y = \dfrac{21,400}{2} = \$10,700$
Therefore, she needs to deposit $10,700 into the account earning 6% interest and the answer is C.

Chapter 14 Exercises

1. Solve the system of equations:
   $5x - 2y = 3$
   $3x + 4y = 20$

2. Solve the system of equations:
   $2x - y = 5$
   $-8x + 4y = 12$

3. Edward is 4 less than twice Jacob's age. The sum of their ages is 44. How old is Edward?

4. The sum of two numbers is 65. One number is 5 more than three times the other number. Find the smaller number.

5. A scoop of ice cream contains 60 more than twice the number of calories in a cup of frozen yogurt. Together, they have a total of 1,650 calories. How many calories are there in a scoop of ice cream?
   A. 530 calories B. 795 calories C. 855 calories D. 1,120 calories

6. Solve the system of equations:
   $y - 2x = 3$
   $8x + 12 = 4y$

7. Solve the system of equations:
   $5y = x^2 + 15$
   $y - x = 3$

8. A candy store creates a mixture of jelly beans and gummy bears. The jelly beans sell for $1.20 per lb while the gummy bears sell for $0.80 and the mixture is worth $0.95 per lb. If they created 20 lbs of the mixture, how many lbs of jelly beans did they add to the mixture?
   A. 2.5 lbs B. 7.5 lbs C. 12.5 lbs D. 17.5 lbs

9. Solve the system of equations:
   $y - x = 1$
   $2x + 3y = -12$

10. Which of the following systems of equations is represented by the graph?
    A. $2x - 3y = 6$
       $y - 2x = -3$
    B. $3y - 2x = 6$
       $y + 2x = 3$
    C. $y - 2x = 3$
       $3y + 2x = -6$
    D. $y - 2x = -3$
       $3y + 2x = 6$

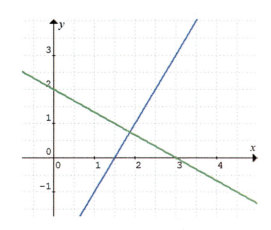

Solutions

1. The elimination method would work well here since no coefficient is equal to 1.
   $2(5x - 2y = 3) \Rightarrow 10x - 4y = 6$
   Add the two equations
   $10x - 4y = 6$
   $3x + 4y = 20$
   _____
   $13x = 26$

   $x = \dfrac{26}{13} = 2$

   Substitute $x = 2$ into either original equation
   $3(2) + 4y = 20$
   $6 + 4y = 20$
   $4y = 14$
   $y = \dfrac{14}{4} = \dfrac{7}{2}$
   Solution is $\left(2, \dfrac{7}{2}\right)$

   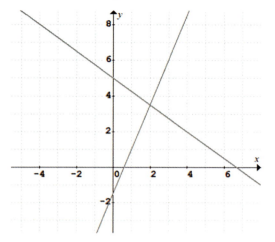

2. The elimination method would work best here.
   $4(2x - y = 5) \Rightarrow 8x - 4y = 20$
   Add the two equations
   $8x - 4y = 20$
   $-8x + 4y = 12$
   _____
   $0 = 32$

   There is no solution. The lines are parallel and inconsistent.

   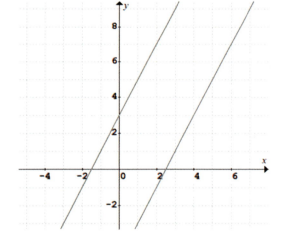

3. Let Edward's age $= x$
   Let Jacob's age $= y$
   Formulate equations based on statements
   $x = 2y - 4$
   $x + y = 44$
   Substitute the 1st equation into the 2nd equation
   $(2y - 4) + y = 44$
   $3y - 4 = 44$
   $3y = 48$
   $y = \dfrac{48}{3} = 16$
   Substitute $y = 16$ into the first equation
   $x = 2(16) - 4 = 32 - 4 = 28$
   Therefore Edward is 28 years old.

4. Let numbers be x and y

   Formulate equations based on statements

   $x + y = 65$

   $y = 3x + 5$

   Substitute the second equation into the first equation

   $x + (3x + 5) = 65$

   $4x = 60$

   $x = \dfrac{60}{4} = 15$

   Substitute $x = 15$ into the second equation

   $y = 3(15) + 5$

   $y = 45 + 5 = 50$

   Therefore the smaller number is 15.

5. **Answer: [D]**   Let the number of calories in a cup of yogurt $= x$

   Let the number of calories in a scoop of ice cream $= y$

   Since the number of calories in a scoop of ice cream is 60 more than twice the number of calories in a cup of yogurt:

   $y = 2x + 60$

   The total number of calories in both equals 1650. So:

   $x + y = 1650$

   Substitute the first equation into the second equation.

   $x + y = 1650$

   $x + (2x + 60) = 1650$

   $3x + 60 = 1650$

   $3x = 1590$

   $x = \dfrac{1590}{3} = 530$

   Substitute into the first equation.

   $y = 2x + 60$

   $y = 2(530) + 60 = 1060 + 60 = 1120$

   There are 1, 120 calories in a scoop of ice cream.

6. We can solve using substitution because the coefficient of y in the first equation is 1. But if we look at the coefficients of x and y we can tell that they may be multiples of each other. Rearrange the equations into standard form to properly inspect.

   $-2x + y = 3$

   $8x - 4y = -12$

   Now we can clearly see that a_2, b_2 & c_2 are factors of a_1, b_1 & c_1 by a factor of -4. Therefore, the equations are 2 forms of the same line and there are infinitely many solutions.

7. In this problem, we have a parabola (quadratic equation) and a line. Since the line has variable of coefficient equal to 1, we can use substitution.
$y - x = 3$
$y = x + 3$
Substitute into the quadratic equation.
$5y = x^2 + 15$
$5(x + 3) = x^2 + 15$
$5x + 15 = x^2 + 15$
$x^2 - 5x = 0$
$x(x - 5) = 0$
$x = 0 \quad x = 5$
Now substitute the values of x into either of the two original equations to find the values of y.

When $x = 0$ \qquad When $x = 5$
$y = x + 3$ \qquad\quad $y = x + 3$
$\ \ = 0 + 3 = 3$ \qquad $\ \ = 5 + 3 = 8$
$(0, 3)$ \qquad\qquad\quad $(5, 8)$

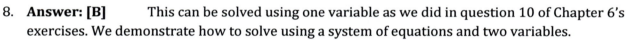

There are therefore two solutions as shown on the sketch of the graphs.

8. **Answer: [B]**   This can be solved using one variable as we did in question 10 of Chapter 6's exercises. We demonstrate how to solve using a system of equations and two variables.
Let number of lbs of jelly beans $= x$
Let number of lbs of gummy bears $= y$
There are 20 lbs of the mix so:
$x + y = 20$
Create the 2nd equation based on the value of the jelly beans, gummy bears and the mixture.
$1.20x + 0.80y = 0.95(20)$
So that we don't have to work with decimals, multiply the entire equation by 100.
$120x + 80y = 95(20)$
$120x + 80y = 1900$
Since we are looking for the number of lbs of jelly beans (x) we solve the first equation for y and substitute into the second equation.
$x + y = 20$
$y = 20 - x$
$120x + 80y = 1900$
$120x + 80(20 - x) = 1900$
$120x + 1600 - 80x = 1900$
$40x = 300$
$x = \dfrac{300}{40} = \dfrac{30}{4} = 7.5$ *lbs of jelly beans*

9. Solve the first equation for y
$y = x + 1$
Substitute into the second equation

$2x + 3y = -12$
$2x + 3(x + 1) = -12$
$2x + 3x + 3 = -12$
$5x = -15$
$x = \dfrac{-15}{5} = -3$

Find the value of y by substituting into either of the two original equations.
$y = x + 1$
$y = -3 + 1 = -2$
There is one solution to the system at $(-3, -2)$.

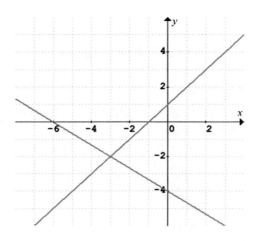

10. **Answer: [D]** The method that we have found to work for most students is to identify the x and y intercepts and formulate the equations using the slope and y-intercept.

$(0, -3)$ and $(1.5, 0)$
$m = \dfrac{0 - (-3)}{1.5 - 0} = \dfrac{3}{1.5} = 2$
Therefore:
$y = mx + b$
$y = 2x - 3$
$y - 2x = -3$

$(0, 2)$ and $(3, 0)$
$m = \dfrac{0 - 2}{3 - 0} = -\dfrac{2}{3}$
Therefore:
$y = mx + b$
$y = -\dfrac{2}{3}x + 2$
$3\left(y = -\dfrac{2}{3}x + 2\right)$
$3y = -2x + 6$
$3y + 2x = 6$

The two equations are $y - 2x = -3$ and $3y + 2x = 6$ and the answer is D.

Chapter 15 – Inverse & Composite Functions   **177**

Chapter 15 | Inverse & Composite Functions

Inverse Functions
- -

For a function $f(a) = b$ the inverse function has the property that $f^{-1}(b) = a$. This means that the domain of the function $f(x)$ becomes the range of the inverse function and the range of $f(x)$ becomes the domain of the inverse function.

If $f(3) = 7$ then $f^{-1}(7) = 3$

To find the inverse of a function:
1. Replace $f(x)$ with y if necessary
2. Interchange x and y
3. Solve for y
4. Replace y with $f^{-1}(x)$

These steps can also be used to find $f(x)$ starting with $f^{-1}(x)$.

Example 15.1:
If $f(x) = x^2 + 5$ then $f^{-1}(21) = ?$

1. Replace $f(x)$ with y
   $y = x^2 + 5$

2. Interchange x and y
   $x = y^2 + 5$

3. Solve for y
   $y^2 = x - 5$
   $y = \sqrt{x - 5}$

4. Replace y with $f^{-1}(x)$
   $f^{-1}(x) = \sqrt{x - 5}$

   To complete the question, substitute $x = 21$ into the equation of the inverse function.
   $f^{-1}(21) = \sqrt{21 - 5}$

   $f^{-1}(21) = \sqrt{16}$

   $f^{-1}(21) = 4$

178   Prep for Success: Florida's PERT Math Study Guide

Example 14.2:

If $h(x) = \dfrac{2x}{x+1}$ then $h^{-1}(x) = ?$

1. Replace $h(x)$ with y

$$y = \dfrac{2x}{x+1}$$

2. Interchange x and y

$$x = \dfrac{2y}{y+1}$$

3. Solve for y

Since the right side is a rational, it is best to cross-multiply

$$\dfrac{x}{1} = \dfrac{2y}{y+1}$$

$$x(y+1) = 2y$$

$$xy + x = 2y$$

Move the terms containing y to one side
$$2y - xy = x$$

Factor y from the left-hand side of the equation
$$y(2 - x) = x$$

$$y = \dfrac{x}{2-x}$$

4. Replace y with $h^{-1}(x)$

$$h^{-1}(x) = \dfrac{x}{2-x}$$

Example 15.3:

If $f^{-1}(x) = \dfrac{x+2}{3}$ and $f^{-1}(x)$ is the inverse of the function $f(x)$, then $f(-4) = ?$

We were provided with the inverse function $f^{-1}(x)$ and asked to find the value of the original function $f(x)$ at -4. The long way to solve this problem would be to find the inverse of $f^{-1}(x)$ to get $f(x)$ then substitute for $x = -4$.

There is however an easier method. Comparing the relationship $f(a) = b$ and $f^{-1}(b) = a$ to the given information:

$$f(a) = \mathbf{b} \qquad\qquad f^{-1}(\mathbf{b}) = a$$
$$f(-4) = ? \qquad\qquad f^{-1}(x) = \dfrac{x+2}{3}$$

From the comparison, we see that we are looking for the value of b. We also see the $b = x$ so that solving for x would give b. Finally, $a = -4$ and $a = \frac{x+2}{3}$ so we can set the two equal to each other and solve for x.

Chapter 15 – Inverse & Composite Functions   **179**

We demonstrate both methods below:

Method 1:

1. Replace $f^{-1}(x)$ with y

    $$y = \frac{x+2}{3}$$

2. Interchange x and y

    $$x = \frac{y+2}{3}$$

3. Solve for y

    $$\frac{x}{1} = \frac{y+2}{3}$$

    Cross multiply
    $$3x = y + 2$$
    $$y = 3x - 2$$

4. Replace y with $f(x)$
    $$f(x) = 3x - 2$$

    Now solve for $f(-4)$
    $$f(-4) = 3(-4) - 2 = -12 - 2 = -14$$

Method 2:

Set and solve
$$\frac{x+2}{3} = -4$$

$$\frac{x+2}{3} = \frac{-4}{1}$$

Cross multiply
$$x + 2 = -4(3)$$
$$x + 2 = -12$$
$$x = -12 - 2 = -14 \text{ therefore } f(-4) = -14$$

Composite Functions

The composite of two functions $f(x)$ and $g(x)$ is defined by:

$$(f \circ g)(x) = f\big(g(x)\big)$$

This means that we substitute all instances of x in the function $f(x)$ by the function $g(x)$.
Always work from right to left.

$(h \circ g)(x)$ means substitute $g(x)$ into $h(x)$
$(f \circ g \circ h)(x)$ means substitute $h(x)$ into $g(x)$ then substitute the result into $f(x)$

180 Prep for Success: Florida's PERT Math Study Guide

Example 15.4:

Find $(g \circ f \circ h)(x)$ if $f(x) = x + 3$, $g(x) = \dfrac{3}{2-x}$ and $h(x) = 4x - 7$

$$(g \circ f \circ h)(x) = g\left(f\big(h(x)\big)\right)$$

$$f(x) = x + 3$$

$$f\big(h(x)\big) = (4x - 7) + 3$$

$$= 4x - 4$$

$$g\left(f\big(h(x)\big)\right) = \frac{3}{2 - (4x - 4)} = \frac{3}{2 - 4x + 4} = \frac{3}{6 - 4x}$$

Example 15.5:

If $h(x) = \dfrac{x+3}{5-x}$ and $g(x) = \dfrac{3}{x}$, then $h\big(g(2)\big) =$

First find $g(2)$

$$g(x) = \frac{3}{x}$$

$$g(2) = \frac{3}{2}$$

Therefore:

$$h(g(2)) = \frac{\frac{3}{2} + 3}{5 - \frac{3}{2}}$$

Simplify the numerator and denominator

$$= \frac{\dfrac{3}{2} + \dfrac{3}{1} \cdot \dfrac{2}{2}}{\dfrac{2}{2} \cdot \dfrac{5}{1} - \dfrac{3}{2}} = \frac{\dfrac{3}{2} + \dfrac{6}{2}}{\dfrac{10}{2} - \dfrac{3}{2}} = \frac{\dfrac{9}{2}}{\dfrac{7}{2}}$$

$$= \frac{9}{2} \times \frac{2}{7} = \frac{9}{7}$$

Rule of Inverses

If two functions $f(x)$ and $g(x)$ are inverses then $f(g(x)) = g(f(x)) = x$

Example 15.6:

Given that $f(x) = x^3 + 5$ and $g(x) = \sqrt[3]{x - 5}$ verify that $f(x)$ and $g(x)$ are inverse functions.

$$f\big(g(x)\big) = \left(\sqrt[3]{x - 5}\right)^3 + 5$$

$$= (x - 5) + 5 = x$$

Therefore $f(x)$ and $g(x)$ are inverse functions.

Chapter 15 Exercises

1. Given the table of values for the functions $g(x)$ and $h(x)$, find the value of $g(h(-1))$.

x	-2	-1	0	1	2	3	4
$g(x)$	-1	2	13	7	9	-5	8
$h(x)$	0	4	-1	6	-2	-9	-11

2. From the graph, what is the value of $g(f(-1))$?

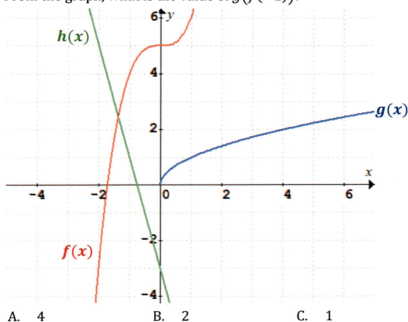

   A. 4   B. 2   C. 1   D. Does not exist

3. What is the domain of $(f/g)(x)$ given that $f(x) = x^2 - 9$ and $g(x) = \sqrt{2-x}$?
   A. $\{x | x \neq -3, -2, 3\}$ B. $\{x | x < 2\}$ C. $(-\infty, 2]$ D. All Real Numbers

4. Given that $f(x) = x - 4$ and $g(x) = 3x^2 - 2x + 11$, find $g(f(x))$

5. Given that $f(x) = x + 2$ and $g(x) = 4 - x^2$, find $f(g(-3))$

6. Find $(f - g)(x)$ if $f(x) = x^3 + 5x - 13$ and $g(x) = x^2 + 23x - 8$

7. Given $g(x) = \dfrac{2}{x}$ and $f(x) = \dfrac{3}{x^2 - 1}$, find $f(g(x))$

8. Find the inverse of the function $f(x) = \dfrac{1 - 3x}{x + 4}$

9. If $f^{-1}(x) = \dfrac{5x}{x + 2}$ and $f^{-1}(x)$ is the inverse of the function $f(x)$ then $f(-3) = ?$

182 Prep for Success: Florida's PERT Math Study Guide

10. Which of the following is the inverse of the function $f(x) = \dfrac{2x}{5-x}$?

A. $f^{-1}(x) = \dfrac{x+5}{2x}$ B. $f^{-1}(x) = \dfrac{5x}{2+x}$ C. $f^{-1}(x) = \dfrac{5-x}{2x}$ D. $f^{-1}(x) = \dfrac{5x}{x-2}$

Solutions

1. First determine the value of $h(x)$ when $x = -1$ using the third row of the table.
   $h(-1) = 4$
   So $g\big(h(-1)\big)$ becomes $g(4)$
   Now determine the value of $g(x)$ when $x = 4$ using the second row of the table.
   $g(4) = 8$
   Therefore $g\big(h(-1)\big) = 8$

2. **Answer: [B]**   First read off the value of $f(x)$ when $x = -1$ using the red graph.
   $f(-1) = 4$
   So $g\big(f(-1)\big)$ becomes $g(4)$
   Now read off the value of $g(x)$ when $x = 4$ using the blue graph.
   $g(4) = 2$
   Therefore $g\big(f(-1)\big) = 2$ and the answer is B.

3. **Answer: [B]**   The domain of the composite function will depend on the domain of the two original functions.
   $$\left(\frac{f}{g}\right)(x) = \frac{x^2 - 9}{\sqrt{2-x}}$$
   Since $f(x)$ is a polynomial, its domain is all real numbers.
   $g(x)$ is an even root so the expression under the radical cannot evaluate to a negative number. Furthermore, since it is in the denominator, $g(x) \neq 0$ because that would result in $(f/g)(x)$ being undefined. Therefore:
   $2 - x > 0$
   $-x > -2$
   $x < 2$
   The domain is therefore $\{x | x < 2\}$. Note that the answer $(-\infty, 2]$ is incorrect because it includes the value $x = 2$ in the domain and this would make the denominator equal to 0.

4. Replace x in $g(x)$ with $f(x)$
   $$\begin{aligned} g\big(f(x)\big) &= 3(x-4)^2 - 2(x-4) + 11 \\ &= 3(x^2 - 8x + 16) - 2x + 8 + 11 \\ &= 3x^2 - 24x + 48 - 2x + 8 + 11 \\ &= 3x^2 - 26x + 67 \end{aligned}$$

5. First find the value of $g(-3)$

$$g(x) = 4 - x^2$$
$$g(-3) = 4 - (-3)^2 = 4 - 9 = -5$$
So $f(g(-3))$ becomes $f(-5)$
$$f(x) = x + 2$$
$$f(-5) = -5 + 2 = -3$$
Therefore $g(f(-5)) = -3$

6. To find $(f - g)(x)$ subtract $g(x)$ from $f(x)$
$$(f - g)(x) = (x^3 + 5x - 13) - (x^2 + 23x - 8)$$
$$= x^3 + 5x - 13 - x^2 - 23x + 8$$
$$= x^3 - x^2 - 18x - 5$$

7. Replace x in $f(x)$ with $g(x)$ then simplify
$$f(g(x)) = \frac{3}{\left(\frac{2}{x}\right)^2 - 1} = \frac{3}{\frac{4}{x^2} - 1}$$

Find the LCD and combine the terms in the denominator
$$= \frac{3}{\frac{4}{x^2} - \frac{1}{1} \cdot \frac{x^2}{x^2}} = \frac{3}{\frac{4 - x^2}{x^2}}$$

Flip the rational in the denominator and multiply.
$$= \frac{3x^2}{4 - x^2}$$

8. To find the inverse, first interchange x and y
$$y = \frac{1 - 3x}{x + 4}$$

$$x = \frac{1 - 3y}{y + 4}$$

Set x over 1 then cross multiply
$$x(y + 4) = 1 - 3y$$
$$xy + 4x = 1 - 3y$$
Move all terms containing y to one side of the equation
$$xy + 3y = 1 - 4x$$
Factor then solve for y

$$y(x + 3) = 1 - 4x$$

$$y = \frac{1 - 4x}{x + 3}$$

$$f^{-1}(x) = \frac{1 - 4x}{x + 3}$$

184   Prep for Success: Florida's PERT Math Study Guide

9. Using the method learned in Example 15.3, set $f^{-1}(x) = -3$

$$\frac{5x}{x+2} = \frac{-3}{1}$$

$$-3(x+2) = 5x$$
$$-3x - 6 = 5x$$
$$-8x = 6$$

$$x = -\frac{6}{8} = -\frac{3}{4}$$

Therefore $f(-3) = -\frac{3}{4}$

10. **Answer: [B]**   To find the inverse, first interchange x and y

$$y = \frac{2x}{5-x}$$

$$x = \frac{2y}{5-y}$$

Set x over 1 then cross multiply
$$x(5-y) = 2y$$
$$5x - xy = 2y$$
Move all terms containing y to one side of the equation
$$2y + xy = 5x$$
Factor then solve for y
$$y(2+x) = 5x$$

$$y = \frac{5x}{2+x}$$

$$f^{-1}(x) = \frac{5x}{2+x}$$

Chapter 16 — Geometric Reasoning

Geometric Reasoning

Chapter 6 introduced some basic geometric formulas. The same basic formulas apply here except that the problems and diagrams are more complex and will involve more in-depth thinking. Some new formulas are also introduced.

Example 16.1:
A triangular plot of land has one side equal to twice the length of the shortest side and the third side 20 feet longer than the shortest side. If the perimeter of the plot is 200 feet, what is the length of the longest side?

Let the shortest side $= x$
Second side $= 2x$
Third side $= x + 20$

The perimeter of a triangle is the sum of all the sides.
$x + 2x + x + 20 = 200$

Now that we have created the formula, proceed with solving for x.
$4x + 20 = 200$
$4x = 180$
$x = \dfrac{180}{4} = 45$

We have just found the length of the shortest side but the question has asked for the length of the longest side. Substitute the value of x into the expressions representing the three sides of the equation to determine which is the longest.

Shortest side: $x = 45$
Second side: $2x = 2(45) = 90$
Third side: $x + 20 = 45 + 20 = 65$

Therefore the longest side of the plot of land is 90 feet long.

Example 16.2:
A wooden rectangular frame measures 6 inches by 8 inches as shown in the diagram to the right. If the distance between the outer edge of the frame and the glass insert is x and the perimeter of the glass is 20 inches, what is the area of the wooden part of the frame?

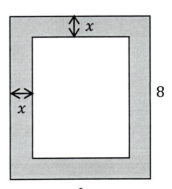

To find the area of the wooden frame, we need to subtract the area of the glass from the area of the outer rectangle. The dimensions of the outer rectangle are given so the area can be calculated easily. For the area of the glass insert, we need the dimensions which were not given. We can, however, create expressions based on x to represent the length and width.
Width $= 6 - x - x = 6 - 2x$ Length $= 8 - x - x = 8 - 2x$

The question provides us with the perimeter of the rectangular glass insert. The equation to find the perimeter of a rectangle is $P = 2l + 2w$

Using the expressions for the length and width of the glass insert, the equation becomes:
$P = 2(8 - 2x) + 2(6 - 2x) = 20$
Solve for x
$2(8 - 2x) + 2(6 - 2x) = 20$
$16 - 4x + 12 - 4x = 20$
$-8x + 28 = 20$
$-8x = -8$
$x = \dfrac{-8}{-8} = 1$

The dimensions of the glass are:
Width $= 6 - 2x = 6 - 2(1) = 4$ Length $= 8 - 2x = 8 - 2(1) = 6$

Area of the outer rectangle:
$A = l \times w = 8 \times 6 = 48 \ inches^2$
Area of the glass insert:
$A = l \times w = 6 \times 4 = 24 \ inches^2$

Therefore the area of the wooden frame:
$48 - 24 = 24 \ inches^2$

Example 16.3:
Six squares are placed inside a rectangle as shown in the diagram. If the perimeter of the rectangle is 25 inches, what is its area?

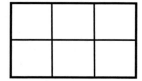

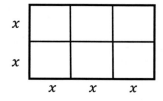

Remembering that the lengths of the sides of a square are equal:
Let the length of one side $= x$

We can now find expressions for the dimensions of the rectangle from the diagram.

Length of rectangle = 3x
Width of rectangle = 2x
$P = 2l + 2w = 25$
$2(3x) + 2(2x) = 25$
$6x + 4x = 25$
$10x = 25$
$x = \dfrac{25}{10} = 2.5 \; inches$

Therefore:
Length of rectangle = $3x = 3(2.5) = 7.5 \; inches$
Width of rectangle = $2x = 2(2.5) = 5 \; inches$
Now that we know the length and width of the rectangle, we can find its area.
$Area = l \times w = 7.5 \times 5 = 37.5 \; inches^2$

Let's review the different types of triangles and add some more information about them.

Types of Triangles

Triangle   **Property**

Isosceles

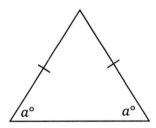

Two congruent (equal) sides.
The base angles are equal.

Equilateral

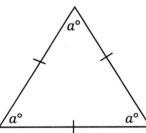

All three sides congruent.
All three angles equal.

Scalene

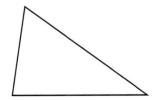

No sides equal.
No angles equal.

The sum of the angles in a triangle is 180°. The sum of any two sides of a triangle is greater than the third. The notation for angle is ∠ so ∠A is read "angle A."

Example 16.4:
Find the value of m in the isosceles triangle below.

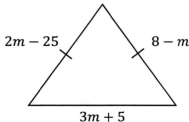

Set the congruent sides equal and solve
$2m - 25 = 8 - m$
$2m + m = 8 + 25$
$3m = 33$
$m = \dfrac{33}{3} = 11$

Pythagorean Theorem

Right-angled triangles are triangles in which one of the angles is a 90° angle.

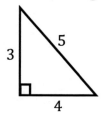

They use a special formula called the **Pythagorean Theorem**.
In a right-angled triangle where the hypotenuse (longest side) is c:
$a^2 + b^2 = c^2$
This only applies to right-angled triangles.

Here are a few right-angled triangles that can be memorized to shorten your calculation time.

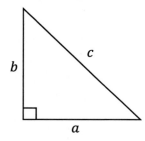 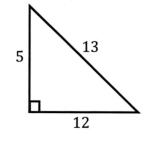 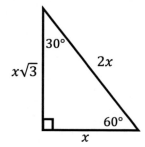 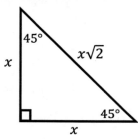

Example 16.5:
A rectangular kiddie's pool measures 4 meters long by 3 meters wide. Sandra's daughter decides to paddle the diameter of the pool. How far did she paddle?

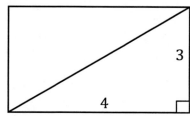

Sketch a diagram of the pool and the diameter. Since all four corners of the rectangle measure 90°, the diameter will dissect the rectangle into two right-angled triangles.

To find the length of the diameter, we can therefore use the Pythagorean Theorem or the first special triangle on the previous page.

Let $a = 3$ and $b = 4$

$3^2 + 4^2 = c^2$

$c^2 = 9 + 16 = 25$

$c = \sqrt{25} = 5 \; meters$

Types of Angles

Angle	Property	Angle	Property
Acute Angles 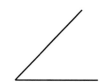	Less than 90°	Supplementary Angles	Straight Angle divided in 2 therefore supplementary angles add to 180° $\angle a + \angle b = 180°$
Right Angles	Equal to 90°	Complementary Angles 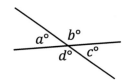	Right angle divided in 2 therefore complementary angles add to 90° $\angle a + \angle b = 90°$
Obtuse Angles	Greater than 90°	Vertical Angles 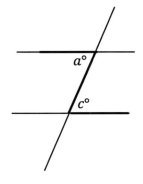	When two lines cross to form an X, the angles opposite each other are equal. $\angle a = \angle c$ and $\angle b = \angle d$
Straight Angles	Sit on a straight line and is equal to 180°	Transverse Angles	Formed when a line intersects two parallel lines (symbol ||). The intersecting line creates vertical angles with each parallel line. It also creates some other equal angles. The easiest way to identify these transverse angles is to picture the Z. $\angle a = \angle c$

Example 16.6:
What is the value of ∠a?

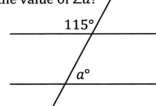

Using the property of transverse angles we see that ∠a is supplementary to 115°.

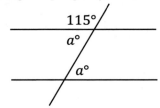

$$115° + a = 180°$$
$$a = 180° - 115°$$
$$a = 65°$$

Polygons and Their Properties

Polygons are identified by the number of sides they have. Common polygons:

Pentagon – Five sides	Octagon – Eight sides
Hexagon – Six sides	Nonagon – Nine sides
Heptagon – Seven sides	Decagon – Ten sides

The sum of the exterior angles in a polygon equals 360°
The sum of the interior angles is $180°(n-2)$ where n is the number of sides of the polygon.
If the polygon is a regular polygon (all sides equal), it has these additional properties:

$$\text{Value of an exterior angle} = \frac{360°}{n}$$
$$\text{Value of an interior angle} = \frac{180°(n-2)}{n}$$

Example 16.7:
What is the value of the interior angle in a regular nonagon?

A regular nonagon has 9 equal sides.
$n = 9$

$$\text{Value of an interior angle} = \frac{180°(n-2)}{n}$$
$$= \frac{180°(9-2)}{9}$$
$$= \frac{180°(7)}{9}$$
$$= 20°(7) = 140°$$

Chapter 16 Exercises

1. The perimeter of a triangle is 240ft. If the length of the sides are $2x + 3$, $x - 4$ and $3x - 17$, what is the length of the shortest side?

2. The expressions in the diagram below represent the values of the 3 angles in the triangle. What is the value of the smallest angle?

    A. 29° B. 40° C. 53° D. 87°

    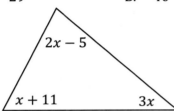

3. What is the value of a in the figure below?

    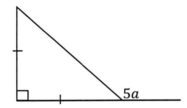

4. In the figure below, the base of the triangle is 4 cm, the height is 6 cm and $AE = \frac{1}{4} AC$. What is the area of the shaded region?

    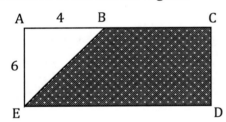

5. Given that the expressions represent the vertical angles, what is the value of y in the figure below?

    A. 20° B. 32° C. 120° D. 180°

    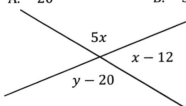

6. Which of the following cannot be the lengths of the sides of a triangle?

    A. 3, 4, 5 B. 4, 6, 12 C. 5, 12, 13 D. 6, 8, 10

192   Prep for Success: Florida's PERT Math Study Guide

7. A baseball diamond is actually a square rotated 45°. What is the shortest distance between home base and 2nd base if the perimeter of the field is 320 feet?
   A. $160\sqrt{2}\ feet$ B. $80\sqrt{2}\ feet$ C. $120\ feet$ D. $160\ feet$

8. Which of the following units cannot be used to describe the amount of water in a tank?
   A.   Gallons   B.   Cubic Inches   C.   Meters   D.   Liters

9. Peter must drive 5 miles south then 12 miles east from his house to work. If the city plans on constructing a freeway at an estimated $2.50 per foot that would provide a straight line of access from his house to his job, how much would it cost the city to construct the freeway?

10. A square baseball stadium is expanded to make room for more stadium seats. If the stadium becomes twice as long on each side, then the area of the expanded stadium is now how many times the area of the original stadium?
    A.   2   B.   4   C.   8   D.   16

Solutions

1. Perimeter $= (2x + 3) + (x - 4) + (3x - 17) = 240$
   $2x + x + 3x + 3 - 4 - 17 = 240$
   $6x - 18 = 240$
   $6x = 258$
   $x = \dfrac{258}{6} = 43$
   This is the value of x but we were asked to find the length of the shortest side. Substitute for x in each of the expressions to determine which is the shortest.
   $side\ 1 = 2(43) + 3 = 86 + 3 = 89$
   $side\ 2 = 43 - 4 = 39$
   $side\ 3 = 3(43) - 17 = 129 - 17 = 112$
   Therefore the shortest side is 39 feet long.

2. **Answer: [B]**   Since the sum of the angles in a triangle is 180°:
   $(2x - 5) + (3x) + (x + 11) = 180°$
   $2x + 3x + x - 5 + 11 = 180$
   $6x = 174$
   $x = \dfrac{174}{6} = 29$
   This is the value of x but we were asked to find the value of the smallest angle. Substitute for x in each of the expressions to determine which is the smallest.
   $1st\ angle = 2x - 5 = 2(29) - 5 = 58° - 5 = 53°$
   $2nd\ angle = 3x = 3(29) = 87°$
   $3rd\ angle = x + 11 = 29 + 11 = 40°$
   The smallest angle is 40°.

3. Note that the triangle is isosceles and therefore the base angles are equal.
Let base angle $= x$

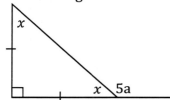

The triangle is a right-angled triangle and so the third angle is $90°$
$x + x + 90° = 180°$
$2x + 90° = 180°$
$2x = 90°$
$x = 45°$

Note that x and $5a$ are supplementary
$5a + 45° = 180°$
$5a = 135°$
$a = \dfrac{135°}{5} = 27°$

4. Area of Shaded Area = Area of rectangle − Area of triangle
Area of triangle $= \dfrac{1}{2}bh = \dfrac{1}{2}(4)(6) = \dfrac{1}{2}(24) = 12 \ cm^2$
$\dfrac{1}{4}AC = AE$
$AC = 4AE = 4(6) = 24 \ cm$
Area of rectangle $= l \times w = AC \times AE = 24 \times 6 = 144 \ cm^2$
Area of Shaded Area $= 144 - 12 = 132 \ cm^2$

5. **Answer: [D]** Using the straight angle property:
$5x + (x - 12) = 180°$
$6x - 12 = 180°$
$6x = 192°$
$x = \dfrac{192°}{6} = 32°$
Now use the vertical angle property
$y - 20 = 5x$
$y - 20 = 5(32) = 160$
$y = 160 + 20 = 180°$

6. **Answer: [B]** Remember that the length of the longest side cannot be longer than the sum of the lengths of the 2 shorter sides.
$3 + 4 > 5$ $4 + 6 < 12$
$5 + 12 > 13$ $6 + 8 > 10$
Therefore 4, 6, and 12 cannot be the lengths of the sides of a triangle.

7. **Answer: [B]**   Do a sketch of the baseball field.

The distance between home base and 2nd base is actually the diagonal of the square. Recall from Chapter 6 that the length of the diagonal of a square can be found using the length of its side. We were given the perimeter of the field so we can find the length of one side.

$Perimeter = 4s = 320$

$s = \frac{320}{4} = 80 \; feet$

$diagonal = s\sqrt{2} = 80\sqrt{2} \; feet$

If you forgot the formula for finding the diagonal of a square, you can use the fact that the diagonal divides the square into 2 right-angled triangles and use the Pythagorean Theorem.

$diagonal^2 = s^2 + s^2$
$ = 80^2 + 80^2$
$ = 6400 + 6400 = 12,800$

$diagonal = \sqrt{12800} = \sqrt{6400 \times 2} = 80\sqrt{2} \; feet$

8. **Answer: [C]**   The amount of liquid a container can hold is represented by its capacity or volume. Gallons, cubic inches and liters are all three dimensional units of volume while "meters" is a one dimensional unit of length. Therefore the answer is C.

9.   Draw a sketch of the problem

We can see that a right-angled triangle is formed. Use the Pythagorean Theorem to solve for the length of the freeway or realize that this is one of the special right-angled triangles on page 188.

The freeway is therefore 13 miles long.

We were given price per foot so we need to convert miles to feet.

$1 \; mile = 5280 \; feet$

$13 \; miles = 5280 \times 13 = 68,640 \; feet$

$Cost \; to \; build \; freeway = cost \; per \; foot \times number \; of \; feet$
$ = \$2.50 \times 68,640 = \$171,600$

10. **Answer: [B]**   The two stadiums are similar but note that we weren't given the dimensions of the original stadium because we don't need them.

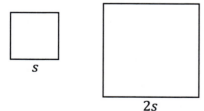

Area of the original stadium $= side^2 = s^2$

New side = twice length of old side $= 2s$

New area $= (2s)^2 = 4s^2$

Comparing the original area and the new area, we see that the area has increased by a factor of 4 so the expanded stadium is 4 times as large as the original stadium.

Final Thoughts

Now that you have completed the topics presented in the study guide portion, it is time to move on to the workbook portion to test your knowledge. In the Appendix at the end of the book, you will find a formula sheet that you can use while completing the exercises in the workbook but you should strive to replicate exam conditions as much as possible.

Worksheets are much longer than the actual Postsecondary Education Readiness Test so you may complete them a little at a time. Ensure that you have a clear and quiet workspace and allocate ample time for practice. Avoid using calculators since most institutions do not allow the use of calculators on the exam. A calculator may be presented onscreen for any questions requiring more complicated or time consuming calculations.

Good Luck!

About this Workbook

The problems presented in the following worksheets are designed to reinforce key concepts necessary to pass the Postsecondary Education Readiness Test.

Track 2 Success makes no guarantee that the problems on the PERT will exactly mirror the ones presented here as each institution is different and as such may use different question formats. It would be impossible to present every type of question within the scope of this book. Instead, we focus on teaching the basic and advanced concepts crucial to conquering the test and obtaining a good score.

In most institutions, the Postsecondary Education Readiness Test is untimed so that you may spend as much time as you need with each question. However, since the real test is adaptive you will not be able to revisit a question once answered and the next presented question will depend on whether or not you got the previous question correct. Therefore, the question difficulty adjusts as you progress through the test. In the following worksheets, easier questions are presented at the beginning of the worksheet and the more difficult ones at the end. The answer sheet and detailed solutions to each worksheet appear at the end of the worksheet.

The bubble sheets are perforated to facilitate easy removal and extra bubble sheets are included at the end of the workbook.

For additional assistance please contact our support department directly at support@track2success.com or submit a support ticket on our website.

198   Prep for Success: Florida's PERT Math Study Guide

Track 2 Success
"Dedicated to helping you stay on the track to academic success"

Postsecondary Education Readiness Test (PERT) – Worksheet #1 Bubble Sheet

1. A B C D
2. A B C D
3. A B C D
4. A B C D
5. A B C D
6. A B C D
7. A B C D
8. A B C D
9. A B C D
10. A B C D
11. A B C D
12. A B C D
13. A B C D
14. A B C D
15. A B C D
16. A B C D
17. A B C D
18. A B C D
19. A B C D
20. A B C D
21. A B C D
22. A B C D
23. A B C D
24. A B C D
25. A B C D

26. A B C D
27. A B C D
28. A B C D
29. A B C D
30. A B C D
31. A B C D
32. A B C D
33. A B C D
34. A B C D
35. A B C D
36. A B C D
37. A B C D
38. A B C D
39. A B C D
40. A B C D
41. A B C D
42. A B C D
43. A B C D
44. A B C D
45. A B C D
46. A B C D
47. A B C D
48. A B C D
49. A B C D
50. A B C D

51. A B C D
52. A B C D
53. A B C D
54. A B C D
55. A B C D
56. A B C D
57. A B C D
58. A B C D
59. A B C D
60. A B C D
61. A B C D
62. A B C D
63. A B C D
64. A B C D
65. A B C D
66. A B C D
67. A B C D
68. A B C D
69. A B C D
70. A B C D
71. A B C D
72. A B C D
73. A B C D
74. A B C D
75. A B C D

The contents of this worksheet are copyrighted. Disclosure or reproduction of any portion herein without express written consent of Track 2 Success © is strictly prohibited.

Track 2 Success
"Dedicated to helping you stay on the track to academic success"

Worksheet #1

The questions contained in this worksheet cover all the topics presented in the study guide section and are ordered by increasing difficulty from Arithmetic through College Level Math.

Since calculators are not generally allowed while taking the Postsecondary Education Readiness Test, use of calculators should be limited while completing this worksheet so as to replicate and properly prepare for exam conditions.

1. $\dfrac{7}{12} + \dfrac{3}{4} =$

   A. $\dfrac{19}{2}$ B. $\dfrac{10}{3}$ C. $\dfrac{5}{6}$ D. $\dfrac{4}{3}$

2. $3\dfrac{1}{3} - 1\dfrac{1}{5} =$

   A. $2\dfrac{2}{15}$ B. $2\dfrac{8}{15}$ C. $4\dfrac{2}{15}$ D. $4\dfrac{8}{15}$

3. $\dfrac{2}{3} \times \dfrac{9}{4} =$

   A. $\dfrac{11}{7}$ B. $\dfrac{4}{9}$ C. $\dfrac{11}{12}$ D. $1\dfrac{1}{2}$

4. $10 \div \dfrac{5}{9} =$

   A. $\dfrac{2}{9}$ B. $\dfrac{5}{9}$ C. $5\dfrac{5}{9}$ D. 18

5. What is 20% of 80?
   A. 10 B. 25 C. 16 D. 20

6. 18 is what percent of 90?
   A. 15% B. 20% C. 18% D. 5%

7. 150 is 20% of what number?
   A. 700 B. 500 C. 600 D. 750

The contents of this worksheet are copyrighted. Disclosure or reproduction of any portion herein without express written consent of Track 2 Success © is strictly prohibited.

8. If the price of a blouse was decreased from $120 to $90, by what percent was the price decreased?

A. 25%   **B.** 30%   **C.** 40%   **D.** 45%

9. What is 40% of 50% of 1200?

A. 240   **B.** 300   **C.** 480   **D.** 600

10. Based on last year's sales, Olivia estimates that she will sell 30 boxes of chocolate bars at this year's state fair. After 4 hrs, she had sold 1, 200 bars of chocolate. If each box contains 25 bars, what was the percentage error of her estimate?

A. 25%   **B.** 37.5%   **C.** 40%   **D.** 62.5%

11. $0.015 + 0.239 =$

A. 0.234   **B.** 0.235   **C.** 0.244   **D.** 0.254

12. $-12.815 - 2.012 =$

A. 10.803   **B.** 14.827   **C.** -14.827 **D.** -10.803

13. $1.02 \times (-3.5) =$

A. -0.357 **B.** -3.57 **C.** 35.7   **D.** -35.7

14. $-25.2 \div 0.4 =$

A. 6.3   **B.** -6.3 **C.** 6.03   **D.** -63

15. Which of the following is equal to $\dfrac{0.000021 \times 240}{0.0014}$?

A. 0.036   **B.** 0.36   **C.** 3.6   **D.** 36

16. A class of 8 students scored an average of 63 on their last math test. What is the sum of their scores?

A. 90   **B.** 483   **C.** 504   **D.** 561

17. Which of the following is listed in order from least to greatest?

A. $\sqrt{2}$, π, $-|4|$, 7

B. $-|-2|$, $\sqrt{3}$, π, $|-5|$

C. $|-8|$, $2\sqrt{3}$, π, 0.5

D. $|-4|$, $\sqrt{25}$, $|6|$, π

18. Amy allocates ⅓ of her monthly budget to rent, ⅙ to food and utilities and ⅛ for transport and entertainment. What fraction of Amy's income is left for savings?

A. $\dfrac{3}{8}$ B. $\dfrac{5}{8}$ C. $\dfrac{10}{13}$ D. $\dfrac{1}{4}$

19. A truck traveled a distance of 435 miles at 29 miles per hour. How many hours did the truck travel?

A. 9   B. 15   C. 30   D. 12,615

20. Find the average of 12, 21, 17 and 6.

A. 13   B. 14   C. 15   D. 16

21. Hair grows at an average of 0.6 inches per month. If you did not cut your hair for 2 years, how many inches would your hair have grown?

A. 0.72   B. 1.44   C. 7.2   D. 14.4

22. What is the greatest number of concert tickets that a school can purchase if the tickets cost $46 each and the school has a budget of $1,541?

A. 33   B. 34   C. 42   D. 48

23. Which of the following is NOT a true statement?

A. $2.5^2 = 6.25$ B. $\dfrac{2}{3} \div \dfrac{7}{9} = \dfrac{6}{7}$ C. $\sqrt{0.09} = 0.03$ D. $4.20 \leq 4.200$

24. A person throwing darts can hit the bull's-eye four times in every 9 attempts. How many times will he hit the bull's-eye in 36 attempts?

A. 12   B. 14   C. 16   D. 20

25. Find the area of the following figure:

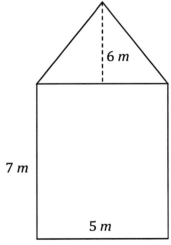

A. $26\ m^2$ B. $36\ m^2$ C. $50\ m^2$ D. $65\ m^2$

26. A party bag contains 36 ounces of potato chips. If one serving is ⅔ of an ounce and each person gets one serving, then how many people get chips?

   A. 54   B. 39   C. 26   D. 24

27. Find the values of x and y in the figure below.

   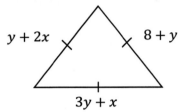

   A. $x = 2, y = 4$ B. $x = 4, y = 0$ C. $x = 5, y = 1$ D. $x = 4, y = 2$

28. Using the table below, what would be the minimum monthly payment required if you have a balance of $250,000?

Balance	$0 – $500	$500.01 – $5000	$5000.01 – $100,000	$100,000 & up
Monthly Payment	Full Balance	$500	$500 + 20% of balance	$500 + 2% of balance over $100,000

   A. $3000   B. $3500   C. $5000   D. $5500

29. A large cube is made up of smaller cubes as shown in the diagram. If the entire top row and right column of cubes are removed, what percentage of small cubes remains?

   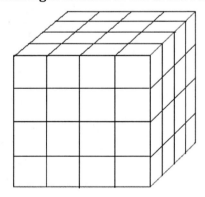

   A. 25%   B. 43.75%   C. 56.25%   D. 75%

30. Evaluate: $10 - 3 \times 4 + 9^2 + (5 - 1) \times 6$

   A. -59 B. 40   C. 103   D. 133

31. Factor: $15xy + 5x^2$

A. $5y(3x + x^2)$

B. $5x(3y + x)$

C. $(15x + 5)(y + x^2)$

D. $5xy(3 + x)$

32. Simplify: $(3x^2 + 5x + 2) + (x^2 - 4x - 8)$

A. $4x^2 + 9x - 6$

B. $2x^2 - x + 6$

C. $2x^2 + 9x - 6$

D. $4x^2 + x - 6$

33. Simplify: $(4x - 1)(3x + 7)$

A. $12x^2 + 25x - 7$

B. $12x^2 - 7$

C. $7x - 6$

D. $12x^2 + 25x + 7$

34. Simplify: $(12x^2y^4)\left(\frac{1}{2}x^5y\right)$

A. $24x^{10}y^4$

B. $24x^7y^5$

C. $6x^3y^3$

D. $6x^7y^5$

35. Find the midpoint of the points: $(-3, 7)$ and $(11, 19)$

A. $(7, 6)$

B. $(4, 13)$

C. $(8, 26)$

D. $(14, 12)$

36. Simplify: $\dfrac{x^2 + x - 12}{x^2 - 16}$

A. $x + 4$

B. $x - 3$

C. $\dfrac{x - 3}{x - 4}$

D. $\dfrac{x - 12}{-16}$

37. If $x^2 + 5x - 24 = 0$, then $x =$

A. -8 and 3

B. -3 and 8

C. 19 and 24

D. -24 and -29

38. What is the value of the expression $(x^2 + 5x - 3)$ when $x = -2$?

A. 12

B. 4

C. -9

D. -17

39. $5\sqrt{27} - 2\sqrt{3} =$

A. $3\sqrt{24}$

B. $13\sqrt{3}$

C. $6\sqrt{6}$

D. $7\sqrt{30}$

40. The sum of two numbers is 27. One number is 3 more than twice the other number. Which of the following represents the equation of the sum of the two numbers?

A. $n(3 + 2n) = 27$

B. $n + (3n + 2) = 27$

C. $n + (2n + 3) = 27$

D. $n(3n + 2) = 27$

41. If $-5(3x + 2) = -3x + 14$, then $x =$
   A. -2
   B. $\dfrac{2}{9}$
   C. 2
   D. 24

42. If $3x + 11 \leq 6x + 8$, then
   A. $x \leq 1$
   B. $x \leq -1$
   C. $x \geq 1$
   D. $x \geq -1$

43. Solve for g in the equation $3g = 4g - 1h - 2$
   A. $h - 1$
   B. $h + 2$
   C. $\dfrac{(2-h)}{3h+2}$
   D. $\dfrac{h-2}{-2-3h}$

44. If $\dfrac{2}{3} - \dfrac{5}{6}x \geq -1$, then
   A. $x \geq \dfrac{2}{5}$
   B. $x \leq -\dfrac{2}{5}$
   C. $x \geq 2$
   D. $x \leq 2$

45. In the figure below, both circles have the same center. The radius of the smaller circle is r and the radius of the larger circle is 4 units larger than the radius of the smaller circle. Which of the following represents the area of the shaded region?

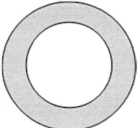

   A. $4\pi r^2$
   B. 16π
   C. $\pi(4r)^2 - 11r^2$
   D. $\pi(r+4)^2 - \pi r^2$

46. $f(x) = \dfrac{3}{x^3 - x}$
   Which of the following is the domain of the function above?
   A. $(-1, 1)$
   C. $\{x : x \neq -1, 0, 1\}$
   B. $\{x : x \neq 0, 1\}$
   D. $(-\infty, \infty)$

47. Which of the following is an equivalent statement?
   A. $(x^3)^2 = x^5$
   C. $x^{12}(x^3) = x^9$
   B. $x^3(x^3) = x^6$
   D. $x^2(x^4 - 3) = x^4 - 3x^2$

48. What is the value of the expression $\dfrac{5x - 4y}{2z}$ when $x = 2$, $y = -3$, and $z = -1$?
   A. 5
   B. -7
   C. -12
   D. -13

49. What are the coordinates of the y-intercept in the following graph?

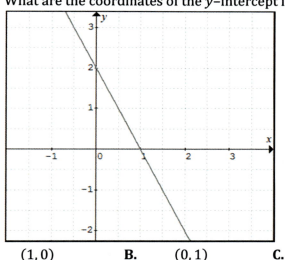

A. (1, 0)   B. (0, 1)   C. (0, 2)   D. (2, 0)

50. Susan gave her two children $82.50 to share. If her daughter took $42 more than her son, write an expression that could be used to determine how much money her son received.

A. $x + 42 = 82.50$ C. $42 = 82.50 - x$
B. $2x + 42 = 82.50$ D. $x = 82.50 - 42 + x$

51. Find the x-intercept of the equation: $3y - 5x = 15$

A. $(0, -3)$ B. $(-3, 0)$ C. $(0, 5)$ D. $(5, 0)$

52. Which of the following graphs shows the area represented by:
$-2 \leq y \leq 5$ and $-3 \leq x \leq 0$

A.
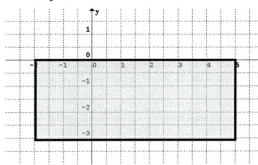

C.
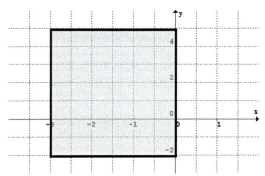

B.
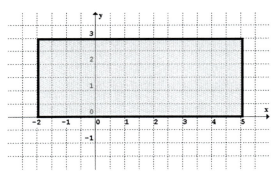

D.
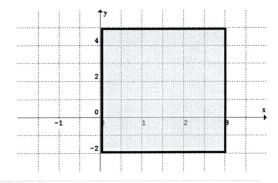

53. Convert to standard form: 1.28×10^{-5}

A. 0.00000128   B. 0.0000128   C. 128,000   D. 12,800,000

---

54. Simplify: $3^{7/2} + 3^{1/2}$

A. 12   B. $21\sqrt{3}$ C. $28\sqrt{3}$ D. 81

---

55. Simplify: $\sqrt[4]{24} \cdot \sqrt[4]{54}$

A. 81   B. 6   C. $\sqrt[4]{30}$ D. Cannot be simplified.

---

56. $\dfrac{2}{3 - \sqrt{5}} =$

A. $2(3 + \sqrt{5})$ B. $\dfrac{3 + \sqrt{5}}{2}$ C. $\dfrac{3 + \sqrt{5}}{4}$ D. $\dfrac{3 + \sqrt{5}}{8}$

---

57. Simplify: $\dfrac{x^2 + 2x - 3}{x^2 + 8x + 16} \cdot \dfrac{9x + 36}{3x - 3}$

A. $\dfrac{9(x + 3)}{x - 4}$ B. $\dfrac{3(x + 3)}{x + 4}$ C. $\dfrac{3(x - 3)}{x + 4}$ D. $\dfrac{x + 3}{x - 1}$

---

58. Simplify: $\dfrac{8}{x - 2} - \dfrac{1}{x}$

A. $\dfrac{7x + 2}{x(x - 2)}$ B. $\dfrac{7}{x(x - 2)}$ C. $\dfrac{-8}{x(x - 2)}$ D. $\dfrac{-7}{2}$

---

59. Factor: $8a^3 + b^3$

A. $(2a - b)(a^2 + 2ab + b^2)$ C. $(2a + b)(4a^2 - 2ab + b^2)$

B. $(2a + b)(a^2 + ab + b^2)$ D. $(2a + b)(a^2 + 2ab + b^2)$

---

60. Expand: $2x(x + 4)^2$

A. $2x^3 + 32x$ C. $2x^3 + 16x^2 + 32x$

B. $2x^3 - 32x$ D. $2x^3 + 8x^2 + 16x$

---

61. If $\dfrac{2}{x + 4} = \dfrac{10}{3(x - 3)}$, then $x =$

A. $-\dfrac{29}{2}$ B. $\dfrac{29}{2}$ C. $-\dfrac{7}{4}$ D. $\dfrac{7}{4}$

---

62. Solve the equation: $-9(m + 8) = m - 12$

A. 2   B. 6   C. -2 D. -6

63. If $|2m - 3| \leq 9$, then
A. $-3 \geq m \geq 6$ B. $-3 \leq m \leq 6$ C. $m \leq 6$ D. No Solution

64. Solve the inequality: $7 - |2x + 8| \geq 1$
A. $x \leq -7$ or $x \geq -1$ C. $-7 \leq x \leq -1$
B. $-7 \leq x \leq 7$ D. No Solution

65. If $2(x^2 + x) = 12$, then $x =$
A. -3 only B. 2 only C. -3 and 2 D. -2 and 3

66. Find the real roots of the following: $3x^2 + 2x + 5 = 0$
A. $x = \dfrac{-1 \pm i\sqrt{7}}{3}$ C. $x = -\dfrac{5}{3}$ and 1
B. $x = \dfrac{-1 \pm i\sqrt{14}}{3}$ D. No Real Roots

67. Solve the system of equations:
$y = 2x + 1$
$2y + 3x = 9$
A. $(1, 3)$ C. The graphs are dependent
B. $(7, 15)$ D. No Solution

68. Write an equation to represent the graph of the line below.

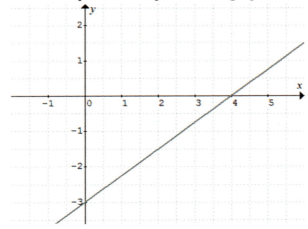

A. $3x - 4y = 12$ B. $4x - 3y = 12$ C. $4x = -3y$ D. $y = \dfrac{3}{4}x$

69. Which of the following represents a line parallel to the graph $4y + 3x = 8$?
A. $4y = -3x - 12$ C. $3y - 4x = 5$
B. $4y = 3x - 7$ D. $3y = 4x - 2$

70. Which of the following could be an equation of the graph below?

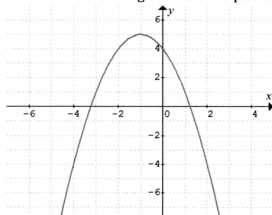

- **A.** $y = -(x+1)^2 + 5$
- **B.** $y = (x+1)^2 - 5$
- **C.** $y = -(x+1)^2 - 5$
- **D.** $y = (x-1)^2 + 5$

---

71. If $f(x) = 2x + 3$ and $g(x) = \dfrac{x+2}{5x-4}$, then $g(f(x)) =$

- **A.** $\dfrac{2x+3}{10x+15}$
- **B.** $\dfrac{2x+5}{10x+11}$
- **C.** $\dfrac{17x-8}{5x-4}$
- **D.** $\dfrac{10x-10}{5x-4}$

---

72. $f(x) = 3x + 4$ and f^{-1} is the inverse of f, then $f^{-1}(-5) =$?

- **A.** -3
- **B.** -5
- **C.** -11
- **D.** 9

---

73. Which of the following can be factored in the form $(x+b)^2$, where b is an integer?

- **A.** $x^2 - 4$
- **B.** $x^2 + 2$
- **C.** $x^2 + 10x + 25$
- **D.** $x^2 + 10x + 100$

---

74. Express $\dfrac{7+3i}{4i}$ in the form $a + bi$.

- **A.** $\dfrac{3}{4} - \dfrac{7}{4i}$
- **B.** $\dfrac{3}{4} + \dfrac{7}{4i}$
- **C.** $\dfrac{3}{4} + \dfrac{7}{4}i$
- **D.** $\dfrac{3}{4} - \dfrac{7}{4}i$

---

75. If $5(a - 2) = 3(2a + 4)$, then:

- **A.** $a = -2$
- **B.** $a = 2$
- **C.** $a = 22$
- **D.** $a = -22$

◆·········· END WORKSHEET ··········◆

Postsecondary Education Readiness Test (PERT) – Worksheet #1 Answer Sheet

1. D	26. A	51. B
2. A	27. D	52. C
3. D	28. B	53. B
4. D	29. C	54. C
5. C	30. C	55. B
6. B	31. B	56. B
7. D	32. D	57. B
8. A	33. A	58. A
9. A	34. D	59. C
10. B	35. B	60. C
11. D	36. C	61. A
12. C	37. A	62. D
13. B	38. C	63. B
14. D	39. B	64. C
15. C	40. C	65. C
16. C	41. A	66. D
17. B	42. C	67. A
18. A	43. B	68. A
19. B	44. D	69. A
20. B	45. D	70. A
21. D	46. C	71. B
22. A	47. B	72. A
23. C	48. C	73. C
24. C	49. C	74. D
25. C	50. B	75. D

The contents of this worksheet are copyrighted. Disclosure or reproduction of any portion herein without express written consent of Track 2 Success © is strictly prohibited.

Prep for Success: Florida's PERT Math Study Guide

Track 2 Success
"Dedicated to helping you stay on the track to academic success"

Worksheet #1 Solutions

1. **D**

   $\dfrac{7}{12} + \dfrac{3}{4} =$

   $= \dfrac{7}{12} + \dfrac{3(3)}{4(3)} = \dfrac{7}{12} + \dfrac{9}{12} = \dfrac{16}{12} = \dfrac{4}{3}$

2. **A**

   $3\dfrac{1}{3} - 1\dfrac{1}{5} =$

   $= \dfrac{10}{3} - \dfrac{6}{5} = \dfrac{(5)10}{(5)3} - \dfrac{6(3)}{5(3)} = \dfrac{50}{15} - \dfrac{18}{15} = \dfrac{32}{15} = 2\dfrac{2}{15}$

3. **D**

   $\dfrac{2}{3} \times \dfrac{9}{4} =$

   $= \dfrac{18}{12} = \dfrac{3}{2} = 1\dfrac{1}{2}$

4. **D**

   $10 \div \dfrac{5}{9} =$

   $= \dfrac{10}{1} \times \dfrac{9}{5} = \dfrac{90}{5} = 18$

5. **C**

   What is 20% of 80?
   This is the simplest type of percentage problem. Multiply the number by the decimal representation of the percentage.
   20% of 80
   $= 20\% \times 80 = 0.20 \times 80 = 16$

6. **B**

   18 is what percent of 90?

   $\dfrac{is}{of} = \dfrac{\%}{100}$

   $\dfrac{18}{90} = \dfrac{x}{100}$

   $(90)(x) = (18)(100)$

   $90x = 1800$

   $x = \dfrac{1800}{90} = 20\%$

The contents of this worksheet are copyrighted. Disclosure or reproduction of any portion herein without express written consent of Track 2 Success © is strictly prohibited.

7. **D**

150 is 20% of what number?

$$\frac{is}{of} = \frac{\%}{100}$$

$$\frac{150}{x} = \frac{20}{100}$$

$(20)(x) = (150)(100)$

$20x = 15000$

$$x = \frac{15000}{20} = 750$$

8. **A**

If the price of a blouse was decreased from \$120 to \$90, by what percent was the price decreased?

$\$120 - \$90 = \$30$

$$percentage\ decrease = \frac{decrease\ amount}{original\ cost} \times 100\%$$

$$= \frac{30}{120} \times 100\% = \frac{1}{4} \times 100\% = 25\%$$

9. **A**

What is 40% of 50% of 1200?

Find 50% of 1200 first, and then find 40% of that result.

$0.5 \times 1200 = 600$

$0.4 \times 600 = 240$

10. **B**

Based on last year's sales, Olivia estimates that she will sell 30 boxes of chocolate bars at this year's state fair. After 4 hrs, she had sold 1,200 bars of chocolate. If each box contains 25 bars, what was the percentage error of her estimate?

Since all the answers are positive, we will take the absolute value of the percentage error. Therefore, we will use the equation:

$$Percentage\ error = \frac{|estimated\ value - actual\ value|}{actual\ value} \times 100\%$$

We already know that she actually sold 1, 200 bars. We need to find out how many bars she estimated that she would sell.

$$Estimated\ \#\ bars = estimated\ \#\ boxes \times \#\ bars\ per\ box$$

$$= 30\ boxes \times 25\ bars = 750\ bars$$

Therefore:

$$Percentage\ error = \frac{|750 - 1200|}{1200} \times 100\% = \frac{450}{1200} \times 100\% = \frac{450}{12}\% = 37.5\%$$

11. **D**

$0.015 + 0.239 =$

Line up the decimal points and add.

$$\begin{array}{r} 0.015 \\ +\quad 0.239 \\ \hline 0.254 \end{array}$$

12. **C**

$-12.815 - 2.012 =$

The signs are the same so line up the decimal points and add.

$$\begin{array}{r} -12.815 \\ +\quad -2.012 \\ \hline -14.827 \end{array}$$

13. **B**

$1.02 \times (-3.5) =$

There are 3 decimal places in the two decimals combined. Multiply the numbers without the decimal points.

$$\begin{array}{r} 102 \\ \times\quad 35 \\ \hline 510 \\ +\quad 3060 \\ \hline 3570 \end{array}$$

The answer will be negative with the decimal point moved 3 places left to become -3.570

14. **D**

$-25.2 \div 0.4 =$

Move the decimal point one place to the right in both the divisor and the dividend. Divide the numbers using whichever method you choose.

$$= -252 \div 4 = -\frac{252}{4} = -63$$

15. **C**

Which of the following is equal to $\dfrac{0.000021 \times 240}{0.0014}$?

Converting to scientific notation would be the easiest method but to make it even easier, keep the coefficients as whole numbers as follows:

$$\frac{21 \times 10^{-6} \times 24 \times 10^1}{14 \times 10^{-4}}$$

Group the coefficients and the powers of 10 and then use the law #4 of exponents to simplify.

$$= \frac{21 \times 24 \times 10^{-6} \times 10^1}{14 \times 10^{-4}} = \frac{21 \times 24 \times 10^{-6+1}}{14 \times 10^{-4}} = \frac{21 \times 24}{14} \times \frac{10^{-5}}{10^{-4}}$$

Using law #5 of exponents:

$$= \frac{21 \times 24}{14} \times 10^{-5+(-4)} = \frac{3 \times 24}{2} \times 10^{-1} = 3 \times 12 \times 10^{-1}$$

$$= 36 \times 10^{-1}$$

$$= 3.6$$

16. C

A class of 8 students scored an average of 63 on their last math test. What is the sum of their scores?

To find the average, we divide the total class score by the number of students so to find the sum of their scores, multiply the number of students by the average.

$$average\ score = \frac{sum\ of\ scores}{\#\ students}$$

$$sum\ of\ scores = \#\ students \times average\ score$$

$$= 8 \times 63 = 504$$

17. B

Which of the following is listed in order from least to greatest?

A. $\sqrt{2}$, π, $-|4|$, 7

B. $-|-2|$, $\sqrt{3}$, π, $|-5|$

C. $|-8|$, $2\sqrt{3}$, π, 0.5

D. $|-4|$, $\sqrt{25}$, $|6|$, π

Given that $\sqrt{2} = 1.4\ldots$, $\sqrt{3} = 1.7\ldots$ and $\pi = 3.14\ldots$ let's rewrite the answer choices.

A. $1.4,\ 3.14,\ -4,\ 7$

B. $-2, 1.7,\ 3.14,\ 5$

C. $8,\ 3.4,\ 3.14,\ 0.5$

D. $4,\ 5,\ 6,\ 3.14$

Now that the answer choices have been simplified, we see that B is the only one listed in order from least to greatest.

18. A

Amy allocates ⅓ of her monthly budget to rent, ⅙ to food and utilities and ⅛ for transport and entertainment. What fraction of Amy's income is left for savings?

First determine the fraction that is being spent by adding the fractions..

$$\frac{1}{3} + \frac{1}{6} + \frac{1}{8} =$$

$$LCD = 48$$

$$\frac{(16)\,1}{(16)\,3} + \frac{(8)\,1}{(8)\,6} + \frac{(6)\,1}{(6)\,8} =$$

$$\frac{16}{48} + \frac{8}{48} + \frac{6}{48} = \frac{30}{48} = \frac{5}{8}$$

Subtract this from 1 to find the fraction left for savings.

$$1 - \frac{5}{8} = \frac{8}{8} - \frac{5}{8} = \frac{3}{8}$$

19. B

A truck traveled a distance of 435 miles at 29 miles per hour. How many hours did the truck travel?

Recall the formula for distance:

$$Distance = rate \times time$$

$$435 = 29 \times t$$

$$t = \frac{435}{29} = 15 \ hours$$

20. B

Find the average of 12, 21, 17 and 6

Find the sum then divide by 4 since there are 4 numbers.

$$\frac{12 + 21 + 17 + 6}{4} = \frac{56}{4} = 14$$

21. D

Hair grows at an average of 0.6 inches per month. If you did not cut your hair for 2 years, how many inches would your hair have grown?

This question is best solved using a direct proportion. Notice that time is given in both months and years so we need to convert years to months before setting up the proportion.

$2 \ years = 2 \times 12 = 24 \ months$

$$\frac{0.6 \ inches}{1 \ month} = \frac{x \ inches}{24 \ months}$$

Cross multiply and solve.

$(1)(x) = (0.6)(24)$

$x = 14.4 \ inches$

22. A

What is the greatest number of concert tickets that a school can purchase if the tickets cost $46 each and the school has a budget of $1,541?

Let x represent the number of tickets.

$\$46x = \$1541$

$$x = \frac{1541}{46} = 33.5$$

We can't purchase half a ticket so the greatest number of tickets that can be purchased is 33.

23. C

Which of the following is NOT a true statement?

$2.5^2 = 2.5 \times 2.5$

There are 2 decimal places combined. Multiplying 25 by 25 gives 625. Moving the decimal point 2 places left gives 6.25 so this is a true statement.

$$\frac{2}{3} \div \frac{7}{9} = \frac{6}{7}$$

Take the reciprocal of the second fraction on the left side of the equation and multiply.

$$\frac{2}{3} \times \frac{9}{7} = \frac{18}{21} = \frac{6}{7}$$

This is therefore a true statement.

$\sqrt{0.09} = 0.03$

Find the factors of 0.09

$\sqrt{0.09} = \sqrt{0.3 \times 0.3} = 0.3 \neq 0.03$

This statement is therefore false and the answer is C.

Track 2 Success © 2014

24. C

A person throwing darts can hit the bull's-eye four times in every 9 attempts. How many times will he hit the bull's-eye in 36 attempts?

This can be easily solved by using a direct proportion.

$$\frac{4 \text{ hits}}{9 \text{ attempts}} = \frac{x \text{ hits}}{36 \text{ attempts}}$$

Cross multiply and solve.

$(9)(x) = (4)(36)$

$x = \frac{(4)(36)}{9} = 16$

25. C

Find the area of the following figure:

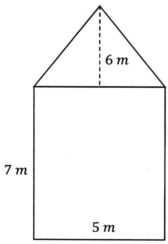

The figure is comprised of a triangle and a rectangle so find the area of the two shapes independently.

Triangle:

The base of the triangle is the width of the rectangle.

$A = \frac{1}{2}bh = \frac{1}{2}(5)(6) = \frac{1}{2}(30)$

$A = 15$ *square meters*

Rectangle:

$A = l \times \omega$

$A = 7 \times 5 = 35$ *square meters*

Total area $= 15 + 35 = 50$ square meters.

26. A

A party bag contains 36 ounces of potato chips. If one serving is ⅔ of an ounce and each person gets one serving, then how many people get chips?

Simply divide the total number of ounces by the number of ounces per serving

$36 \div \frac{2}{3} = 36 \times \frac{3}{2} = \frac{108}{2} = 54$ people

27. D

Find the values of x and y in the figure below.

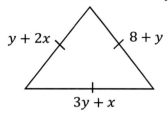

The sides are all equal because this is an equilateral triangle.

$y + 2x = 8 + y$ $3y + x = 8 + y$
$y + 2x - y = 8$ $2y = 8 - x = 8 - 4 = 4$
$2x = 8$ $y = \dfrac{4}{2} = 2$
$x = \dfrac{8}{2} = 4$

28. **B**

Using the table below, what would be the minimum monthly payment required if you have a balance of $250,000?

Balance	$0 – $500	$500.01 – $5000	$5000.01 – $100,000	$100,000 & up
Monthly Payment	Full Balance	$500	$500 + 20% of balance	$500 + 2% of balance over $100,000

A balance of $250, 000 falls in the last column.
First find the amount over $100, 000
$$250,000 - 100,000 = 150,000$$
Find 2% of 150, 000
$$\dfrac{2}{100} \times 150,000 = \dfrac{300,000}{100} = 3,000$$
Now add this to $500
$$\$3,000 + \$500 = \$3,500$$

29. **C**

A large cube is made up of smaller cubes as shown in the diagram. If the entire top row and right column of cubes are removed, what percentage of small cubes remains?

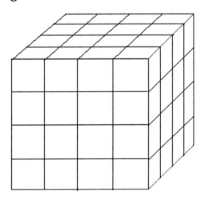

The large cube is made up of 64 smaller cubes (4 cubes high × 4 cubes wide × 4 cubes deep).

If the top row and right column are removed (28 small cubes) we would be left with:

$64 - 28 = 36 \; cubes$

$$percentage\;left = \frac{cubes\;left}{original\;\#\;cubes} \times 100$$

$$= \frac{36}{64} \times 100\% = \frac{9}{16} \times 100\%$$

$$= 9 \times \frac{1}{16} \times 100\% = (9)(0.0625)(100) = (9)(6.25) = 56.25\%$$

30. C

Evaluate: $\quad 10 - 3 \times 4 + 9^2 + (5 - 1) \times 6$

Using PEMDAS, complete the operations in correct order.

$= 10 - 3 \times 4 + 9^2 + \mathbf{4} \times 6$

$= 10 - 3 \times 4 + \mathbf{81} + 4 \times 6$

$= 10 - \mathbf{12} + 81 + 4 \times 6$

$= 10 - 12 + 81 + \mathbf{24}$

$= \mathbf{-2} + 81 + 24$

$= \mathbf{79} + 24 = 103$

31. B

Factor: $15xy + 5x^2$

Greatest common factor is $5x$

$5x(3y + x)$

32. D

Simplify: $\quad (3x^2 + 5x + 2) + (x^2 - 4x - 8)$

This is an addition with nothing to distribute. Just drop the parentheses and add like terms.

$= 3x^2 + x^2 + 5x - 4x + 2 - 8$

$= 4x^2 + x - 6$

33. A

Simplify: $\quad (4x - 1)(3x + 7)$

FOIL

$= 12x^2 + 28x - 3x - 7$

$= 12x^2 + 25x - 7$

34. D

Simplify: $\quad (12x^2y^4)\left(\frac{1}{2}x^5y\right)$

$12 \cdot \dfrac{1}{2} \cdot x^2 \cdot x^5 \cdot y^4 \cdot y = 6x^{2+5} \cdot y^{4+1} = 6x^7y^5$

35. B

Find the midpoint of the points: $\quad (-3, 7)$ and $(11, 19)$

$$M = \left(\frac{x_1 + x_2}{2}, \frac{y_1 + y_2}{2}\right) = \left(\frac{-3 + 11}{2}, \frac{7 + 19}{2}\right) = \left(\frac{8}{2}, \frac{26}{2}\right) = (4, 13)$$

36. C

Simplify: $\dfrac{x^2 + x - 12}{x^2 - 16}$

Factor both numerator and denominator

$\dfrac{(x + 4)(x - 3)}{(x + 4)(x - 4)}$

Eliminate any factors common to both the numerator and denominator

$\dfrac{x - 3}{x - 4}$

37. A

If $x^2 + 5x - 24 = 0$, then $x =$

Factor then set each factor equal to zero and solve

$(x - 3)(x + 8) = 0$

$x - 3 = 0 \qquad\qquad x + 8 = 0$

$x = 3 \qquad\qquad\quad x = -8$

38. C

What is the value of the expression $(x^2 + 5x - 3)$ when $x = -2$

Substitute -2 for the variable x

$x^2 + 5x - 3$

$= (-2)^2 + 5(-2) - 3 = 4 - 10 - 3 = -9$

39. B

$5\sqrt{27} - 2\sqrt{3} =$

Simplify the radicands if possible

$\sqrt{27} = \sqrt{3 \times 3 \times 3} = 3\sqrt{3}$

Therefore:

$5\sqrt{27} - 2\sqrt{3} = 5(3\sqrt{3}) - 2\sqrt{3} = 15\sqrt{3} - 2\sqrt{3}$

Now that the radicands match, combine the terms

$13\sqrt{3}$

40. C

The sum of two numbers is 27. One number is 3 more than twice the other number. Which of the following represents the equation of the sum of the two numbers?

Let one of the numbers be n

First number $= n$

The second number is 3 more than twice the other number

Second number $= 2n + 3$

The sum of both numbers is 27

$n + (2n + 3) = 27$

41. A

If $-5(3x + 2) = -3x + 14$, then $x =$

$-5(3x + 2) = -3x + 14$

$-15x - 10 = -3x + 14$

$$-15x + 3x = 14 + 10$$
$$-12x = 24$$
$$x = \frac{24}{-12} = -2$$

42. C

If $3x + 11 \leq 6x + 8$, then
$$3x + 11 \leq 6x + 8$$
$$3x - 6x \leq 8 - 11$$
$$-3x \leq -3$$
The inequality sign is reversed when dividing or multiplying by a negative number.
$$x \geq \frac{-3}{-3}$$
$$x \geq 1$$

43. B

Solve for g in the equation $3g = 4g - 1h - 2$
Move all terms containing g to one side of the equation
$$3g - 4g = -h - 2$$
$$-g = -h - 2$$
$$g = -(-h - 2) = h + 2$$

44. D

If $\frac{2}{3} - \frac{5}{6}x \geq -1$, then
$$\frac{2}{3} - \frac{5}{6}x \geq -1$$

$$-\frac{5}{6}x \geq -1 - \frac{2}{3}$$

$$-\frac{5}{6}x \geq -\frac{1(3)}{1(3)} - \frac{2}{3}$$

$$-\frac{5}{6}x \geq -\frac{3}{3} - \frac{2}{3}$$

$$-\frac{5}{6}x \geq -\frac{5}{3}$$

The inequality sign is reversed when dividing or multiplying by a negative number.
$$\left(-\frac{6}{5}\right)\left(-\frac{5}{6}\right)x \leq -\frac{5}{3}\left(-\frac{6}{5}\right)$$

$$x \leq \frac{30}{15}$$

$$x \leq 2$$

45. D

In the figure below, both circles have the same center. The radius of the smaller circle is r and the radius of the larger circle is 4 units larger than the radius of the smaller circle.

Which of the following represents the area of the shaded region?

The formula for calculating the area of a circle is $A = \pi r^2$
Radius of smaller circle $= r$
Radius of larger circle $= r + 4$
Area of smaller circle:
$$A = \pi(r)^2$$
Area of larger circle:
$$A = \pi(r + 4)^2$$
Area of shaded region = Area of larger circle − Area of smaller circle
$$A = \pi(r + 4)^2 - \pi r^2$$

46. **C**

The denominator of the function cannot equal 0. Set the domain equal to 0 to find the excluded values of x.
$x^3 - x = 0$
$x(x^2 - 1) = 0$
$x(x - 1)(x + 1) = 0$

Therefore the excluded values are $x = -1, 0$ and 1 and the answer is C.

47. **B**

Which of the following is an equivalent statement?
$x^{(3)(2)} = x^6 \neq x^5$ false
$x^3(x^3) = x^{3+3} = x^6$ true
$x^{12}(x^3) = x^{12+3} = x^{15} \neq x^9$ false
$x^2(x^4 - 3) = (x^2)(x^4) + (x^2)(-3) = x^{2+4} - 3x^2 = x^6 - 3x^2 \neq x^4 - 3x^2$ false

48. **C**

What is the value of the expression $\dfrac{5x - 4y}{2z}$ when $x = 2, y = -3,$ and $z = -1$?

Substitute the values of the variables into the expression and simplify

$$\dfrac{5x - 4y}{2z}$$
$$= \dfrac{5(2) - 4(-3)}{2(-1)}$$
$$= \dfrac{10 + 12}{-2}$$
$$= \dfrac{24}{-2}$$
$$= -12$$

49. C

What are the coordinates of the y-intercept in the following graph?

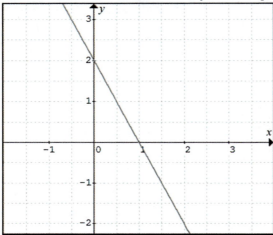

The y-intercept is the point where the graph intersects the y-axis $(0, 2)$

50. B

Susan gave her two children $82.50 to share. If her daughter took $42 more than her son, write an expression that could be used to determine how much money her son received.

Let the amount of money taken by her son $= x$
Amount of money taken by her daughter $= x + 42$
$(x) + (x + 42) = 82.50$
$2x + 42 = 82.50$
If we had to solve:
$2x = 40.50$
$x = \dfrac{40.50}{2}$
$x = 20.25$
Her son took $20.25

51. B

Find the x-intercept of the equation: $3y - 5x = 15$
To find the x-intercept, set $y = 0$ and solve for x
$3y - 5x = 15$
$3(0) - 5x = 15$
$-5x = 15$
$x = \dfrac{15}{-5}$
$x = -3$ or $(-3, 0)$

52. C

Which of the following graphs shows the area represented by $-2 \leq y \leq 5$ and $-3 \leq x \leq 0$?

The graph will be bounded between −2 and 5 on the y-axis and −3 and 0 on the x-axis.

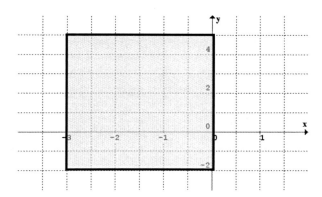

53. B

Convert to standard form: 1.28×10^{-5}

The negative exponent indicates that we need to move the decimal point 5 places to the left

0.0000128

54. C

Simplify: $3^{7/2} + 3^{1/2}$

First find the common factor which is the term with the lowest exponent. Divide each term by the common factor to find the remaining factor.

$$\frac{3^{7/2}}{3^{1/2}} = 3^{7/2 - 1/2} = 3^{6/2} = 3^3$$

$$\frac{3^{1/2}}{3^{1/2}} = 1$$

Therefore:

$3^{7/2} + 3^{1/2} = 3^{1/2}(3^3 + 1)$

$3^{1/2}(27 + 1)$

$3^{1/2}(28)$

Remember that a fractional exponent indicates a root.

$28\sqrt{3}$

55. B

Simplify: $\sqrt[4]{24} \cdot \sqrt[4]{54}$

$\sqrt[4]{24} = \sqrt[4]{2 \times 2 \times 2 \times 3}$

$\sqrt[4]{54} = \sqrt[4]{2 \times 3 \times 3 \times 3}$

So:

$$\begin{aligned}\sqrt[4]{24} \cdot \sqrt[4]{54} &= \sqrt[4]{24 \times 54} \\ &= \sqrt[4]{(2 \times 2 \times 2 \times 3) \times (2 \times 3 \times 3 \times 3)} \\ &= \sqrt[4]{2 \times 2 \times 2 \times 2 \times 3 \times 3 \times 3 \times 3} = 2 \times 3 = 6\end{aligned}$$

56. B

$$\frac{2}{3 - \sqrt{5}}$$

Rationalize the denominator (multiply by the conjugate)

$$\frac{2}{3-\sqrt{5}}\cdot\frac{3+\sqrt{5}}{3+\sqrt{5}}$$

Multiply and simplify

$$\frac{2(3+\sqrt{5})}{(3-\sqrt{5})(3+\sqrt{5})}$$

$$\frac{6+2\sqrt{5}}{9+3\sqrt{5}-3\sqrt{5}-5}$$

$$\frac{6+2\sqrt{5}}{9-5}$$

$$\frac{6+2\sqrt{5}}{4}$$

Reduce by a factor of 2

$$\frac{3+\sqrt{5}}{2}$$

57. B

Simplify: $\dfrac{x^2+2x-3}{x^2+8x+16}\cdot\dfrac{9x+36}{3x-3}$

Factor as much as possible

$$\frac{(x+3)(x-1)}{(x+4)(x+4)}\cdot\frac{9(x+4)}{3(x-1)}$$

Multiply straight across and eliminate factors common to both the numerator and denominator

$$\frac{9(x+3)(x-1)(x+4)}{3(x+4)(x+4)(x-1)}=\frac{3(x+3)}{x+4}$$

58. A

Simplify: $\dfrac{8}{x-2}-\dfrac{1}{x}$

LCD: $x(x-2)$

$$\frac{(x)}{(x)}\cdot\frac{8}{x-2}-\frac{1}{x}\cdot\frac{(x-2)}{(x-2)}$$

$$\frac{8x}{x(x-2)}-\frac{x-2}{x(x-2)}$$

$$\frac{8x-(x-2)}{x(x-2)}=\frac{7x+2}{x(x-2)}$$

59. C

Factor: $8a^3+b^3$

Use the sum of cubes formula to factor.

$$8a^3+b^3$$

$$(2a)^3+(b)^3$$

$$(2a+b)[(2a)^2-(2a)(b)+(b)^2]$$

$$(2a+b)(4a^2-2ab+b^2)$$

60. C

Expand: $2x(x + 4)^2$
FOIL $(x + 4)^2$ first
$2x(x + 4)(x + 4)$
$2x(x^2 + 4x + 4x + 16)$
$2x(x^2 + 8x + 16)$
Now distribute
$2x^3 + 16x^2 + 32x$

61. A

If $\dfrac{2}{x + 4} = \dfrac{10}{3(x - 3)}$, then $x =$
The simplest method would be to cross multiply.
$2[3(x - 3)] = 10(x + 4)$
$6(x - 3) = 10(x + 4)$
$6x - 18 = 10x + 40$
$6x - 10x = 40 + 18$
$-4x = 58$
$x = \dfrac{58}{-4} = -\dfrac{29}{2}$

62. D

Solve the equation: $-9(m + 8) = m - 12$
$-9(m + 8) = m - 12$
$-9m - 72 = m - 12$
$-9m - m = -12 + 72$
$-10m = 60$
$m = \dfrac{60}{-10} = -6$

63. B

If $|2m - 3| \leq 9$, then
$-9 \leq 2m - 3 \leq 9$
$+3 \qquad +3 \quad +3$

$-6 \leq 2m \leq 12$
$-\dfrac{6}{2} \leq \dfrac{2m}{2} \leq \dfrac{12}{2}$
$-3 \leq m \leq 6$

64. C

Solve the inequality: $7 - |2x + 8| \geq 1$
$7 - |2x + 8| \geq 1$
Isolate the absolute value.
$-|2x + 8| \geq 1 - 7$
$-|2x + 8| \geq -6$

228   Prep for Success: Florida's PERT Math Study Guide

Multiply both sides by –1 and change the direction of the inequality sign.

$|2x + 8| \leq 6$

Remove the absolute value sign and set the expression between -6 and 6 with the inequality signs separating the 3 terms.

$-6 \leq 2x + 8 \leq 6$

$-8 \qquad -8 \quad -8$

$-14 \leq 2x \leq -2$

$\dfrac{-14}{2} \leq \dfrac{2x}{2} \leq \dfrac{-2}{2}$

$-7 \leq x \leq -1$

65.   **C**

If $2(x^2 + x) = 12$, then $x =$

$2(x^2 + x) = 12$

$2x^2 + 2x = 12$

$2x^2 + 2x - 12 = 0$

$2(x^2 + x - 6) = 0$

$2(x + 3)(x - 2) = 0$

$x + 3 = 0 \qquad\qquad x - 2 = 0$

$x = -3 \qquad\qquad\quad x = 2$

66.   **D**

Find the real roots of the following: $3x^2 + 2x + 5 = 0$

The expression cannot be factored so use the quadratic formula

$x = \dfrac{-b \pm \sqrt{b^2 - 4ac}}{2a}$

$a = 3, \;\; b = 2, \;\; c = 5$

$x = \dfrac{-2 \pm \sqrt{2^2 - 4(3)(5)}}{2(3)}$

$x = \dfrac{-2 \pm \sqrt{4 - 60}}{6}$

$x = \dfrac{-2 \pm \sqrt{-56}}{6}$

The solution involves the square root of a negative number which will yield only imaginary roots. The equation therefore has no real roots.

67.   **A**

Solve the system of equations:

$y = 2x + 1$

$2y + 3x = 9$

One of the equations contains a variable of coefficient 1 so substitution would be fastest. Substitute the first equation into the second equation.

$2(2x + 1) + 3x = 9$

$4x + 2 + 3x = 9$

$7x + 2 = 9$

Track 2 Success, Inc © 2014   All Rights Reserved

$7x = 7$

$x = \dfrac{7}{7} = 1$

Substitute this value into the first equation and solve for y

$y = 2(1) + 1 = 2 + 1 = 3$

Therefore the solution is $(1, 3)$

68. A

Write an equation to represent the graph of the line below.

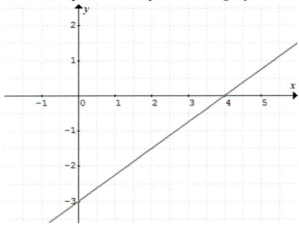

Pick two points on the graph. We select the x-intercept and the y-intercept.

$(0, -3)$ and $(4, 0)$

(x_1, y_1) and (x_2, y_2)

Now find the slope.

$m = \dfrac{y_2 - y_1}{x_2 - x_1} = \dfrac{0 - (-3)}{4 - 0} = \dfrac{3}{4}$

Use the slope and a point in the point-slope formula to find the equation of the line

$y - y_1 = m(x - x_1)$

$y - (-3) = \dfrac{3}{4}(x - 0)$

$y + 3 = \dfrac{3}{4}x$

$y = \dfrac{3}{4}x - 3$

$4\left(y = \dfrac{3}{4}x - 3\right)$

$4y = 3x - 12$

$4y - 3x = -12$

Compare to the answers and see that this is equivalent to choice A (multiply the entire equation by -1).

69. A

Which of the following represents a line parallel to the graph of $4y + 3x = 8$?

Parallel lines have the same slope so solve for y to find the slope of the given line

$4y + 3x = 8$

$4y = -3x + 8$

$$y = -\frac{3}{4}x + \frac{8}{4}$$
Slope: $-\frac{3}{4}$

Starting with choice A, find the slope of each answer choice until you find one that has the given slope.
$$4y = -3x - 12$$
$$y = -\frac{3}{4}x - \frac{12}{4}$$
This line has the required slope.

70. **A**

Which of the following could be an equation of the graph below?

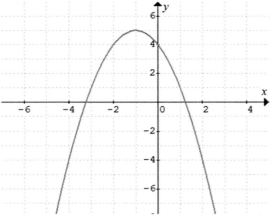

The graph is quadratic equation with a maximum point so the leading coefficient has to be negative.
In addition the vertex occurs at $(-1, 5)$
Eliminate answers choices B and D because the leading coefficients are positive.
Remember that the vertex is (h, k) in the equation $y = a(x - h)^2 + k$
The vertex of answer choice A is $(-1, 5)$
The vertex of answer choice C is $(-1, -5)$
The correct answer is therefore A.

71. **B**

If $f(x) = 2x + 3$ and $g(x) = \frac{x + 2}{5x - 4}$, then $g(f(x)) =$

Substitute x for $f(x)$ in $g(x)$
$$g(f(x)) = \frac{(2x + 3) + 2}{5(2x + 3) - 4}$$
$$g(f(x)) = \frac{2x + 3 + 2}{10x + 15 - 4}$$
$$g(f(x)) = \frac{2x + 5}{10x + 11}$$

72. **A**

$f(x) = 3x + 4$ and f^{-1} is the inverse of f, then $f^{-1}(-5) =?$

By the rule of inverses:

$3x + 4 = -5$

$3x = -9$

$x = \dfrac{-9}{3} = -3$

73. C

Which of the following can be factored in the form $(x + b)^2$, where b is an integer?

Answer choice A

$\quad x^2 - 4$

$\quad$ This is a difference of squares and factors into $(x - 2)(x + 2)$

Answer choice B

$\quad x^2 + 2$

$\quad$ This is not factorable

Answer choice C

$\quad x^2 + 10x + 25$

$\quad$ This factors into: $(x + 5)(x + 5) = (x + 5)^2$

Therefore the correct answer is C.

74. D

Express $\dfrac{7 + 3i}{4i}$ in the form $a + bi$.

Multiply by the conjugate of the denominator

$\dfrac{7 + 3i}{4i} \cdot \dfrac{-4i}{-4i}$

$= \dfrac{-4i(7 + 3i)}{-16i^2} = \dfrac{-28i - 12i^2}{-16(-1)}$

$= \dfrac{-28i - 12(-1)}{16}$

$= \dfrac{12 - 28i}{16} = \dfrac{12}{16} - \dfrac{28}{16}i = \dfrac{3}{4} - \dfrac{7}{4}i$

75. D

If $5(a - 2) = 3(2a + 4),$ then:

Distribute and solve

$5a - 10 = 6a + 12$

$5a - 6a = 12 + 10$

$-a = 22$

$a = -22$

Postsecondary Education Readiness Test (PERT) – Worksheet #2 Bubble Sheet

1. A B C D
2. A B C D
3. A B C D
4. A B C D
5. A B C D
6. A B C D
7. A B C D
8. A B C D
9. A B C D
10. A B C D
11. A B C D
12. A B C D
13. A B C D
14. A B C D
15. A B C D
16. A B C D
17. A B C D
18. A B C D
19. A B C D
20. A B C D
21. A B C D
22. A B C D
23. A B C D
24. A B C D
25. A B C D

26. A B C D
27. A B C D
28. A B C D
29. A B C D
30. A B C D
31. A B C D
32. A B C D
33. A B C D
34. A B C D
35. A B C D
36. A B C D
37. A B C D
38. A B C D
39. A B C D
40. A B C D
41. A B C D
42. A B C D
43. A B C D
44. A B C D
45. A B C D
46. A B C D
47. A B C D
48. A B C D
49. A B C D
50. A B C D

51. A B C D
52. A B C D
53. A B C D
54. A B C D
55. A B C D
56. A B C D
57. A B C D
58. A B C D
59. A B C D
60. A B C D
61. A B C D
62. A B C D
63. A B C D
64. A B C D
65. A B C D
66. A B C D
67. A B C D
68. A B C D
69. A B C D
70. A B C D
71. A B C D
72. A B C D
73. A B C D
74. A B C D
75. A B C D

The contents of this worksheet are copyrighted. Disclosure or reproduction of any portion herein without express written consent of Track 2 Success © is strictly prohibited.

Track 2 Success
"Dedicated to helping you stay on the track to academic success"

Worksheet #2

The questions contained in this worksheet cover all the topics presented in the study guide section and are ordered by increasing difficulty from Arithmetic through College Level Math.

Since calculators are not generally allowed while taking the Postsecondary Education Readiness Test, use of calculators should be limited while completing this worksheet so as to replicate and properly prepare for exam conditions.

1. $1\frac{2}{3} + 2\frac{1}{2} =$

   A. $3\frac{1}{6}$ B. $4\frac{1}{6}$ C. $3\frac{3}{5}$ D. $4\frac{2}{5}$

2. $4\frac{1}{8} - 2\frac{1}{4} =$

   A. $1\frac{3}{8}$ B. $1\frac{7}{8}$ C. $-2\frac{1}{8}$ D. $2\frac{1}{8}$

3. $1\frac{4}{9} \times 1\frac{1}{2} =$

   A. $1\frac{2}{9}$ B. $2\frac{1}{6}$ C. $2\frac{5}{18}$ D. $\frac{26}{27}$

4. $\frac{3}{7} \div 1\frac{1}{14} =$

   A. $\frac{45}{98}$ B. $2\frac{1}{5}$ C. $\frac{2}{5}$ D. $3\frac{3}{14}$

5. What is $\frac{1}{2}$% of 100?

   A. 5   B. 50   C. 0.05   D. 0.5

6. 55 is what percent of 125?
   A. 60%   B. 55%.   C. 48%   D. 44%

7. 18 is 30% of what number?
   A. 54   B. 60   C. 36   D. 40

The contents of this worksheet are copyrighted. Disclosure or reproduction of any portion herein without express written consent of Track 2 Success © is strictly prohibited.

8. The Electricity Company advises that you could save 12% on your electricity bill by installing a tankless water heater. If Susan's bill was $125 last month, how much would she have saved by using a tankless water heater?

   A. $10   B. $15   C. $18   D. $22

9. Which of the following is an irrational number?

   A. $-\dfrac{2}{3}$ B. $\dfrac{2\sqrt{27}}{\sqrt{3}}$ C. $\sqrt{3}$ D. $\sqrt[3]{-8}$

10. $\dfrac{13}{20} =$

    A. 65%   B. 6.5%   C. 0.65%   D. 0.13

11. $3.125 + 1.728 =$

    A. 4.843   B. 4.853   C. 5.843   D. 5.853

12. $5 - 3.135 =$

    A. 2.865   B. 1.865   C. −2.865   D. 3.865

13. $-0.042 \times (-4) =$

    A. 0.168   B. −0.168   C. 0.4168   D. 4.168

14. $-0.036 \div (-0.048) =$

    A. 0.75   B. 7.5   C. −7.5   D. 75

15. Henry bought 4.5 pounds of candy at the grocery store. If each pound of candy costs $3.92, what is a reasonable estimate of the amount he paid for the candy?

    A. $12.00   B. $15.00   C. $16.00   D. $20.00

16. Which of the following is least?

    A. 0.325   B. $0.\overline{32}$ C. $\dfrac{4}{11}$ D. 3.9×10^{-2}

17. Dale and John each own one half of the stock in an oil company. If Dale decides to sell ⅜ of his stock to Earl, what fraction of the oil company does Earl now own?

    A. $\dfrac{3}{8}$ B. $\dfrac{5}{8}$ C. $\dfrac{1}{4}$ D. $\dfrac{3}{16}$

18. A car needs to travel 275 miles in 5 hours. How fast should the car be driven?

A. 45 *mph*   **B.** 50 *mph*   **C.** 55 *mph*   **D.** 65 *mph*

19. Tim scored 92, 87, 56, 89 and 94 on his five English Exams. What was Tim's average score?

A. 76.2   **B.** 79.0   **C.** 83.6   **D.** 85.0

20. How many 4ft long pieces of fencing are needed to surround a rectangular garden that is 16ft wide and 32ft long?

A. 12   **B.** 24   **C.** 32   **D.** 96

21. 15 students took Mrs. Brown's math test and the results are presented in the table below. Find the median score.

# of Students	Test Score
4	72
2	85
1	48
5	60
3	92

A. 48   **B.** 60   **C.** 72   **D.** 92

22. A watch loses 10 secs every hour. How many minutes will the watch lose in a week?

A. 4   **B.** 28   **C.** 120   **D.** 240

23. Kevin borrowed $48,816 from the bank to purchase a house five years ago. If he has repaid ⅝ of the loan, how much money does he still owe the bank?

A. $6,102   **B.** $18,306   **C.** $21,276   **D.** $30,510

24. Find the area in square cm of a rectangle that has a width of 12 cm and a length of 16.2 cm.

A. 192.2   **B.** 194.4   **C.** 196.8   **D.** 204.2

25. The Dodge family is planning a vacation to Lake Tahoe. If the distance from their house to the lake is 300 miles and their van gets 18 miles per gallon, approximately how many gallons does the Dodge family need to make the trip and back?

A. 16 *gallons*   **B.** 17 *gallons*   **C.** 32 *gallons*   **D.** 34 *gallons*

26. Write the decimal 0.56 as a fraction in its lowest terms.

A. $\dfrac{7}{25}$ **B.** $\dfrac{14}{25}$ **C.** $\dfrac{7}{125}$ **D.** $\dfrac{14}{250}$

27. Find the perimeter of the following figure if the length is three times the width.

8

A. 24 **B.** 48 **C.** 56 **D.** 64

28. Evaluate: $|-15 - (-4)|$
A. 11 **B.** -19 **C.** 19 **D.** -11

29. Which of the following is a linear factor of: $3x^2 + 16x + 5$
A. $3x + 1$ **B.** $3x - 1$ **C.** $x - 5$ **D.** $3x + 5$

30. Simplify: $(x^3 + 4x^2 - 5) + (2x^2 - x - 6)$
A. $3x^2 + 3x - 11$ **C.** $3x^2 + 3x + 1$
B. $x^3 + 6x^2 - x - 11$ **D.** $x^3 + 2x^2 - 6x - 6$

31. Simplify: $(x + 2y)(3x - y)$
A. $4x + y$ **C.** $3x^2 + 4xy + y^2$
B. $3x^2 - 2y^2$ **D.** $3x^2 + 5xy - 2y^2$

32. Simplify: $\dfrac{(2x^3)^2(3y^4)^3}{(x^3y)^4}$
A. $\dfrac{6y^3}{x^2}$ **B.** $\dfrac{6y^8}{x^3}$ **C.** $\dfrac{108y^8}{x^6}$ **D.** $\dfrac{108y^3}{x^2}$

33. Simplify: $\dfrac{x^2 - x - 12}{x^2 + 5x + 6}$
A. $\dfrac{x-4}{x+2}$ **B.** -2 **C.** $4x - 2$ **D.** $\dfrac{x+3}{x+2}$

34. If $y^2 - 5y = 0$, then $y =$
A. -5 **B.** 0 **C.** -5 and 0 **D.** 0 and 5

35. What percentage of the diagram is not shaded?

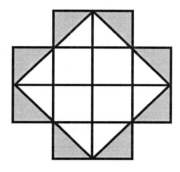

A. 25% B. $33\frac{1}{3}\%$ C. $66\frac{2}{3}\%$ D. 75%

36. $8\sqrt{5} + \sqrt{20} =$

A. $10\sqrt{5}$ B. $8\sqrt{5}$ C. $9\sqrt{15}$ D. 13

37. If M pounds of grapes are purchased at the market for 49 cents per lb and N lbs of oranges are purchased for 50 cents per lb, how much was spent at the market (in cents)?

A. $0.49M + 0.5N$
B. $M + N$
C. $99(M + N)$
D. $49M + 50N$

38. Given that $LM \parallel OP$, find the value of a and b.

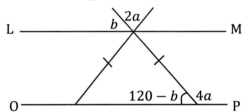

A. $a = 20, b = 20$
B. $a = 25, b = 40$
C. $a = 30, b = 60$
D. $a = 35, b = 80$

39. Given $g(x) = 5 + 2x - x^2$, find $g(5)$

A. -10 B. 20 C. -40 D. 60

40. If $3x - 4(x - 2) = 0$, then $x =$

A. -8 B. 8 C. $-\frac{8}{7}$ D. 6

41. Solve: $2(7x - 3) \leq 12x + 16$

A. $x \leq 11$ B. $x \leq -5$ C. $x \geq -5$ D. $x \leq 12$

42. Solve for c in the equation $\dfrac{3}{a} - \dfrac{2}{b} = \dfrac{1}{c}$

A. ab **B.** $\dfrac{1}{ab}$ **C.** $\dfrac{3b-2a}{ab}$ **D.** $\dfrac{ab}{3b-2a}$

43. Simplify: $-9 + \dfrac{1}{3} - 4 \times \dfrac{1}{2} \div 2^2 + \dfrac{5}{6}$

A. $-8\dfrac{1}{3}$ **B.** $-5\dfrac{1}{3}$ **C.** $-10\dfrac{1}{4}$ **D.** $-4\dfrac{1}{3}$

44. Samuel wants to enclose his rectangular pool with protective fencing. If the length of the pool is 25 feet and the perimeter of the pool is 120 feet, write an equation that can be used to solve for the width of the pool.

A. $2x + 25 = 120$ **C.** $2x = 95$

B. $2x + 50 = 120$ **D.** $120 - x = 50$

45. Given the equation $3y - 4x = 12$, find the missing value in the ordered pair $(-2, \_\_)$

A. $-\dfrac{3}{2}$ **B.** $\dfrac{4}{3}$ **C.** $-\dfrac{1}{2}$ **D.** 12

46. Which of the following is NOT between -1 and 2?

A. $-\dfrac{9}{8}$ **B.** $\dfrac{11}{7}$ **C.** $\dfrac{3}{2}$ **D.** $-\dfrac{1}{5}$

47. Simplify the expression: $3a - 5a^2 + 4b - 6a \div 3 - 12b$

A. $a - 5a^2 - 8b$ **C.** $5a^2 - 3a$

B. $\dfrac{-5a^2 - 3a - 8b}{3}$ **D.** $-(5a^2 + 8b)$

48. Write $\sqrt{\sqrt[3]{x^4}}$ using a single exponent.

A. $x^{\frac{3}{2}}$ **B.** x^6 **C.** $x^{\frac{2}{3}}$ **D.** $\dfrac{1}{x^5}$

49. The simple interest formula is given by the equation $I = P \times R \times T$. Find the amount of money P that should be saved in order to receive \$395 in interest after 2 years in an account that earns 10% interest annually.

A. \$1,975 **B.** \$2,425 **C.** \$3,950 **D.** \$7,900

50. Find the y-intercept of the equation: $y = 2(x - 4)$

A. $(0, -4)$ **B.** $(-4, 0)$ **C.** $(0, -8)$ **D.** $(-8, 0)$

51. Graph the line: $4y - 2x = 8$

A.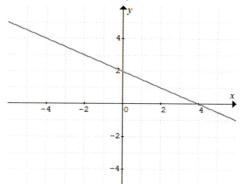

C.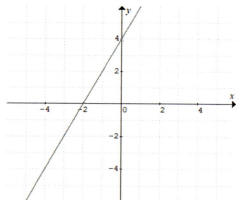

B.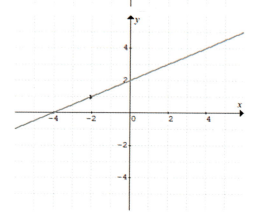

D.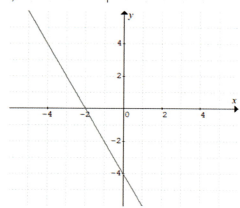

52. Solve for x: $-2 < 3x - 5 < 11$

A. $1 < x < 2$

B. $-\frac{7}{3} < x < 2$

C. $-\frac{7}{3} < x < \frac{16}{3}$

D. $1 < x < \frac{16}{3}$

53. $$\frac{\left(\frac{4}{3}\right)^{-2} \cdot 12^{-1}}{2^{-3}}$$

What is the value of the expression above?

A. $-\frac{3}{8}$ B. $\frac{8}{3}$ C. $-\frac{2}{3}$ D. $\frac{3}{8}$

54. Simplify: $5\sqrt[3]{81} - \sqrt[3]{24}$

A. $4\sqrt[3]{57}$ B. $4\sqrt[3]{3}$ C. $6\sqrt[3]{3}$ D. $13\sqrt[3]{3}$

55. A boat travelling upstream against the river's current of 12 mph makes the trip in 2 hrs. The return trip downstream takes 45 mins. What is the boat's speed in still water?

A. 16.5 mph B. 17.4 mph C. 26.4 mph D. 28.5 mph

56. Simplify: $\dfrac{x^2 + 5x + 6}{x^2 + 6x + 5} \div \dfrac{x^2 - x - 6}{x^2 + 2x - 15}$

A. $\dfrac{x+1}{x+3}$ **B.** $\dfrac{x+1}{x-3}$ **C.** $\dfrac{x+3}{x+1}$ **D.** $\dfrac{x-3}{x+1}$

---

57. Simplify by rationalizing the denominator: $\dfrac{4\sqrt{2}}{\sqrt{3} - \sqrt{2}}$

A. $4\sqrt{6}$ **B.** $4\sqrt{6} + 8$ **C.** $4\sqrt{6} - 8$ **D.** $\dfrac{4}{\sqrt{3}}$

---

58. Factor: $2y^4 - 6y^3 - 8y^2 + 24y$

A. $2y(y-3)(y-2)(y+2)$ **C.** $2y(y+3)(y^2-4)$

B. $2y(y-3)(y+4)$ **D.** $-2y(y-3)(y-2)(y+2)$

---

59. Expand: $(x-3)^3$

A. $x^3 - 9x^2 + 27x - 27$ **C.** $x^3 - 27$

B. $x^3 + 9x^2 - 27x + 27$ **D.** $x^3 - 9x^2 - 27x + 27$

---

60. What is the domain of $\dfrac{f}{g}(x)$ if $f(x) = |x+3|$ and $g(x) = \sqrt[4]{6-x}$?

A. $(-\infty, \infty)$ **C.** $(-\infty, -3) \cup (6, \infty)$

B. $(3, 6]$ **D.** $(-\infty, 6)$

---

61. Solve the equation: $\dfrac{-3}{2(x+4)} + \dfrac{5}{(x-4)} = \dfrac{-(x+1)}{x^2-16}$

A. -6 **B.** 6 **C.** 10 **D.** -10

---

62. If $|15x - 14| < -8$, then

A. $x < -\dfrac{4}{15}$ **B.** $x > \dfrac{4}{5}$ **C.** $\dfrac{4}{5} < x < -\dfrac{4}{15}$ **D.** No Solution

---

63. Solve the inequality: $\dfrac{x-1}{x+1} > 2$

A. $-3 < x < -1$ **B.** $x > 3$ **C.** $x < -1$ **D.** $-1 < x < 3$

---

64. If $3x^2 + 8x + 5 = 0$, then $x =$

A. -1 and $-\dfrac{5}{3}$ **B.** $-\dfrac{1}{3}$ and -5 **C.** $-\dfrac{5}{3}$ **D.** -5 only

65. Find the solution(s) of the following: $5x^2 - 2x + 1 = 0$

A. $\dfrac{1 \pm 2i}{5}$
B. $\dfrac{1 \pm \sqrt{6}}{5}$
C. $\dfrac{-1 \pm 2i}{5}$
D. $\dfrac{-1 \pm \sqrt{6}}{5}$

66. Solve the system of equations:
$y + 5x = 0$
$y - x^2 = 4$

A. $(1, -5)$ and $(4, -20)$
B. $(-1, 5)$ and $(-4, 20)$
C. $\left(-\dfrac{2}{3}, \dfrac{10}{3}\right)$
D. No Solution

67. If m, x and y are integers such that $x \neq y$ and $\dfrac{3}{m} - \dfrac{2}{x} = \dfrac{1}{y}$ then $m =$

A. $\dfrac{6x + 3y}{xy}$
B. xy
C. $\dfrac{3xy}{x + 2y}$
D. $\dfrac{xy}{3(y + 2x)}$

68. Write an equation in slope–intercept form to represent the graph of the line below.

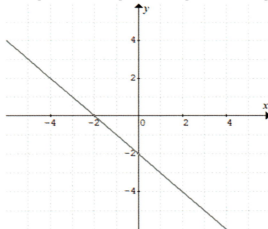

A. $y - x = 2$
B. $y = -2x - 2$
C. $y = x$
D. $y + x = -2$

69. Which of the following represents a line perpendicular to the graph $y = 2x + 5$ and passing through the point $(-3, 1)$?

A. $y = 2x + 7$
B. $y = 2x - 5$
C. $2y + x = -1$
D. $2y + x = 5$

70. If $h(x) = \dfrac{x+1}{x-1}$ and $f(x) = \dfrac{3}{x}$, then $h(f(-2)) =$

A. $-\dfrac{1}{2}$
B. $\dfrac{5}{4}$
C. $\dfrac{1}{2}$
D. $\dfrac{1}{5}$

71. Which of the following could be the graph of the equation $y + x^2 = a$ where $a > 0$?

A.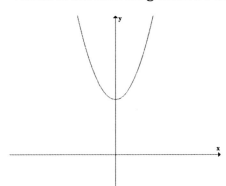

C.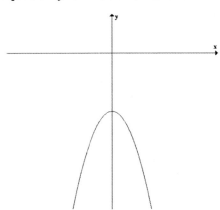

B.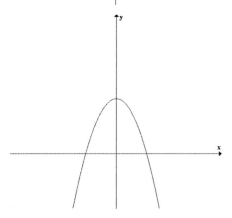

D.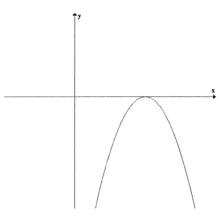

72. What is the inverse of the function $f(x) = \dfrac{1+3x}{5-2x}$?

A. $\dfrac{5x+3}{1-2x}$ B. $\dfrac{5-2x}{1+3x}$ C. $\dfrac{3x-5}{1+2x}$ D. $\dfrac{5x-1}{3+2x}$

73. $\dfrac{2-5i}{1+3i} =$

A. $\dfrac{13}{8} + \dfrac{11}{8}i$ B. $-\dfrac{13}{10} - \dfrac{11}{10}i$ C. $-\dfrac{17}{8} + \dfrac{11}{8}i$ D. $-\dfrac{17}{10} - \dfrac{11}{10}i$

74. John earns a total of $1,260 in interest from two savings accounts earning 6% and 8% respectively. If he deposited a total of $20,000, how much did he invest in the account earning 8%?

A. $17,000   B. $3,000   C. $1,500   D. $1,200

75. Which of the following is a solution to the equation $\sqrt{3x-5} + 7 = x + 4$?

A. $x = -1$ B. $x = 2$ C. $x = 7$ D. $x = 9$

◆·················· END WORKSHEET ··················◆

Postsecondary Education Readiness Test (PERT) – Worksheet #2 Answer Sheet

1. B	26. B	51. B
2. B	27. D	52. D
3. B	28. A	53. D
4. C	29. A	54. D
5. D	30. B	55. C
6. D	31. D	56. C
7. B	32. C	57. B
8. B	33. A	58. A
9. C	34. D	59. A
10. A	35. C	60. D
11. B	36. A	61. A
12. B	37. D	62. D
13. A	38. C	63. A
14. A	39. A	64. A
15. D	40. B	65. A
16. D	41. A	66. B
17. D	42. D	67. C
18. C	43. A	68. D
19. C	44. B	69. C
20. B	45. B	70. D
21. C	46. A	71. B
22. B	47. A	72. D
23. B	48. C	73. B
24. B	49. A	74. B
25. D	50. C	75. C

The contents of this worksheet are copyrighted. Disclosure or reproduction of any portion herein without express written consent of Track 2 Success © is strictly prohibited.

Track 2 Success

"Dedicated to helping you stay on the track to academic success"

Worksheet #2 Solutions

1. **B**

   $1\frac{2}{3} + 2\frac{1}{2} =$

   First convert the mixed numbers to improper fractions

   $\frac{5}{3} + \frac{5}{2}$

   LCD: 6

   $\frac{(2)5}{(2)3} + \frac{5(3)}{2(3)} = \frac{10}{6} + \frac{15}{6} = \frac{25}{6} = 4\frac{1}{6}$

2. **B**

   $4\frac{1}{8} - 2\frac{1}{4} =$

   First convert the mixed numbers to improper fractions

   $\frac{33}{8} - \frac{9}{4}$

   LCD: 8

   $\frac{33}{8} - \frac{9(2)}{4(2)} = \frac{33}{8} - \frac{18}{8} = \frac{15}{8} = 1\frac{7}{8}$

3. **B**

   $1\frac{4}{9} \times 1\frac{1}{2} =$

   First convert the mixed numbers to improper fractions

   $\frac{13}{9} \times \frac{3}{2}$

   Multiply straight across

   $\frac{39}{18} = 2\frac{3}{18} = 2\frac{1}{6}$

4. **C**

   $\frac{3}{7} \div 1\frac{1}{14} =$

   First convert the mixed number to an improper fraction

   $\frac{3}{7} \div \frac{15}{14}$

   Take the reciprocal of the second fraction then change the operator to multiplication.

   $\frac{3}{7} \times \frac{14}{15}$

   Multiply straight across

The contents of this worksheet are copyrighted. Disclosure or reproduction of any portion herein without express written consent of Track 2 Success © is strictly prohibited.

$$\frac{3(14)}{7(15)}$$

Reduce

$$\frac{(14)(3)}{(7)(15)} = \frac{2}{1} \cdot \frac{1}{5} = \frac{2}{5}$$

5. **D**

   What is $\frac{1}{2}$% of 100?

   $\frac{1}{2}$% $= 0.5\% = 0.005$

   Now simply multiply the number by the decimal representation of the percentage.

   $\frac{1}{2}$% of $100 = 0.005 \times 100 = 0.5$

6. **D**

   55 is what percent of 125?

   $$\frac{is}{of} = \frac{\%}{100}$$

   $$\frac{55}{125} = \frac{x}{100}$$

   $$125x = (55)(100)$$

   $$x = \frac{5500}{125} = 44\%$$

7. **B**

   18 is 30% of what number?

   $$\frac{is}{of} = \frac{\%}{100}$$

   $$\frac{18}{x} = \frac{30}{100}$$

   $$30x = (18)(100)$$

   $$x = \frac{1800}{30} = 60$$

8. **B**

   The Electricity Company advises that you could save 12% on your electricity bill by installing a tankless water heater. If Susan's bill was $125 last month, how much would she have saved by using a tankless water heater?

   The problem is essentially asking "what is 12% of $125".

   12% of $125 = 0.12 \times 125 = \$15$

9. **C**

   An irrational number is one that cannot be expressed as a fraction. Answer choice A is already a fraction, so eliminate it.

$$\frac{2\sqrt{27}}{\sqrt{3}} = 2\sqrt{\frac{27}{3}} = 2\sqrt{9} = 6$$

$$\sqrt{3} = 1.73\ldots$$
$$\sqrt[3]{-8} = \sqrt[3]{-2 \times -2 \times -2} = -2$$

Therefore the answer is clearly C because $\sqrt{3}$ is an infinite decimal.

10. **A**

$$\frac{13}{20} =$$

Convert to a decimal either by long division or by using the table of common fractions on page 20 then convert to a percentage by moving the decimal point two places to the right.

$$\frac{13}{20} = 13 \times \frac{1}{20} = 13 \times 0.05 = 0.65 = 65\%$$

11. **B**

$$3.125 + 1.728 =$$

Line up decimal points and add

$$
\begin{array}{r}
3.125 \\
+\quad 1.728 \\
\hline
4.853
\end{array}
$$

12. **B**

$$5 - 3.135 =$$

Add the decimal point and three place holding zeros after the 5.

$$5.000 - 3.135$$

Line up decimal points and subtract from right to left

$$
\begin{array}{r}
5.000 \\
-\quad 3.135 \\
\hline
1.865
\end{array}
$$

13. **A**

$$-0.042 \times (-4) =$$

Remove the decimal point and multiply

$$
\begin{array}{r}
-42 \\
\times \quad -4 \\
\hline
168
\end{array}
$$

There were 3 decimal places in the two original numbers combined so move the decimal point in the answer three places to the left

$$0.168$$

14. **A**

$$-0.036 \div (-0.048) =$$

Move the decimal point three places to the right in each number to convert to whole numbers.

$$-36 \div -48$$

$$\frac{-36}{-48} = \frac{3}{4} = 0.75$$

15. **D**

Henry bought 4.5 pounds of candy at the grocery store. If each pound of candy costs $3.92, what is a reasonable estimate of the amount he paid for the candy?

Round the numbers to the nearest whole number and multiply.

$5 \times \$4 = \$20$

16. **D**

Convert all the choices to decimals. Write out any recurring decimals then compare.

$0.\overline{32} = 0.323232\ldots$

$$\frac{4}{11} = 4 \times \frac{1}{11} = 4 \times 0.\overline{09} = 0.\overline{36} = 0.363636\ldots$$

$3.9 \times 10^{-2} = 0.039$

Comparing the four decimals:

0.325

0.323232 ...

0.363636 ...

0.039

We can therefore see that 3.9×10^{-2} is the smallest.

17. **D**

Dale and John each own one half of the stock in an oil company. If Dale decides to sell ⅜ of his stock to Earl, what fraction of the oil company does Earl now own?

We need to find $\frac{3}{8}$ of $\frac{1}{2}$

$$\frac{3}{8} \times \frac{1}{2} = \frac{3}{16}$$

18. **C**

A car needs to travel 275 miles in 5 hours. How fast should the car be driven?

$Distance = rate \times time$

$275 = r \times 5$

$$r = \frac{275}{5} = 55 \; mph$$

19. **C**

Tim scored 92, 87, 56, 89 and 94 on his five English Exams. What was Tim's average score?

Add all the scores and divide by 5

$$\frac{92 + 87 + 56 + 89 + 94}{5} = \frac{418}{5} = 83.6$$

20. **B**

How many 4ft long pieces of fencing are needed to surround a rectangular garden that is 16ft wide and 32ft long?

First find the perimeter

Track 2 Success © 2014

$$P = 2l \times 2w$$
$$P = 2(32) + 2(16)$$
$$P = 64 + 32 = 96$$

Since each piece is 4 ft long, divide by 4 to determine how many pieces are needed.

$$\frac{96}{4} = 24 \ pieces$$

21. C

15 students took Mrs. Brown's math test and the results are presented in the table below. Find the median score.

# of Students	Test Score
4	72
2	85
1	48
5	60
3	92

There are $4 + 2 + 1 + 5 + 3 = 15$ students in the class.

Since there are an odd number of students, the median is the middle or 8th score. Reorganize the table so that the scores are either in ascending or descending order.

# of Students	Test Score
1	48
5	60
4	72
2	85
3	92

Start at the top row and add the number of students in each row.

1 + 5 = 6

1 + 5 + 4 = 10

Therefore the 8th student is in the third row and the median score is 72.

22. B

A watch loses 10 secs every hour. How many minutes will the watch lose in a week?
We can use a direct proportion to solve this problem. Notice however that we have seconds per hour and minutes per week which do not match. If we convert 1 week to hours, we can use a proportion to determine how many seconds the watch will lose in a week.

$$1 \ week = 1 \times 7 days \times 24 \ hours = 168 \ hours$$
$$\frac{10 \ seconds}{1 \ hours} = \frac{x \ seconds}{168 \ hours}$$

Cross multiply and solve

$$(1)(x) = (10)(168)$$
$$x = 1680 \ seconds \ lost \ in \ 1 \ week$$

The question asks for the solution in minutes so convert seconds to minutes. There are 60 seconds in one minute.

$$\frac{1680}{60} = 28 \ minutes$$

23. B

Kevin borrowed $48, 816 from the bank to purchase a house five years ago. If he has repaid ⅝ of the loan, how much money does he still owe the bank?

Since he repaid $\frac{5}{8}$ he still owes $\frac{3}{8}$

$\frac{3}{8} \times \frac{\$48816}{1} = \frac{146448}{8} = \$18,306$

24. B

Find the area in square cm of a rectangle that has a width of 12 cm and a length of 16.2 cm.
$A = l \times w$
$A = 16.2 \times 12$
Multiply using the method for multiplying decimals.
$A = 194.4 \; square \; cm$

25. D

The Dodge family is planning a vacation to Lake Tahoe. If the distance from their house to the lake is 300 miles and their van gets 18 miles per gallon, approximately how many gallons does the Dodge family need to make the trip and back?
The trip there and back covers 300 + 300 = 600 miles. Divide miles to be travelled by miles per gallon to determine the number of gallons needed.
$\frac{600}{18} = 33\frac{6}{18} = 33\frac{1}{3}$
Rounding this up to the nearest gallon gives 34 gallons.

26. B

Write the decimal 0.56 as a fraction in its lowest terms.
There are two decimal places so drop the decimal point, place the number over 100 and simplify.
$0.56 = \frac{56}{100} = \frac{14}{25}$

27. D

Find the perimeter of the following figure if the length is three times the width.

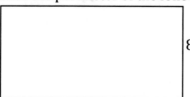

Length = 3 × width = 3 × 8 = 24
$P = 2l \times 2w$
$P = 2(24) + 2(8) = 48 + 16 = 64$

28. A

Evaluate: $|-15 - (-4)|$

Simplify the expression within the absolute value first.

$|-15 + 4|$

$|-11|$

The absolute value sign makes the answer positive

$|-11| = 11$

29. A

Which of the following is a linear factor of: $3x^2 + 16x + 5$

Factor the quadratic to determine its linear factors and choose the one that appears in the answer choices.

$3x^2 + 16x + 5 = (3x + 1)(x + 5)$

30. B

Simplify: $(x^3 + 4x^2 - 5) + (2x^2 - x - 6)$

We are dealing with addition and there is nothing to distribute so drop the parentheses and combine like terms.

$= x^3 + 4x^2 - 5 + 2x^2 - x - 6$

$= x^3 + 4x^2 + 2x^2 - x - 5 - 6$

$= x^3 + 6x^2 - x - 11$

31. D

Simplify: $(x + 2y)(3x - y)$

FOIL

$(x)(3x) + (x)(-y) + (2y)(3x) + (2y)(-y)$

$3x^2 - xy + 6xy - 2y^2$

Combine like terms

$3x^2 + 5xy - 2y^2$

32. C

Simplify: $\dfrac{(2x^3)^2(3y^4)^3}{(x^3y)^4}$

Using law #1 of exponents

$\dfrac{2^2(x^3)^2 \cdot 3^3(y^4)^3}{(x^3)^4 y^4}$

$\dfrac{4x^6 \cdot 27y^{12}}{x^{12}y^4}$

Using law #5 of exponents

$108x^{6-12}y^{12-4}$

$108x^{-6}y^8$

Finally, apply law #6 of exponents

$\dfrac{108y^8}{x^6}$

33. A

Simplify: $\dfrac{x^2 - x - 12}{x^2 + 5x + 6}$

Factor and simplify

$\dfrac{(x-4)(x+3)}{(x+3)(x+2)} = \dfrac{x-4}{x+2}$

34. D

If $y^2 - 5y = 0$, then $y =$

Factor, set factors equal to zero and solve

$y^2 - 5y = 0$

$y(y - 5) = 0$

$y = 0 \qquad y - 5 = 0$

$\qquad\qquad\quad y = 5$

35. C

What percentage of the diagram is not shaded?

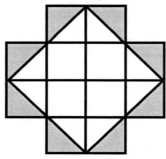

Count the number of triangles in the entire shape and then count the number of shaded triangles (note that the squares count as 2 triangles).

There are 24 triangles in total and 8 of them are shaded so 16 are not shaded.

Percentage not shaded $= \dfrac{not\ shaded}{total} \times 100 = \dfrac{16}{24} \times 100 = \dfrac{2}{3} \times 100 = 66\dfrac{2}{3}\%$

36. A

$8\sqrt{5} + \sqrt{20} =$

Reduce $\sqrt{20}$

$\sqrt{20} = \sqrt{4 \times 5} = 2\sqrt{5}$

Therefore:

$8\sqrt{5} + \sqrt{20} = 8\sqrt{5} + 2\sqrt{5} = 10\sqrt{5}$

37. D

If M pounds of grapes are purchased at the market for 49 cents per pound and N pounds of oranges are purchased for 50 cents per pound, how much was spent at the market (in cents)?

Cost of grapes $=$ *pounds* $\times$ *cost per pound* $= M \times 49 = 49M$

Cost of oranges $=$ *pounds* $\times$ *cost per pound* $= N \times 50 = 50N$

Total $= 49M + 50N$

Note that the decimal points in answer choice A indicate dollar format not cents

38. C

Given that $LM \parallel OP$, find the value of a and b.

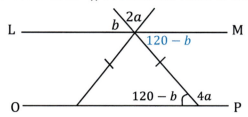

Using transverse angles, we add the value of the angle vertical to b to the diagram in blue. Since vertical angles are equal:
$b = 120 - b$
$2b = 120$
$b = \dfrac{120}{2} = 60$

Using the straight angle property we see that:
$(120 - b) + 4a = 180$
Substitute for b and solve for a.
$(120 - 60) + 4a = 180$
$60 + 4a = 180$
$4a = 120$
$a = \dfrac{120}{4} = 30$

39. A

Given $g(x) = 5 + 2x - x^2$, find $g(5)$
Substitute 5 for x
$g(5) = 5 + 2(5) - (5)^2$
$g(5) = 5 + 10 - 25 = -10$

40. B

If $3x - 4(x - 2) = 0$, then $x =$
$3x - 4(x - 2) = 0$
$3x - 4x + 8 = 0$
$-x + 8 = 0$
$-x = -8$
$x = 8$

41. A

Solve: $2(7x - 3) \leq 12x + 16$
$2(7x - 3) \leq 12x + 16$
$14x - 6 \leq 12x + 16$
$14x - 12x \leq 16 + 6$
$2x \leq 22$
$x \leq \dfrac{22}{2}$
$x \leq 11$

42. D

Solve for c in the equation $\dfrac{3}{a} - \dfrac{2}{b} = \dfrac{1}{c}$

LCD of left side: ab

$$\frac{(b)3}{(b)a} - \frac{2(a)}{b(a)} = \frac{1}{c}$$

$$\frac{3b}{ab} - \frac{2a}{ab} = \frac{1}{c}$$

$$\frac{3b - 2a}{ab} = \frac{1}{c}$$

Cross multiply

$$c(3b - 2a) = ab$$

$$c = \frac{ab}{3b - 2a}$$

43. A

Simplify: $-9 + \frac{1}{3} - 4 \times \frac{1}{2} \div 2^2 + \frac{5}{6}$

Applying PEMDAS

$$= -9 + \frac{1}{3} - 4 \times \frac{1}{2} \div 4 + \frac{5}{6}$$

$$= -9 + \frac{1}{3} - 2 \div 4 + \frac{5}{6}$$

$$= -9 + \frac{1}{3} - \frac{1}{2} + \frac{5}{6}$$

$$= -\frac{26}{3} - \frac{1}{2} + \frac{5}{6}$$

$$= -\frac{55}{6} + \frac{5}{6}$$

$$= -\frac{50}{6} = -8\frac{1}{3}$$

44. B

Samuel wants to enclose his rectangular pool with protective fencing. If the length of the pool is 25 feet and the perimeter of the pool is 120 feet, write an equation that can be used to solve for the width of the pool.

Let width $= x$

$$P = 2l + 2w$$

$$120 = 2(25) + 2(x)$$

$$2x + 50 = 120$$

45. B

Given the equation $3y - 4x = 12$, find the missing value in the ordered pair $(-2, \_\_)$

Substitute -2 into the equation for x

$$3y - 4x = 12$$

$$3y - 4(-2) = 12$$

$$3y + 8 = 12$$

$$3y = 4$$

$$y = \frac{4}{3}$$

Prep for Success: Florida's PERT Math Study Guide   257

46. **A**

Which of the following is NOT between -1 and 2?

$$-\frac{9}{8}, \frac{11}{7}, \frac{3}{2}, -\frac{1}{5}$$

$-\frac{1}{5} = -0.2$

$\frac{3}{2} = 1.5$

$\frac{11}{7} = 1\frac{4}{7}$

$-\frac{9}{8} = -1\frac{1}{8}$

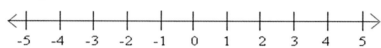

Comparing the answer choices to the number line above we see that the only one that does not fall between -1 and 2 is A.

47. **A**

Simplify the expression: $3a - 5a^2 + 4b - 6a \div 3 - 12b$

Using PEMDAS

$3a - 5a^2 + 4b - 6a \div 3 - 12b$
$= 3a - 5a^2 + 4b - 2a - 12b$

Combine like terms

$= 3a - 2a - 5a^2 + 4b - 12b$
$= a - 5a^2 - 8b$

48. **C**

Using law #7 of exponents we can convert the roots to fractional exponents.

$$\sqrt{\sqrt[3]{x^4}} = \sqrt{(x^4)^{\frac{1}{3}}} = \left[(x^4)^{\frac{1}{3}}\right]^{\frac{1}{2}}$$

Now combine using law #1 of exponents.

$$\left[(x^4)^{\frac{1}{3}}\right]^{\frac{1}{2}} = \left[x^{\frac{4}{3}}\right]^{\frac{1}{2}} = x^{\left(\frac{4}{3}\right)\left(\frac{1}{2}\right)} = x^{\frac{2}{3}}$$

49. **A**

The simple interest formula is given by the equation $I = P \times R \times T$. Find the amount of money P that should be saved in order to receive \$395 in interest after 2 years in an account that earns 10% interest annually.

$I = P \times R \times T$
$395 = P \times 0.10 \times 2$
$P \times 0.2 = 395$
$P = \dfrac{395}{0.2} = \dfrac{3950}{2}$
$P = \$1,975$

50. C

Find the y-intercept of the equation: $y = 2(x - 4)$

Substitute $x = 0$

$y = 2(0 - 4)$

$y = 2(-4)$

$y = -8$

$(0, -8)$

51. B

Graph the line: $4y - 2x = 8$

$x - intercept$

$4(0) - 2x = 8$

$-2x = 8$

$x = -\dfrac{8}{2}$

$x = -4$

$y - intercept$

$4y - 2(0) = 8$

$4y = 8$

$y = \dfrac{8}{4} = 2$

Plot the x and y intercepts then connect the points

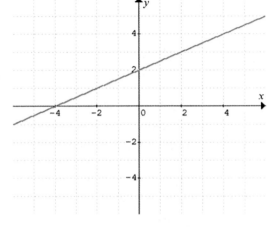

52. D

Solve for x: $-2 < 3x - 5 < 11$

$-2 < 3x - 5 < 11$

$+5 \quad\quad +5 \quad +5$

$3 < 3x < 16$

$\dfrac{3}{3} < \dfrac{3x}{3} < \dfrac{16}{3}$

$1 < x < \dfrac{16}{3}$

53. D

$\dfrac{\left(\dfrac{4}{3}\right)^{-2} \cdot 12^{-1}}{2^{-3}}$

What is the value of the expression above?

Use rule #6 of exponents then simplify

$\dfrac{\left(\dfrac{4}{3}\right)^{-2} \cdot 12^{-1}}{2^{-3}} = \dfrac{\left(\dfrac{3}{4}\right)^{2} \cdot \dfrac{1}{12}}{\dfrac{1}{2^3}} = \dfrac{\dfrac{9}{16} \cdot \dfrac{1}{12}}{\dfrac{1}{8}}$

$= \dfrac{9}{16} \cdot \dfrac{1}{12} \cdot \dfrac{8}{1} = \dfrac{3 \times 3 \times 2 \times 4}{4 \times 4 \times 3 \times 4} = \dfrac{3 \times 2}{4 \times 4}$

$= \dfrac{6}{16} = \dfrac{3}{8}$

54. D

Simplify: $5\sqrt[3]{81} - \sqrt[3]{24}$

$5\sqrt[3]{81} - \sqrt[3]{24}$

$5\sqrt[3]{3 \times 3 \times 3 \times 3} - \sqrt[3]{2 \times 2 \times 2 \times 3}$

$5 \cdot 3\sqrt[3]{3} - 2\sqrt[3]{3}$

$15\sqrt[3]{3} - 2\sqrt[3]{3} = 13\sqrt[3]{3}$

55. C

A boat travelling upstream against the river's current of 12 mph makes the trip in 2 hrs. The return trip downstream takes 45 mins. What is the boat's speed in still water?

When the boat travels upstream against the current, the rate of the river flowing downstream slows the boat down. So the boat's speed upstream is its speed alone (in still water) minus the speed of the river going against it. Therefore:

$rate\ of\ boat = (x - 12)\ mph$

When the boat travels downstream with the current, the rate of the river flowing in the same direction speeds the boat up. So the boat's speed downstream is its speed alone plus the speed of the river going with it. Therefore:

$rate\ of\ boat = (x + 12)\ mph$

Since the rate is in mph, we need both times in hrs.

$45\ mins = \dfrac{45}{60} = 0.75\ hrs$

Create a table relating the rate, distance and time.

Let distance travelled $= d$

	Distance	Rate	Time
Upstream	d	$x - 12$	2 hrs
Downstream	d	$x + 12$	0.75 hrs

$distance = rate \times time$

$d = (x - 12)(2) \quad and \quad d = (x + 12)(0.75)$

Since the distance upstream and downstream are equal:

$(x - 12)(2) = (x + 12)(0.75)$

$2x - 24 = 0.75x + 9$

$1.25x = 33$

$x = \dfrac{33}{1.25} = 26.4\ mph$

56. C

Simplify: $\dfrac{x^2 + 5x + 6}{x^2 + 6x + 5} \div \dfrac{x^2 - x - 6}{x^2 + 2x - 15}$

Factor, take the reciprocal of the second fraction and multiply

$\dfrac{x^2 + 5x + 6}{x^2 + 6x + 5} \div \dfrac{x^2 - x - 6}{x^2 + 2x - 15}$

$\dfrac{(x + 3)(x + 2)}{(x + 5)(x + 1)} \div \dfrac{(x - 3)(x + 2)}{(x + 5)(x - 3)}$

$$\frac{(x+3)(x+2)}{(x+5)(x+1)} \times \frac{(x+5)(x-3)}{(x-3)(x+2)}$$

Eliminate factors common to both the numerator and the denominator

$$\frac{(x+3)(x+2)(x+5)(x-3)}{(x+5)(x+1)(x-3)(x+2)} = \frac{(x+3)}{(x+1)}$$

57. B

Simplify by rationalizing the denominator: $\dfrac{4\sqrt{2}}{\sqrt{3}-\sqrt{2}}$

Multiply by the conjugate of the denominator

$$\frac{4\sqrt{2}}{\sqrt{3}-\sqrt{2}} \cdot \frac{\sqrt{3}+\sqrt{2}}{\sqrt{3}+\sqrt{2}}$$

$$\frac{(4\sqrt{2})(\sqrt{3})+(4\sqrt{2})(\sqrt{2})}{(\sqrt{3})(\sqrt{3})+(\sqrt{3})(\sqrt{2})-(\sqrt{2})(\sqrt{3})-(\sqrt{2})(\sqrt{2})}$$

$$\frac{4\sqrt{6}+4(2)}{3-2} = 4\sqrt{6}+8$$

58. A

Factor: $2y^4 - 6y^3 - 8y^2 + 24y$

Factor by grouping

$2y^4 - 6y^3 - 8y^2 + 24y$

$2y^3(y-3) - 8y(y-3)$

$(2y^3 - 8y)(y-3)$

$2y(y^2 - 4)(y-3)$

$2y(y+2)(y-2)(y-3)$

59. A

Expand: $(x-3)^3$

$(x-3)(x-3)(x-3)$

FOIL first two factors

$(x^2 - 3x - 3x + 9)(x-3)$

$(x^2 - 6x + 9)(x-3)$

Now multiply these two factors

$x^2(x-3) - 6x(x-3) + 9(x-3)$

$x^3 - 3x^2 - 6x^2 + 18x + 9x - 27$

$x^3 - 9x^2 + 27x - 27$

60. D

$$\frac{f}{g}(x) = \frac{|x+3|}{\sqrt[4]{6-x}}$$

The domain of $f(x)$ is $(-\infty, \infty)$ or all real numbers so we focus on $g(x)$ which is an even root. We cannot take the even root of a negative number and we combine this with the fact that the denominator cannot equal 0.

$6 - x > 0$

$-x > -6$

$x < 6$

Domain: $(-\infty, 6)$

61. A

Solve the equation: $\dfrac{-3}{2(x+4)} + \dfrac{5}{(x-4)} = \dfrac{-(x+1)}{x^2 - 16}$

$\dfrac{-3}{2(x+4)} + \dfrac{5}{(x-4)} = \dfrac{-(x+1)}{(x+4)(x-4)}$

LCD: $2(x+4)(x-4)$

$\dfrac{-3(x-4)}{2(x+4)(x-4)} + \dfrac{5\big(2(x+4)\big)}{(x-4)\big(2(x+4)\big)} = \dfrac{-(x+1)(2)}{(x+4)(x-4)(2)}$

Since denominators are now the same, solve the equation formed by numerators

$-3(x-4) + 5\big(2(x+4)\big) = -(x+1)(2)$

$-3x + 12 + 10x + 40 = -2x - 2$

$7x + 52 = -2x - 2$

$7x + 2x = -2 - 52$

$9x = -54$

$x = -\dfrac{54}{9} = -6$

62. D

If $|15x - 14| < -8$, then

There is no solution because an absolute value cannot be less than –8

63. A

Solve the inequality: $\dfrac{x-1}{x+1} > 2$

$\dfrac{x-1}{x+1} - \dfrac{2}{1} > 0$

$\dfrac{x-1}{x+1} - \dfrac{2(x+1)}{1(x+1)} > 0$

$\dfrac{x-1-2x-2}{x+1} > 0$

$\dfrac{-x-3}{x+1} > 0$

Find the critical points

$-x - 3 = 0 \qquad\qquad x + 1 = 0$

$x = -3 \qquad\qquad\quad x = -1$

Place the critical values on a number line and choose test points between them.

Substitute the test points into the original inequality to see which satisfy the inequality.

262   Prep for Success: Florida's PERT Math Study Guide

When $x = -4$

$$\frac{-4 - 1}{-4 + 1} > 2$$

$$\frac{-5}{-3} > 2$$

$$\frac{5}{3} > 2$$

False

When $x = -2$

$$\frac{-2 - 1}{-2 + 1} > 2$$

$$\frac{-3}{-1} > 2$$

$$3 > 2$$

True

When $x = 0$

$$\frac{0 - 1}{0 + 1} > 2$$

$$\frac{-1}{1} > 2$$

$$-1 > 2$$

False

Therefore the solution is:

$$-3 < x < -1$$

64.   **A**

If $3x^2 + 8x + 5 = 0$, then $x =$

Factor if possible

$(3x + 5)(x + 1) = 0$

$3x + 5 = 0 \qquad\qquad x + 1 = 0$

$3x = -5 \qquad\qquad x = -1$

$$x = -\frac{5}{3}$$

65.   **A**

Find the solution(s) of the following: $5x^2 - 2x + 1 = 0$

The quadratic equation cannot be factored so use the quadratic formula

$a = 5, \; b = -2, \; c = 1$

$$x = \frac{-b \pm \sqrt{b^2 - 4ac}}{2a}$$

$$x = \frac{-(-2) \pm \sqrt{(-2)^2 - 4(5)(1)}}{2(5)}$$

$$x = \frac{2 \pm \sqrt{4 - 20}}{10}$$

$$x = \frac{2 \pm \sqrt{-16}}{10} = \frac{2 \pm \sqrt{16i^2}}{10}$$

$$x = \frac{2 \pm 4i}{10} = \frac{1 \pm 2i}{5}$$

66.   **B**

Solve the system of equations:

$y + 5x = 0$

$y - x^2 = 4$

This is a system consisting of a line and a parabola. Solve the linear equation for y.

$y = -5x$

Substitute into the parabola.

$-5x - x^2 = 4$

Move all terms to one side.

$x^2 + 5x + 4 = 0$

Factor to find the value(s) of x

Track 2 Success © 2014   All Rights Reserved

$(x+4)(x+1) = 0$
$x + 4 = 0 \quad\quad x + 1 = 0$
$x = -4 \quad\quad x = -1$
Substitute the values of x into either of the original equations to find the values of y

$x = -4 \quad\quad\quad\quad x = -1$
$y + 5x = 0 \quad\quad\quad y + 5x = 0$
$y + 5(-4) = 0 \quad\quad y + 5(-1) = 0$
$y - 20 = 0 \quad\quad\quad y - 5 = 0$
$y = 20 \quad\quad\quad\quad y = 5$
$(-4, 20) \quad\quad\quad\quad (-1, 5)$

67. C

If m, x and y are integers such that $x \neq y$ and $\dfrac{3}{m} - \dfrac{2}{x} = \dfrac{1}{y}$ then $m =$

Move all terms containing m to one side of the equation and all the other terms to the other side of the equation

$\dfrac{3}{m} = \dfrac{1}{y} + \dfrac{2}{x}$

Find the LCD of the right side and combine the fractions

LCD: xy

$\dfrac{3}{m} = \dfrac{1(x)}{y(x)} + \dfrac{2(y)}{x(y)}$

$\dfrac{3}{m} = \dfrac{x}{xy} + \dfrac{2y}{xy}$

$\dfrac{3}{m} = \dfrac{x + 2y}{xy}$

Cross multiply then solve for m.

$3xy = m(x + 2y)$

$m = \dfrac{3xy}{x + 2y}$

68. D

Write an equation in slope–intercept form to represent the graph of the line below.

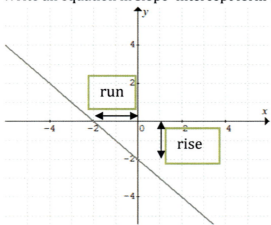

We can pick two points then find the slope and use the point–slope formula to find the equation of the line or we can find the y–intercept then use rise and run to the x–intercept to find the slope.

First find the y–intercept b.

$b = -2$

Next find rise and run from the y–intercept to the x–intercept as shown by the arrows on the graph above.

Rise $= +2$

Run $= -2$

Slope $m = \dfrac{rise}{run} = \dfrac{+2}{-2} = -1$

Now that we have m and b, simply plug into the equation $y = mx + b$

$y = -1x - 2$

69. **C**

Which of the following represents a line perpendicular to the graph of $y = 2x + 5$ and passing through the point $(-3, 1)$?

In the given line, the slope $m = 2$

The slope of the new line will be the negative reciprocal of the current slope

$m = -\dfrac{1}{2}$

Using the new slope and the given point:

$y - y_1 = m(x - x_1)$

$y - 1 = -\dfrac{1}{2}(x - (-3))$

$y - 1 = -\dfrac{1}{2}(x + 3)$

$y - 1 = -\dfrac{1}{2}x - \dfrac{3}{2}$

$y = -\dfrac{1}{2}x - \dfrac{3}{2} + 1$

$y = -\dfrac{1}{2}x - \dfrac{1}{2}$

$2\left(y = -\dfrac{1}{2}x - \dfrac{1}{2}\right)$

$2y = -x - 1$

$2y + x = -1$

70. **D**

If $h(x) = \dfrac{x + 1}{x - 1}$ and $f(x) = \dfrac{3}{x}$, then $h\big(f(-2)\big) =$

$f(-2) = \dfrac{3}{-2}$

$$h(f(-2)) = \frac{-\frac{3}{2}+1}{-\frac{3}{2}-1}$$

$$= \frac{-\frac{1}{2}}{-\frac{5}{2}}$$

$$= -\frac{1}{2} \times -\frac{2}{5} = \frac{1}{5}$$

71. B

Which of the following could be the graph of the equation $y + x^2 = a$ where $a > 0$?
Solve the equation for y
$y = -x^2 + a$
From this we see that the leading coefficient is negative so the graph will have a maximum point or peak. When $x = 0$, we see that the y-intercept occurs at $y = a$. Since a is positive, the y-intercept is positive. The only graph that match these criteria is B

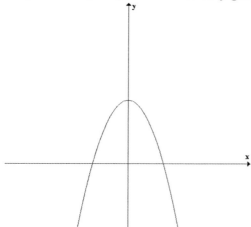

72. D

What is the inverse of the function $f(x) = \frac{1+3x}{5-2x}$

Let $f(x) = y$
$$y = \frac{1+3x}{5-2x}$$
Interchange x and y
$$x = \frac{1+3y}{5-2y}$$
Cross multiply and solve for y
$x(5 - 2y) = 1 + 3y$
$5x - 2xy = 1 + 3y$
$5x - 1 = 3y + 2xy$
$5x - 1 = y(3 + 2x)$

$$y = \frac{5x - 1}{3 + 2x}$$

Replace y with $f^{-1}(x)$

$$f^{-1}(x) = \frac{5x - 1}{3 + 2x}$$

73. B

$$\frac{2 - 5i}{1 + 3i} =$$

Multiply by the conjugate of the denominator

$$\frac{2 - 5i}{1 + 3i} \cdot \frac{1 - 3i}{1 - 3i}$$

FOIL

$$\frac{(2)(1) + (2)(-3i) + (-5i)(1) + (-5i)(-3i)}{(1)(1) + (1)(-3i) + (3i)(1) + (3i)(-3i)}$$

$$\frac{2 - 6i - 5i + 15i^2}{1 - 3i + 3i - 9i^2}$$

$$\frac{2 - 11i + 15(-1)}{1 - 9(-1)}$$

$$\frac{-13 - 11i}{10} = -\frac{13}{10} - \frac{11}{10}i$$

74. B

John earns a total of \$1,260 in interest from two savings accounts earning 6% and 8% respectively. If he deposited a total of \$20,000, how much did he invest in the account earning 8%?

Let the amount deposited into the first account earning 6% $= x$

Let the amount deposited into the second account earning 8% $= y$

$x + y = 20000$

$0.06x + 0.08y = 1260$

Multiply the second equation by 100 to make it easier to work with

$6x + 8y = 126000$

Solve the first equation for either x or y and substitute into the second equation. We solve for x since we are trying to find y

$x + y = 20000$

$x = 20000 - y$

Substitute to find y

$6(20000 - y) + 8y = 126000$

$120000 - 6y + 8y = 126000$

$2y = 126000 - 120000$

$2y = 6000$

$$y = \frac{6000}{2} = \$3,000$$

75. C

Which of the following is a solution to the equation $\sqrt{3x-5}+7=x+4$?

Isolate the radical

$\sqrt{3x-5}=x-3$

Square both sides:

$\left(\sqrt{3x-5}\right)^2=(x-3)^2$

$3x-5=x^2-6x+9$

Move all terms to one side and factor

$x^2-6x+9-3x+5=0$

$x^2-9x+14=0$

$(x-7)(x-2)=0$

$x=7$ and $x=2$

Test answers

$\sqrt{3(7)-5}+7=7+4$ $\sqrt{3(2)-5}+7=7+4$

$\sqrt{21-5}+7=7+4$ $\sqrt{6-5}+7=7+4$

$\sqrt{16}+7=7+4$ $\sqrt{1}+7=7+4$

$4+7=7+4$ True $1+7=7+4$ False

Therefore, only $x=7$ is a solution.

268   Prep for Success: Florida's PERT Math Study Guide

Postsecondary Education Readiness Test (PERT) – Worksheet #3 Bubble Sheet

1. A B C D
2. A B C D
3. A B C D
4. A B C D
5. A B C D
6. A B C D
7. A B C D
8. A B C D
9. A B C D
10. A B C D
11. A B C D
12. A B C D
13. A B C D
14. A B C D
15. A B C D
16. A B C D
17. A B C D
18. A B C D
19. A B C D
20. A B C D
21. A B C D
22. A B C D
23. A B C D
24. A B C D
25. A B C D

26. A B C D
27. A B C D
28. A B C D
29. A B C D
30. A B C D
31. A B C D
32. A B C D
33. A B C D
34. A B C D
35. A B C D
36. A B C D
37. A B C D
38. A B C D
39. A B C D
40. A B C D
41. A B C D
42. A B C D
43. A B C D
44. A B C D
45. A B C D
46. A B C D
47. A B C D
48. A B C D
49. A B C D
50. A B C D

51. A B C D
52. A B C D
53. A B C D
54. A B C D
55. A B C D
56. A B C D
57. A B C D
58. A B C D
59. A B C D
60. A B C D
61. A B C D
62. A B C D
63. A B C D
64. A B C D
65. A B C D
66. A B C D
67. A B C D
68. A B C D
69. A B C D
70. A B C D
71. A B C D
72. A B C D
73. A B C D
74. A B C D
75. A B C D

The contents of this worksheet are copyrighted. Disclosure or reproduction of any portion herein without express written consent of Track 2 Success © is strictly prohibited.

Track 2 Success
"Dedicated to helping you stay on the track to academic success"

Worksheet #3

The questions contained in this worksheet cover all the topics presented in the study guide section and are ordered by increasing difficulty from Arithmetic through College Level Math.

Since calculators are not generally allowed while taking the Postsecondary Education Readiness Test, the use of calculators should be limited while completing this worksheet so as to replicate and properly prepare for exam conditions.

1. $2\frac{1}{3} + \frac{4}{7} =$

   A. $2\frac{4}{21}$
   B. $2\frac{1}{2}$
   C. $2\frac{5}{21}$
   D. $2\frac{19}{21}$

2. $\frac{10}{21} - \frac{1}{7} =$

   A. $\frac{9}{14}$
   B. $\frac{3}{7}$
   C. $\frac{1}{3}$
   D. $\frac{11}{21}$

3. $2\frac{1}{4} \times \frac{1}{3} =$

   A. $\frac{3}{4}$
   B. $2\frac{1}{12}$
   C. $2\frac{7}{12}$
   D. $2\frac{3}{4}$

4. $-2\frac{1}{3} \div 1\frac{2}{7} =$

   A. $1\frac{22}{27}$
   B. $-1\frac{22}{27}$
   C. $2\frac{1}{7}$
   D. -3

5. A basketball team played 60 games and lost 15% of them. How many games did they win?

   A. 9
   B. 15
   C. 45
   D. 51

6. 90 is what percent of 60?

   A. 67%
   B. 110%
   C. 125%
   D. 150%

The contents of this worksheet are copyrighted. Disclosure or reproduction of any portion herein without express written consent of Track 2 Success © is strictly prohibited.

7. 10 is 50% of what number?
A. 20 B. 30 C. 50 D. 15

8. The price of gasoline will increase by 10% next month. If gasoline costs $1.80 per gallon now, what will be the price per gallon next month?
A. $1.88 B. $1.90 C. $1.98 D. $2.00

9. The graph below represents the average rainfall, measured in inches, in Davenport during the year 1997. How much higher is the average rainfall in October than it is in April?

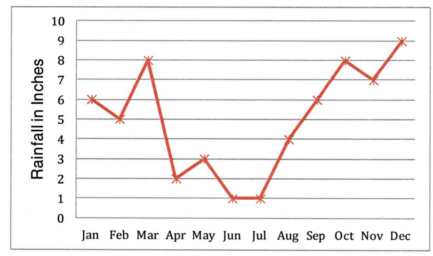

A. 25% B. 75% C. 133% D. 300%

10. If 60 percent of a number is 750, then 10% of the same number is?
A. 45 B. 75 C. 125 D. 250

11. $1.923 + 0.598 =$
A. 1.521 B. 2.421 C. 2.511 D. 2.521

12. $-9.023 + 4.748 =$
A. −4.275 B. 4.275 C. 13.771 D. 13.761

13. $7.2 \times 0.31 =$
A. 2.131 B. 2.232 C. 2.17 D. 22.32

14. $0.016 \div 0.8 =$
A. 0.002 B. 0.02 C. 0.2 D. 2

15. The Lions cheerleading squad held a car wash to raise funds for new uniforms. If they charged $5.00 per car, $7.00 per van and $12.00 per truck, what would be a reasonable estimate of the amount they charged per vehicle?

A. $6.50   **B.** $7.75   **C.** $9.95   **D.** $11.00

16. Which of the following is greatest?

A. $\dfrac{1}{4}$ **B.** $\dfrac{2}{5}$ **C.** $\dfrac{7}{8}$ **D.** $\dfrac{13}{16}$

17. Four people are hired to work together on a project so that the total time spent on the project is equivalent to one person working full time. The first person spends ⅛ of his time, the second spends ⅓ of his time, and the third person spends ½ of his time working on the project. How much of his time should the fourth person spend on the project?

A. $\dfrac{1}{24}$ **B.** $\dfrac{3}{24}$ **C.** $\dfrac{21}{24}$ **D.** $\dfrac{23}{24}$

18. Find the arithmetic mean of the following numbers: 8.2, 4.9 and 6.7

A. 6.6   **B.** 7.1   **C.** 7.4   **D.** 8.1

19. The area of a rectangle is 192 square inches and the length is 16 inches. What is the width of the rectangle?

A. 12   **B.** 13   **C.** 14   **D.** 15

20. Sammy scored 69% on one math exam and 95% on another. What does he need to score on the third exam in order to receive an 85% average?

A. 82%   **B.** 91%   **C.** 93%   **D.** 94%

21. Randy bought a new entertainment system. He paid $975 for the LCD television, $159 for the speakers and $60 for the TV stand. What is his total price after 8% sales tax?

A. $95.52   **B.** $1,194.00   **C.** $1,289.52   **D.** $2,149.20

22. Find the area of a triangle in ft² if its base measures 60 ft and its height is 48 inches.

A. 120   **B.** 24   **C.** 72   **D.** 144

23. Three workers can finish building a barn in 8 days by working together. Working alone, the first worker can build the barn in 24 days while another worker would take 18 days. How long would the third worker need to build the barn alone?

A. 24 days   **B.** 36 days   **C.** 48 days   **D.** 72 days

24. At a boutique, a blouse costs $23.50 and a skirt costs $17.25. Both are on sale at 40% off. What is the price of a blouse and skirt bought together after the discount is applied?

   A. $16.30   **B.** $18.15   **C.** $24.45   **D.** $26.25

25. In 1990, a new house costs $9,850. Today the same house would cost $315,790. How much did the house increase?

   A. $35,940   **B.** $206,940   **C.** $305,940   **D.** $306,940

26. Write the following in decimal notation: Seven and eleven hundredths.

   A. 711   **B.** 7.011   **C.** 7.0011   **D.** 7.11

27. The midnight train has 280 miles left to travel. If the train has traveled ¾ of the way, how long is the total trip?

   A. 280 *miles*   **B.** 840 *miles*   **C.** 1120 *miles*   **D.** 1260 *miles*

28. Evaluate: $\dfrac{4 \cdot 5^2 + 8^2 \div 16}{7 - 12 \div \sqrt{36} + 8}$

   A. $1\dfrac{37}{50}$ **B.** $8\dfrac{12}{13}$ **C.** 8 **D.** 15

29. Which of the following is a linear factor of $x^2 + 5x + 6$?

   A. $x - 2$ **B.** $x - 1$ **C.** $x - 6$ **D.** $x + 3$

30. Simplify: $(3x^2 + x + 1) - (2x^2 - 3x - 5)$

   A. $5x^2 - 2x - 4$ **C.** $x^2 + 4x + 6$

   B. $5x^2 - 2x + 6$ **D.** $x^2 - 2x + 6$

31. Simplify: $(2x - 3)^2$

   A. $4x - 6$ **C.** $4x^2 - 12x + 9$

   B. $4x^2 - 6x + 9$ **D.** $4x^2 - 9$

32. Simplify: $\dfrac{a^{-3}b^4}{a^{-5}b^5}$

   A. $a^2 b$ **B.** $\dfrac{a^2}{b}$ **C.** $a^{-8}b^9$ **D.** $\dfrac{b^9}{a^{-8}}$

33. Simplify: $\dfrac{y^2 + y}{y^2 - 1}$

   A. $-y$ **B.** $\dfrac{y}{y - 1}$ **C.** $y + 1$ **D.** $y - 1$

34. If $6x^2 + 5x = 4$, then $x =$

A. $-\frac{1}{3}$ and 2 B. $\frac{4}{3}$ and $-\frac{1}{2}$ C. $-\frac{4}{3}$ and $\frac{1}{2}$ D. $\frac{1}{3}$ and -2

35. What is the value of the expression $(2x^2 - 3y^2)^2$ when $x = -1$ and $y = 2$?

A. 10 B. -100 C. -20 D. 100

36. Solve the equation: $|-2x + 3| = 15$

A. 6 only B. 9 only C. -6 and 9 D. No Solution

37. The difference between two integers is 8. The sum of the two integers is 30. Which of the following equations is correct given that the larger of the two integers is represented by p?

A. $p + (8p) = 30$
B. $p + (p - 8) = 30$
C. $p - (p + 30) = 8$
D. $p - (p - 30) = 8$

38. If $\frac{x+4}{7} = \frac{4}{9}$, then $x =$

A. $-\frac{1}{4}$ B. $-\frac{8}{9}$ C. $\frac{24}{9}$ D. $\frac{36}{7}$

39. If $5 - \frac{2}{3}x > -7$, then

A. $x < -3$ B. $x > -3$ C. $x > -18$ D. $x < 18$

40. Solve for m in the equation: $5n - 3 = 2n\left(\frac{m}{3}\right)$

A. $\frac{9}{2n + 15}$ B. $\frac{9}{13n}$ C. $\frac{3(5n - 3)}{2n}$ D. $\frac{3}{5 - 6n}$

41. Find the length of AC given the following figure.

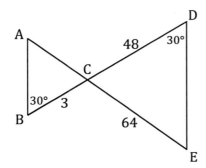

A. 2 B. 4 C. 5 D. 6

42. What is the value of the expression $x^2 - 3x + 5$ when $x = \frac{1}{2}$?

    A. $-4\frac{11}{12}$ B. $3\frac{3}{4}$ C. $-6\frac{1}{4}$ D. $5\frac{1}{12}$

43. The distance between points $(-3, 7)$ and $(x, 12)$ is 13. Find x.

    A. $x = 9$
    B. $x = 15$
    C. $x = -15$ or $x = 9$
    D. $x = -9$ and $x = 15$

44. Sheila has a triangular plot of land with one side three times the length of the shortest side and the third side 10 ft longer than the shortest side. If the perimeter of the plot is 145 ft and the shortest side is represented by x, write an equation that can be used to solve for the lengths of the three sides.

    A. $x + 3x + 10 = 145$
    B. $3(x + 10) = 145$
    C. $x + 3x + x + 10 = 145$
    D. $3x + 10 = 145$

45. What are the coordinates of the x-intercept in the following graph?

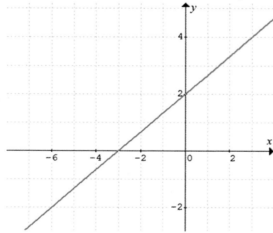

    A. $(2, 0)$ B. $(0, 2)$ C. $(0, -3)$ D. $(-3, 0)$

46. Which one of the following ordered pairs is NOT a solution of the equation: $y - 4x = 5$

    A. $(-3, -7)$ B. $(0.5, 7)$ C. $(1, 8)$ D. $(0, 5)$

47. Solve for p: $\dfrac{p + 5m}{p} = 4$

    A. $4 - \dfrac{5m}{p}$ B. $\dfrac{3p}{5}$ C. $4 - 5m$ D. $\dfrac{5m}{3}$

48. A factory that provides fabric has 8 less than 7 times the number of females as male employees. Let x represent the number of male employees at the factory. Give an expression for the number of females.

A. $8x + 7$ **B.** $7(x - 8)$ **C.** $7x - 8$ **D.** $8 - 7x$

49. If $3x$ is 16 more than the sum of $4y$ and $7x$, then $x + y =$?

A. 0 **B.** 4 **C.** -4 **D.** 12

50. Sharon deposited \$2,400 into a savings account that earns 8% per year in simple interest. How much interest will her account earn in 260 weeks?

A. \$960   **B.** \$1,500   **C.** \$3,840   **D.** \$9,600

51. Find the slope of the equation: $\quad 3y + 2x = 14$

A. $-\dfrac{2}{3}$ **B.** $-\dfrac{3}{2}$ **C.** $\dfrac{2}{3}$ **D.** $\dfrac{3}{2}$

52. Find the y–intercept of the equation: $\quad 5x - 3y = 11$

A. $\left(0, -\dfrac{11}{3}\right)$ **B.** $\left(-\dfrac{11}{3}, 0\right)$ **C.** $\left(0, \dfrac{11}{5}\right)$ **D.** $\left(\dfrac{11}{5}, 0\right)$

53. Simplify: $\quad 2^{-3} \cdot 16^{3/2} \cdot 8^{1/3}$

A. 2^4 **B.** 2^{-4} **C.** 2^8 **D.** 2^{-8}

54. $\sqrt[3]{48} + \sqrt[3]{6} =$

A. $9\sqrt{6}$ **B.** $9\sqrt{2}$ **C.** $3\sqrt[3]{2}$ **D.** $3\sqrt[3]{6}$

55. $\dfrac{3}{\sqrt{7} + 4} =$

A. $\dfrac{4 - \sqrt{7}}{3}$ **B.** $\dfrac{\sqrt{7} - 4}{3}$ **C.** $3(\sqrt{7} - 4)$ **D.** $3(4 - \sqrt{7})$

56. Simplify: $\quad \dfrac{x^2 - 16}{x^2 + 5x + 4} \cdot \dfrac{x^2 - 2x - 3}{-2x^2 + 8x}$

A. $\dfrac{(x + 4)(x - 3)}{2x(x - 4)}$ **C.** $\dfrac{-2x}{x - 3}$

B. $\dfrac{2x}{3 - x}$ **D.** $\dfrac{3 - x}{2x}$

57. Simplify: $\dfrac{1}{2x} - \dfrac{x-2}{x(x+3)} + \dfrac{5}{3x+9}$

A. $\dfrac{7x-3}{6x(x+3)}$ **B.** $\dfrac{19x-3}{6x(x+3)}$ **C.** $\dfrac{4-x}{6x(x+3)}$ **D.** $\dfrac{7}{6x}$

---

58. Factor: $-3x^2 - 6x + 45$

A. $3(x+5)(x+3)$ **C.** $-3(x-5)(x+3)$

B. $-3(x+5)(x-3)$ **D.** $3(x-5)(x-3)$

---

59. Expand: $(x-2)^2(x+5)$

A. $x^3 + 9x^2 - 16x + 20$ **C.** $x^3 + x^2 - 16x + 20$

B. $x^3 + 9x^2 + 16x - 20$ **D.** $x^3 - x^2 + 16x - 20$

---

60. In the xy plane, which of the following linear equations is parallel to the graph of the line $8y - 3x = 16$?

A. $3y = 8x + 3$ **C.** $8x = -3y - 2$

B. $8y = -3x + 5$ **D.** $8y = 3x - 1$

---

61. Solve the equation: $\dfrac{4}{n+2} - \dfrac{6}{2n} = \dfrac{5}{2n+4}$

A. $-\dfrac{2}{3}$ **B.** $\dfrac{2}{3}$ **C.** -4 **D.** 4

---

62. If $-2x + 7 > 21$, then

A. $x > -\dfrac{3}{2}$ **B.** $x < -\dfrac{3}{2}$ **C.** $x < -7$ **D.** $x > -7$

---

63. If $(x-3)(x+5) \geq 0$, then

A. $-5 \leq x \leq 3$ **C.** $x \geq 3$

B. $x < 5$ **D.** $x \leq -5 \text{ or } x \geq 3$

---

64. If $x^2 - 5x = 14$, then $x =$

A. $-7 \text{ and } 2$ **B.** $-2 \text{ and } 7$ **C.** -2 only **D.** -7 only

---

65. Find the roots (real or imaginary) of the equation: $c(c+2) = -2$

A. $-1 \pm 2i$ **B.** $-1 \pm i$ **C.** $-1 \pm 2\sqrt{3}$ **D.** No Roots Exist

66. Solve the system of equations:
$$4x + 2y = -1$$
$$12x + 6y = 5$$

A. $(0, 2)$ **B.** $(0, 4)$ **C.** $(2, -1)$ **D.** No Solution

67. Which of the following linear equations matches the data given in the table below?

x	-3	0	3	6
y	0	-4	-8	-12

A. $3y + 4x = -12$ **B.** $3x + 4y = -12$ **C.** $3y - 4x = 12$ **D.** $3x - 4y = 12$

68. The equation of the line that passes through the point $(5, -7)$ and the origin is:

A. $5x + 7y = 0$ **B.** $5y + 7x = 0$ **C.** $5x - 7y = 0$ **D.** $5y - 7x = 0$

69. If $f(x) = 5x - 1$ and $g(x) = \dfrac{x}{2x - 3}$ then $f \circ g(x) =$

A. $\dfrac{5x - 1}{5(2x - 1)}$ **B.** $\dfrac{3(x + 1)}{2x - 3}$ **C.** $\dfrac{5x - 1}{2x - 3}$ **D.** $\dfrac{5x - 1}{2(5x - 2)}$

70. If $f(x) = 2x^3 - 5x$ then $f(-x) =$

A. $f(x)$ **B.** $-f(x)$ **C.** $-f(-x)$ **D.** 0

71. If $f(x) = 4x^2 - 1$, then $\dfrac{f(x + h) - f(x)}{h}$ is:

A. 4

B. $8x + 4h$

C. $\dfrac{8x + 4}{h}$

D. $\dfrac{-3x^2 + 2xh + h^2}{h}$

72. If $4^{x-1} = 8^{3+x}$, then $x =$

A. 5 **B.** 7 **C.** 11 **D.** -11

73. Write $\dfrac{10i}{1 + 2i}$ in the form $a + bi$

A. $4 + 2i$ **B.** $-4 + 2i$ **C.** $5 + 10i$ **D.** $-\dfrac{20}{3} - \dfrac{10}{3}i$

74. Tickets for a basketball game cost $5 for seniors and $10 for adults. If there are 50 seats available and the basketball team needs to sell $400 in tickets to afford new uniforms, how many adult tickets must they sell?

A. 30 **B.** 20 **C.** 15 **D.** 12

75. Does the graph below represent a function? If so, what is the domain of the function?

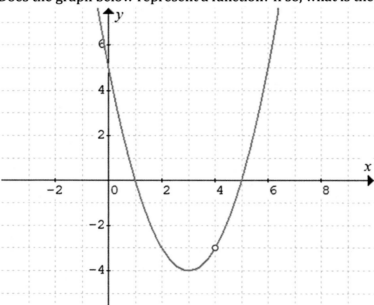

A. No, the graph is not a function

B. Yes; All Real Numbers

C. Yes; $\{x|x \geq -4\}$

D. Yes; $\{x|x \neq 4\}$

••••••••••••••••••• END WORKSHEET •••••••••••••••••••

Track 2 Success
"Dedicated to helping you stay on the track to academic success"

Postsecondary Education Readiness Test (PERT) – Worksheet #3 Answer Sheet

1. D	26. D	51. A
2. C	27. C	52. A
3. A	28. C	53. A
4. B	29. D	54. D
5. D	30. C	55. A
6. D	31. C	56. D
7. A	32. B	57. D
8. C	33. B	58. B
9. D	34. C	59. C
10. C	35. D	60. D
11. D	36. C	61. C
12. A	37. B	62. C
13. B	38. B	63. D
14. B	39. D	64. B
15. B	40. C	65. B
16. C	41. B	66. D
17. A	42. B	67. A
18. A	43. C	68. B
19. A	44. C	69. B
20. B	45. D	70. B
21. C	46. C	71. B
22. A	47. D	72. D
23. B	48. C	73. A
24. C	49. C	74. A
25. C	50. A	75. D

The contents of this worksheet are copyrighted. Disclosure or reproduction of any portion herein without express written consent of Track 2 Success © is strictly prohibited.

Worksheet #3 Solutions

1. **D**

   $2\dfrac{1}{3} + \dfrac{4}{7} =$

   Change the mixed number to an improper fraction

   $\dfrac{7}{3} + \dfrac{4}{7}$

   LCD: 21

   $\dfrac{(7)7}{(7)3} + \dfrac{4(3)}{7(3)} = \dfrac{49}{21} + \dfrac{12}{21} = \dfrac{61}{21} = 2\dfrac{19}{21}$

2. **C**

   $\dfrac{10}{21} - \dfrac{1}{7} =$

   LCD: 21

   $\dfrac{10}{21} - \dfrac{1(3)}{7(3)} = \dfrac{10}{21} - \dfrac{3}{21} = \dfrac{7}{21} = \dfrac{1}{3}$

3. **A**

   $2\dfrac{1}{4} \times \dfrac{1}{3} =$

   Change the mixed number to an improper fraction

   $\dfrac{9}{4} \times \dfrac{1}{3}$

   Multiply straight across and reduce the resulting fraction

   $\dfrac{9}{12} = \dfrac{3}{4}$

4. **B**

   $-2\dfrac{1}{3} \div 1\dfrac{2}{7} =$

   Change the mixed numbers to improper fractions

   $-\dfrac{7}{3} \div \dfrac{9}{7}$

   Take the reciprocal of the second fraction and then multiply straight across

   $-\dfrac{7}{3} \times \dfrac{7}{9} = -\dfrac{49}{27} = -1\dfrac{22}{27}$

5. **D**

   A basketball team played 60 games and lost 15% of them. How many games did they win?
   If they lost 15% of the games then they won 100% − 15% = 85% of the games.

The contents of this worksheet are copyrighted. Disclosure or reproduction of any portion herein without express written consent of Track 2 Success © is strictly prohibited.

The question is basically asking, what is 85% of 60?

Number of games won:

85% of $60 = 0.85 \times 60 = 51$ games

6. **D**

90 is what percent of 60?

$$\frac{is}{of} = \frac{\%}{100}$$

$$\frac{90}{60} = \frac{x}{100}$$

Cross multiply and solve

$$60x = (90)(100)$$

$$60x = 9000$$

$$x = \frac{9000}{60} = 150\%$$

7. **A**

10 is 50% of what number?

$$\frac{is}{of} = \frac{\%}{100}$$

$$\frac{10}{x} = \frac{50}{100}$$

Cross multiply and solve

$$50x = (10)(100)$$

$$50x = 1000$$

$$x = \frac{1000}{50} = 20$$

8. **C**

The price of gasoline will increase by 10% next month. If gasoline costs $1.80 per gallon now, what will be the price per gallon next month?

An increase of 10% gives a new value of 110% of the current price

110% of $1.80 = 1.10 \times \$1.80 = \$1.98$

Cost per gallon next month will be $1.98.

9. **D**

First read off the values corresponding to the months Oct and Apr.

Oct = 8 inches

Apr = 2 inches

Since we are trying to find how much higher the rainfall is in October than April, we subtract the two amounts to find the difference then use the percentage formula. Note that the words "than April" indicate that the basis of the comparison is April so we use April as the original amount.

Difference $= 8 - 2 = 6$

$$\text{Percentage difference} = \frac{difference}{original\ amount} \times 100\% = \frac{6}{2} \times 100\% = 300\%$$

10. **C**

If 60 percent of a number is 750, then 10% of the same number is?

First find the number by using the relationship:

$$\frac{is}{of} = \frac{\%}{100}$$

$$\frac{750}{x} = \frac{60}{100}$$

Cross multiply and solve

$$60x = 75000$$

$$x = \frac{75000}{60} = 1250$$

Next find 10% of this number

$$1250 \times 0.10 = 125$$

11. **D**

$$1.923 + 0.598 =$$

Line up the decimal points and add

$$
\begin{array}{r}
1.923 \\
+\ 0.598 \\
\hline
2.521
\end{array}
$$

12. **A**

$$-9.023 + 4.748 =$$

Since the signs are opposite, line up the decimal points with the larger number above and subtract. Keep the sign of the larger decimal.

$$
\begin{array}{r}
9.023 \\
-\ 4.748 \\
\hline
4.275
\end{array}
$$

Since the larger decimal was negative, the solution is also negative.

$$-4.275$$

13. **B**

$$7.2 \times 0.31 =$$

Multiply 72 by 31

$$72 \times 31 = 2232$$

The two original decimals had 3 decimal places combined so move the decimal point in the answer three places to the left

$$2.232$$

14. **B**

$$0.016 \div 0.8 =$$

Move the decimal point three places to the right in both numbers

$16 \div 800$

$$
\begin{array}{r}
0.02 \\
800\,\overline{\smash)16.00} \\
-\underline{1600} \\
0
\end{array}
$$

Therefore:

$0.016 \div 0.8 = 0.02$

15. B

The Lions cheerleading squad held a car wash to raise funds for new uniforms. If they charged $5.00 per car, $7.00 per van and $12.00 per truck, what would be a reasonable estimate of the amount they charged per vehicle?

The average amount charged is:

$$\frac{5 + 7 + 12}{3} = \frac{24}{3} = 8$$

Choose the amount closest to $8 which is $7.75

16. C

Which of the following is greatest?

$$\frac{1}{4}, \ \frac{2}{5}, \ \frac{7}{8}, \ \frac{13}{16}$$

Since they are all fractions, it may be easier to convert the fractions to the same common denominator and then compare the fractions but you may use the table of common fractions.

LCD: 80

$$\frac{(20)1}{(20)4}, \ \frac{(16)2}{(16)5}, \ \frac{(10)7}{(10)8}, \ \frac{(5)13}{(5)16}$$

$$\frac{20}{80}, \ \frac{32}{80}, \ \frac{70}{80}, \ \frac{65}{80}$$

Now we can clearly see that $\dfrac{7}{8}$ is the largest fraction.

17. A

Four people are hired to work together on a project so that the total time spent on the project is equivalent to one person working full time. The first person spends ⅛ of his time, the second spends ⅓ of his time, and the third person spends ½ of his time working on the project. How much of his time should the fourth person spend on the project?

$$\frac{1}{8} + \frac{1}{3} + \frac{1}{2}$$

LCD: 24

$$\frac{(3)1}{(3)8} + \frac{(8)1}{(8)3} + \frac{(12)1}{(12)2} = \frac{3}{24} + \frac{8}{24} + \frac{12}{24} = \frac{23}{24}$$

The fourth person needs to spend whatever fraction remains to make sure the project is complete (adds to 1):

$$1 - \frac{23}{24} = \frac{24}{24} - \frac{23}{24} = \frac{1}{24}$$

18. A

Find the arithmetic mean of the following numbers: 8.2, 4.9 and 6.7

$$\frac{8.2 + 4.9 + 6.7}{3} = \frac{19.8}{3} = 6.6$$

19. A

The area of a rectangle is 192 square inches and the length is 16 inches. What is the width of the rectangle?

$A = l \times w$

$192 = 16 \times w$

$w = \dfrac{192}{16}$

$w = 12$ inches

20. B

Sammy scored 69% on one math exam and 95% on another. What does he need to score on the third exam in order to receive an 85% average?

Let his grade on third exam $= x$

Average:

$$\frac{69 + 95 + x}{3} = 85$$

$$\frac{164 + x}{3} = \frac{85}{1}$$

Cross multiply and solve

$164 + x = (85)(3)$

$164 + x = 255$

$x = 255 - 164 = 91$

21. C

Randy bought a new entertainment system. He paid $975 for the LCD television, $159 for the speakers and $60 for the TV stand. What is his total price after 8% sales tax?

Total price before tax:

$975 + 159 + 60 = 1194$

The addition of tax is a percentage increase of 8% so we need to find 108% of the total.

108% of $1194 = 1.08 \times 1194 = 1289.52$

Total after tax is $1,289.52.

22. A

Find the area of a triangle in ft² if its base measures 60 ft and its height is 48 inches.

Notice that the base is in feet and the height is in inches so we need to convert one to the other. We need to convert inches to feet because we were asked to find the area in feet².

12 inches = 1 foot

48 inches $= \dfrac{48}{12}$ ft $= 4$ ft

$A = \dfrac{1}{2}bh = \dfrac{1}{2}(60)(4) = 30(4) = 120$ ft^2

23. B

Three workers can finish building a barn in 8 days by working together. Working alone, the first worker can build the barn in 24 days while another worker would take 18 days. How long would the third worker need to build the barn alone?

Let x = third worker's time

$$\frac{1}{24} + \frac{1}{18} + \frac{1}{x} = \frac{1}{8}$$

$$\frac{1}{x} = \frac{1}{8} - \frac{1}{24} - \frac{1}{18}$$

$8 = 2 \times 2 \times 2$

$18 = 2 \times 3 \times 3$

$24 = 2 \times 2 \times 2 \times 3$

$\text{LCD} = 2 \times 2 \times 2 \times 3 \times 3 = 72$

$$\frac{1}{x} = \frac{9}{72} - \frac{3}{72} - \frac{4}{72} = \frac{2}{72} = \frac{1}{36}$$

$x = 36 \ days$

24. C

At a boutique, a blouse costs $23.50 and a skirt costs $17.25. Both are on sale at 40% off. What is the price of a blouse and skirt bought together after the discount is applied?

Total before discount:

$23.50 + $17.25 = $40.75

A discount of 40% means that you will pay 60% of the original cost.

60% of $40.75 = 0.60 × 40.75 = $24.45

25. C

In 1990, a new house costs $9,850. Today the same house would cost $315,790. How much did the house increase?

$315,790 – $9,850 = $305,940

26. D

Write the following in decimal notation: Seven and eleven hundredths.

Seven represents the number before the decimal point and 11 hundredths represents the numbers behind the decimal point.

7.11

27. C

The midnight train has 280 miles left to travel. If the train has traveled ¾ of the way, how long is the total trip?

If the train has travelled $\frac{3}{4}$ of the way then it has $\frac{1}{4}$ left to travel

Let the total distance $= x$

$$\frac{1}{4}x = 280$$

$$x = 280 \cdot \frac{4}{1} = 1120 \ miles$$

28. C

Evaluate: $\dfrac{4 \cdot 5^2 + 8^2 \div 16}{7 - 12 \div \sqrt{36} + 8}$

Using PEMDAS

Simplify Exponents and Radicals.

$= \dfrac{4 \cdot \mathbf{25} + 8^2 \div 16}{7 - 12 \div \sqrt{36} + 8}$

$= \dfrac{4 \cdot 25 + \mathbf{64} \div 16}{7 - 12 \div \sqrt{36} + 8}$

$= \dfrac{4 \cdot 25 + 64 \div 16}{7 - 12 \div \mathbf{6} + 8}$

Do multiplication and division working from left to right.

$= \dfrac{\mathbf{100} + 64 \div 16}{7 - 12 \div 6 + 8}$

$= \dfrac{100 + \mathbf{4}}{7 - 12 \div 6 + 8}$

$= \dfrac{100 + 4}{7 - \mathbf{2} + 8}$

Do addition and subtraction working from left to right.

$= \dfrac{\mathbf{104}}{\mathbf{5} + 8} = \dfrac{104}{\mathbf{13}} = 8$

29. D

Which of the following is a linear factor of $x^2 + 5x + 6$?

Factor the quadratic to find its linear factors

$x^2 + 5x + 6$

$= (x + 3)(x + 2)$

Choose the one that appears in the answer choices

$x + 3$

30. C

Simplify: $(3x^2 + x + 1) - (2x^2 - 3x - 5)$

Distribute the negative sign and combine like terms

$= 3x^2 + x + 1 - 2x^2 + 3x + 5$

$= 3x^2 - 2x^2 + x + 3x + 1 + 5$

$= x^2 + 4x + 6$

31. C

Simplify: $(2x - 3)^2$

FOIL

$= (2x - 3)(2x - 3)$

$= 4x^2 - 6x - 6x + 9$

$= 4x^2 - 12x + 9$

32. B

Simplify: $\dfrac{a^{-3}b^4}{a^{-5}b^5}$

Using law #5 of exponents:

$a^{-3-(-5)}b^{4-5}$

$a^{-3+5}b^{-1}$

$a^2 b^{-1} = \dfrac{a^2}{b}$

33. B

Simplify: $\dfrac{y^2 + y}{y^2 - 1}$

Factor the numerator and denominator and eliminate any common factors.

$\dfrac{y(y+1)}{(y-1)(y+1)} = \dfrac{y}{y-1}$

34. C

If $6x^2 + 5x = 4$, then $x =$

$6x^2 + 5x - 4 = 0$

Factor, set each factor equal to zero and solve.

$(3x + 4)(2x - 1) = 0$

$3x + 4 = 0 \qquad\qquad 2x - 1 = 0$

$3x = -4 \qquad\qquad 2x = 1$

$x = -\dfrac{4}{3} \qquad\qquad x = \dfrac{1}{2}$

35. D

What is the value of the expression $(2x^2 - 3y^2)^2$ when $x = -1$ and $y = 2$?

Substitute the values of the variables and simplify

$(2x^2 - 3y^2)^2$

$= (2(-1)^2 - 3(2)^2)^2$

$= \left(2(1) - 3(4)\right)^2$

$= (2 - 12)^2$

$= (-10)^2$

$= 100$

36. C

Solve the equation: $\qquad |-2x + 3| = 15$

$-2x + 3 = 15 \qquad\qquad -2x + 3 = -15$

$-2x = 12 \qquad\qquad -2x = -18$

$x = \dfrac{12}{-2} = -6 \qquad\qquad x = \dfrac{-18}{-2} = 9$

37. **B**

The difference between two integers is 8. The sum of the two integers is 30. Which of the following equations is correct given that the larger of the two integers is represented by p?

Let the first integer $= p$

Since p is the larger of the two integers and there is a difference of 8 between the two, the second integer will be 8 less than the larger.

Second integer $= p - 8$

The sum of the two integers is 30, therefore:

$p + (p - 8) = 30$

38. **B**

If $\dfrac{x+4}{7} = \dfrac{4}{9}$, then $x =$

Cross multiply and solve

$9(x + 4) = 7(4)$

$9x + 36 = 28$

$9x = -8$

$x = -\dfrac{8}{9}$

39. **D**

If $5 - \dfrac{2}{3}x > -7$, then

$5 - \dfrac{2}{3}x > -7$

$-\dfrac{2}{3}x > -7 - 5$

$-\dfrac{2}{3}x > -12$

Remember to switch the direction of the inequality sign when dividing or multiplying by a negative number.

$x < -12 \cdot -\dfrac{3}{2}$

$x < 18$

40. **C**

Solve for m in the equation: $5n - 3 = 2n\left(\dfrac{m}{3}\right)$

$5n - 3 = 2n\left(\dfrac{m}{3}\right)$

$\dfrac{5n - 3}{1} = \dfrac{2mn}{3}$

Cross multiply and solve.

$3(5n - 3) = 2mn$

$m = \dfrac{3(5n - 3)}{2n}$

41. **B**

Find the length of AC given the following figure.

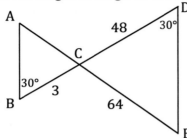

The figure contains 2 similar triangles. We can tell because the angles at C are identical (vertical angles) and ∠B = ∠D therefore ∠A = ∠E.
We can set up a proportion to determine the length of AC.

$$\frac{BC}{CD} = \frac{AC}{CE}$$

$$\frac{3}{48} = \frac{AC}{64}$$

$$AC = \frac{3(64)}{48} = \frac{192}{48} = 4$$

42. **B**

What is the value of the expression $x^2 - 3x + 5$ when $x = \frac{1}{2}$?

Substitute the value of x into the expression

$$\left(\frac{1}{2}\right)^2 - 3\left(\frac{1}{2}\right) + 5$$

$$\frac{1}{4} - \frac{3}{2} + \frac{5}{1}$$

$$\frac{1}{4} - \frac{6}{4} + \frac{20}{4} = \frac{15}{4} = 3\frac{3}{4}$$

43. **C**

The distance between points $(-3, 7)$ and $(x, 12)$ is 13. Find x

$$d = \sqrt{(x_2 - x_1)^2 + (y_2 - y_1)^2}$$

$$\sqrt{(x - (-3))^2 + (12 - 7)^2} = 13$$

Square both sides to eliminate the radical.

$(x + 3)^2 + 5^2 = 13^2$

$(x + 3)^2 = 169 - 25 = 144$

Take the square root of both sides

$x + 3 = \pm\sqrt{144}$

$x = -3 \pm 12$

$x = -15$ or $x = 9$

44. **C**

Sheila has a triangular plot of land with one side three times the length of the shortest side and the third side 10 ft longer than the shortest side. If the perimeter of the plot is 145 ft

and the shortest side is represented by x, write an equation that can be used to solve for the lengths of the three sides.
Shortest side = x
Second side = $3x$
Third side = $x + 10$
The perimeter is the sum of all three sides
$x + 3x + x + 10 = 145$

45. **D**

What are the coordinates of the x-intercept in the following graph?

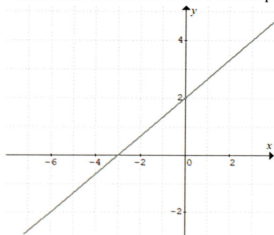

The x-intercept is the point where the graph touches the x-axis with y-coordinate equal to 0.
$(-3, 0)$

46. **C**

Which one of the following ordered pairs is NOT a solution of the equation:
$y - 4x = 5$
Substitute the ordered pairs into the equation until you find the one that does not yield a true statement.

$(-3, -7)$	$(0.5, 7)$	$(1, 8)$
$y - 4x = 5$	$y - 4x = 5$	$y - 4x = 5$
$-7 - 4(-3) = 5$	$7 - 4(0.5) = 5$	$8 - 4(1) = 5$
$-7 + 12 = 5$	$7 - 2 = 5$	$8 - 4 = 5$
True	True	False

Therefore $(1, 8)$ is not a solution of the equation

47. **D**

Solve for p: $\dfrac{p + 5m}{p} = 4$

Cross multiply
$\dfrac{p + 5m}{p} = \dfrac{4}{1}$
$4p = p + 5m$

$$4p - p = 5m$$
$$3p = 5m$$
$$p = \frac{5m}{3}$$

48. C

A factory that provides fabric has 8 less than 7 times the number of females as male employees. Let x represent the number of male employees at the factory. Give an expression for the number of females.

8 less than 7 times $\Rightarrow 7x - 8$ females

49. C

If $3x$ is 16 more than the sum of $4y$ and $7x$, then $x + y =$?

Translate the words into an equation

16 more than the sum of $4y$ and $7x \Rightarrow$ "$4y + 7x + 16$"

$3x$ is $\Rightarrow$ "$= 3x$"

Therefore the equation is:

$$4y + 7x + 16 = 3x$$
$$4y + 7x - 3x = -16$$
$$4y + 4x = -16$$

Since we are looking for $x + y$, factor the 4 from the left side

$$4(y + x) = -16$$
$$x + y = -\frac{16}{4}$$
$$x + y = -4$$

50. A

Sharon deposited \$2,400 into a savings account that earns 8% per year in simple interest. How much interest will her account earn in 260 weeks?

Note that the interest rate is annual but the time was given in weeks. We therefore need to convert weeks to years.

$$\frac{52 \; weeks}{1 \; year} = \frac{260 \; weeks}{x \; years}$$

Cross multiply and solve

$$52x = 260$$
$$x = \frac{260}{52} = 5 \; years$$

Interest earned in 260 weeks:

$$I = P \times R \times T = \$2,400 \times 0.08 \times 5 = \$960$$

51. A

Find the slope of the equation: $3y + 2x = 14$

Solve for y

$$3y + 2x = 14$$
$$3y = -2x + 14$$

$$y = -\frac{2}{3}x + \frac{14}{3}$$

Therefore the slope of the line is:

$$m = -\frac{2}{3}$$

52. A

Find the y–intercept of the equation: $5x - 3y = 11$

Solve for y

$5x - 3y = 11$

$-3y = -5x + 11$

$y = \frac{-5}{-3}x + \frac{11}{-3}$

$y = \frac{5}{3}x - \frac{11}{3}$

Now that the equation is in the form $y = mx + b$ we see that the y–intercept is:

$-\frac{11}{3}$ or $\left(0, -\frac{11}{3}\right)$

53. A

Simplify: $2^{-3} \cdot 16^{3/2} \cdot 8^{1/3}$

Convert all the bases to base 2

$= 2^{-3} \cdot (2^4)^{3/2} \cdot (2^3)^{1/3}$

$= 2^{-3} \cdot 2^{4 \times 3/2} \cdot 2^{3 \times 1/3}$

$= 2^{-3} \cdot 2^{12/2} \cdot 2^{3/3}$

$= 2^{-3} \cdot 2^6 \cdot 2^1$

$= 2^{-3+6+1}$

$= 2^4$

54. D

$\sqrt[3]{48} + \sqrt[3]{6} =$

Simplify the radicands if possible

$\sqrt[3]{48} + \sqrt[3]{6}$

$\sqrt[3]{2 \times 2 \times 2 \times 2 \times 3} + \sqrt[3]{2 \times 3}$

$2\sqrt[3]{2 \times 3} + \sqrt[3]{2 \times 3}$

$2\sqrt[3]{6} + \sqrt[3]{6} = 3\sqrt[3]{6}$

55. A

$\frac{3}{\sqrt{7} + 4} =$

Multiply by the conjugate of the denominator

$\frac{3}{\sqrt{7} + 4} \cdot \frac{\sqrt{7} - 4}{\sqrt{7} - 4}$

FOIL

$$\frac{3\sqrt{7} - 12}{7 - 4\sqrt{7} + 4\sqrt{7} - 16}$$

$$\frac{3\sqrt{7} - 12}{7 - 16}$$

$$\frac{3\sqrt{7} - 12}{-9} \qquad \text{Now Reduce by} - 3$$

$$\frac{-\sqrt{7} + 4}{3} \quad or \quad \frac{4 - \sqrt{7}}{3}$$

56. D

Simplify: $\dfrac{x^2 - 16}{x^2 + 5x + 4} \cdot \dfrac{x^2 - 2x - 3}{-2x^2 + 8x}$

Factor

$$\frac{(x - 4)(x + 4)}{(x + 4)(x + 1)} \cdot \frac{(x - 3)(x + 1)}{-2x(x - 4)}$$

$$\frac{(x - 4)(x + 4)(x - 3)(x + 1)}{-2x(x + 4)(x + 1)(x - 4)}$$

Eliminate factors common to the numerator and the denominator.

$$-\frac{(x - 3)}{2x} = \frac{3 - x}{2x}$$

57. D

Simplify: $\dfrac{1}{2x} - \dfrac{x - 2}{x(x + 3)} + \dfrac{5}{3x + 9}$

$$\frac{1}{2x} - \frac{x - 2}{x(x + 3)} + \frac{5}{3(x + 3)}$$

LCD: $6x(x + 3)$

$$\frac{(3)(x + 3)1}{(3)(x + 3)2x} - \frac{(6)x - 2}{(6)x(x + 3)} + \frac{(2x)5}{(2x)(3)(x + 3)}$$

$$\frac{3x + 9}{6x(x + 3)} - \frac{6x - 12}{6x(x + 3)} + \frac{10x}{6x(x + 3)}$$

$$\frac{3x + 9 - (6x - 12) + 10x}{6x(x + 3)}$$

$$\frac{3x - 6x + 10x + 9 + 12}{6x(x + 3)}$$

$$\frac{7x + 21}{6x(x + 3)}$$

$$\frac{7(x + 3)}{6x(x + 3)} = \frac{7}{6x}$$

58. B

Factor: $-3x^2 - 6x + 45$

$-3(x^2 + 2x - 15)$

$= -3(x + 5)(x - 3)$

59. **C**

Expand: $(x-2)^2(x+5)$

$(x-2)(x-2)(x+5)$

$(x^2-2x-2x+4)(x+5)$

$(x^2-4x+4)(x+5)$

$x^2(x+5)-4x(x+5)+4(x+5)$

$x^3+5x^2-4x^2-20x+4x+20$

$x^3+x^2-16x+20$

60. **D**

Remember that parallel lines have the same slope so we only need to solve the equations for y to determine which choice has the same slope as the original. Since we know that the slope will be a combination of the x and y coefficients, we know that the parallel line will have **exactly** the same coefficients as the original. So as a shortcut we only need to find the equation that fits this criterion. Since all the answer choices have x and y on opposite sides of the equation, we rearrange the original equation to match so that we do less work.

$8y-3x=16$

$8y=3x+16$

Comparing this to the answer choices, the only one that matches is choice D.

61. **C**

Solve the equation: $\dfrac{4}{n+2}-\dfrac{6}{2n}=\dfrac{5}{2n+4}$

$\dfrac{4}{n+2}-\dfrac{6}{2n}=\dfrac{5}{2(n+2)}$

LCD: $2n(n+2)$

$\dfrac{(2n)4}{(2n)n+2}-\dfrac{(n+2)6}{(n+2)2n}=\dfrac{(n)5}{(n)2n+4}$

$\dfrac{8n}{2n(n+2)}-\dfrac{6(n+2)}{2n(n+2)}=\dfrac{5n}{2n(n+2)}$

Since the denominators are now the same, solve the equation formed by the numerators

$8n-6(n+2)=5n$

$8n-6n-12=5n$

$2n-12=5n$

$2n-5n=12$

$-3n=12$

$n=\dfrac{12}{-3}=-4$

62. **C**

If $-2x+7>21,$ then

$-2x+7>21$

$-2x>14$

When dividing or multiplying by a negative number, switch the direction of the inequality sign

$$x < \frac{14}{-2}$$
$$x < -7$$

63. D

If $(x-3)(x+5) \geq 0$, then
Set each factor equal to zero and solve for the critical points
$x - 3 = 0 \qquad x + 5 = 0$
$x = 3 \qquad x = -5$
Place the critical values on a number line and choose test points in the 3 regions formed.

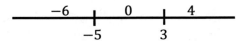

Substitute the test points into original inequality to see which satisfy the inequality

When $x = -6$
$(-6-3)(-6+5) \geq 0$
$(-9)(-1) \geq 0$
$9 \geq 0$
True

When $x = 0$
$(0-3)(0+5) \geq 0$
$(-3)(5) \geq 0$
$-15 \geq 0$
False

When $x = 4$
$(4-3)(4+5) \geq 0$
$(1)(9) \geq 0$
$9 \geq 0$
True

Therefore the solution set is:
$(-\infty, -5] \cup [3, \infty)$

64. B

If $x^2 - 5x = 14$, then $x =$
$x^2 - 5x - 14 = 0$
Factor
$(x-7)(x+2) = 0$
$x - 7 = 0 \qquad x + 2 = 0$
$x = 7 \qquad x = -2$

65. B

Find the roots (real or imaginary) of the equation: $c(c+2) = -2$
$c(c+2) = -2$
$c^2 + 2c + 2 = 0$
This cannot be factored so use the quadratic formula
$a = 1, \ b = 2, \ c = 2$
$$c = \frac{-b \pm \sqrt{b^2 - 4ac}}{2a}$$
$$c = \frac{-2 \pm \sqrt{2^2 - 4(1)(2)}}{2(1)}$$
$$c = \frac{-2 \pm \sqrt{4-8}}{2}$$
$$c = \frac{-2 \pm \sqrt{-4}}{2}$$
$$c = \frac{-2 \pm 2i}{2} = -1 \pm i$$

66. D

Solve the system of equations:

$4x + 2y = -1$

$12x + 6y = 5$

Multiply the first equation by –3

$-3(4x + 2y = -1)$

$-12x - 6y = 3$

Add the two equations

$$-12x - 6y = 3$$
$$\underline{12x + 6y = 5}$$
$$0 = 8$$

Since this is an untrue statement, the lines are parallel and inconsistent so there is no solution.

67. A

Which of the following linear equations matches the data given in the table below?

x	-3	0	3	6
y	0	-4	-8	-12

Choose any of the points in the table and plug into the answer choices to see which is valid. Alternatively: select any two points in the table, find the slope and then find the equation of the line using the point slope formula. Since points in the table may work for more than one equation, the latter method is normally a better approach.

Using $(-3, 0)$ and $(0, -4)$

$$m = \frac{y_2 - y_1}{x_2 - x_1} = \frac{-4 - 0}{0 - (-3)} = -\frac{4}{3}$$

Since the y–intercept $b = -4$, the equation of the line is:

$y = mx + b$

$y = -\dfrac{4}{3}x - 4$

$3\left(y = -\dfrac{4}{3}x - 4\right)$

$3y = -4x - 12$

$3y + 4x = -12$

Therefore the answer is A

68. B

The equation of the line that passes through the point $(5, -7)$ and the origin is:

Using $(5, -7)$ and $(0, 0)$

$$m = \frac{y_2 - y_1}{x_2 - x_1} = \frac{(-7) - 0}{5 - 0} = -\frac{7}{5}$$

$y - y_1 = m(x - x_1)$

$y - 0 = -\dfrac{7}{5}(x - 0)$

$y = -\dfrac{7}{5}x$

300   Prep for Success: Florida's PERT Math Study Guide

$$5y = -7x$$
$$5y + 7x = 0$$

69. B

If $f(x) = 5x - 1$ and $g(x) = \dfrac{x}{2x - 3}$ then $f \circ g(x) =$

Substitute $g(x)$ into $f(x)$

$$f(g(x)) = 5\left(\frac{x}{2x - 3}\right) - 1$$

$$= \frac{5x}{2x - 3} - \frac{1}{1}$$

$$= \frac{5x}{2x - 3} - \frac{(2x - 3)1}{(2x - 3)1}$$

$$= \frac{5x - (2x - 3)}{2x - 3}$$

$$= \frac{3x + 3}{2x - 3} = \frac{3(x + 1)}{2x - 3}$$

70. B

If $f(x) = 2x^3 - 5x$ then $f(-x) =$

$$f(-x) = 2(-x)^3 - 5(-x)$$
$$f(-x) = -2x^3 + 5x$$
$$f(-x) = -(2x^3 - 5x)$$
$$f(-x) = -f(x)$$

71. B

If $f(x) = 4x^2 - 1$, then $\dfrac{f(x + h) - f(x)}{h}$ is:

$$f(x + h) = 4(x + h)^2 - 1$$
$$= 4(x + h)(x + h) - 1$$
$$= 4(x^2 + xh + xh + h^2) - 1$$
$$= 4(x^2 + 2xh + h^2) - 1$$
$$= 4x^2 + 8xh + 4h^2 - 1$$

$$\frac{f(x + h) - f(x)}{h} = \frac{4x^2 + 8xh + 4h^2 - 1 - (4x^2 - 1)}{h}$$

$$= \frac{8xh + 4h^2}{h} = \frac{h(8x + 4h)}{h} = 8x + 4h$$

72. D

If $4^{x-1} = 8^{3+x}$, then $x =$

Convert both sides to base 2

$$(2^2)^{x-1} = (2^3)^{3+x}$$
$$2^{2(x-1)} = 2^{3(3+x)}$$

Since the bases are now the same, set the exponents equal to each other and solve

$$2(x - 1) = 3(3 + x)$$
$$2x - 2 = 9 + 3x$$

Track 2 Success © 2014   All Rights Reserved

$$2x - 3x = 9 + 2$$
$$-x = 11$$
$$x = -11$$

73. A

Write $\dfrac{10i}{1 + 2i}$ in the form $a + bi$

Multiply by the conjugate of the denominator

$$\frac{10i}{1 + 2i} \cdot \frac{1 - 2i}{1 - 2i}$$
$$\frac{10i(1 - 2i)}{1 - 2i + 2i - 4i^2}$$
$$\frac{10i - 20i^2}{1 - 4i^2}$$
$$\frac{10i - 20(-1)}{1 - 4(-1)}$$
$$\frac{20 + 10i}{5}$$
$$4 + 2i$$

74. A

Tickets for a basketball game cost \$5 for seniors and \$10 for adults. If there are 50 seats available and the basketball team needs to sell \$400 in tickets to afford new uniforms, how many adult tickets must they sell?

Let the number of senior tickets $= x$

Let the number of adult tickets $= y$

$$x + y = 50$$
$$5x + 10y = 400$$

Since we are solving for y, solve the first equation for x and substitute into the second equation

$$x = 50 - y$$
$$5(50 - y) + 10y = 400$$
$$250 - 5y + 10y = 400$$
$$5y + 250 = 400$$
$$5y = 150$$
$$y = \frac{150}{5}$$
$$y = 30 \text{ adult tickets}$$

75. D

Does the graph below represent a function? If so, what is the domain of the function?

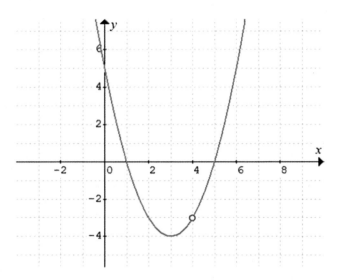

This is the graph of a quadratic function. If we conduct the vertical line test, we find that no value of x maps unto more than one value of y. Therefore, the graph represents a function.

There is an open circle at $x = 4$ which indicates that $x = 4$ is excluded from the domain.

The domain of the function is therefore $\{x | x \neq 4\}$ and the answer is D.

Track 2 Success

"Dedicated to helping you stay on the track to academic success"

Postsecondary Education Readiness Test (PERT) – Worksheet #4 Bubble Sheet

1. Ⓐ Ⓑ Ⓒ Ⓓ
2. Ⓐ Ⓑ Ⓒ Ⓓ
3. Ⓐ Ⓑ Ⓒ Ⓓ
4. Ⓐ Ⓑ Ⓒ Ⓓ
5. Ⓐ Ⓑ Ⓒ Ⓓ
6. Ⓐ Ⓑ Ⓒ Ⓓ
7. Ⓐ Ⓑ Ⓒ Ⓓ
8. Ⓐ Ⓑ Ⓒ Ⓓ
9. Ⓐ Ⓑ Ⓒ Ⓓ
10. Ⓐ Ⓑ Ⓒ Ⓓ
11. Ⓐ Ⓑ Ⓒ Ⓓ
12. Ⓐ Ⓑ Ⓒ Ⓓ
13. Ⓐ Ⓑ Ⓒ Ⓓ
14. Ⓐ Ⓑ Ⓒ Ⓓ
15. Ⓐ Ⓑ Ⓒ Ⓓ
16. Ⓐ Ⓑ Ⓒ Ⓓ
17. Ⓐ Ⓑ Ⓒ Ⓓ
18. Ⓐ Ⓑ Ⓒ Ⓓ
19. Ⓐ Ⓑ Ⓒ Ⓓ
20. Ⓐ Ⓑ Ⓒ Ⓓ
21. Ⓐ Ⓑ Ⓒ Ⓓ
22. Ⓐ Ⓑ Ⓒ Ⓓ
23. Ⓐ Ⓑ Ⓒ Ⓓ
24. Ⓐ Ⓑ Ⓒ Ⓓ
25. Ⓐ Ⓑ Ⓒ Ⓓ
26. Ⓐ Ⓑ Ⓒ Ⓓ
27. Ⓐ Ⓑ Ⓒ Ⓓ
28. Ⓐ Ⓑ Ⓒ Ⓓ
29. Ⓐ Ⓑ Ⓒ Ⓓ
30. Ⓐ Ⓑ Ⓒ Ⓓ
31. Ⓐ Ⓑ Ⓒ Ⓓ
32. Ⓐ Ⓑ Ⓒ Ⓓ
33. Ⓐ Ⓑ Ⓒ Ⓓ
34. Ⓐ Ⓑ Ⓒ Ⓓ
35. Ⓐ Ⓑ Ⓒ Ⓓ
36. Ⓐ Ⓑ Ⓒ Ⓓ
37. Ⓐ Ⓑ Ⓒ Ⓓ
38. Ⓐ Ⓑ Ⓒ Ⓓ
39. Ⓐ Ⓑ Ⓒ Ⓓ
40. Ⓐ Ⓑ Ⓒ Ⓓ
41. Ⓐ Ⓑ Ⓒ Ⓓ
42. Ⓐ Ⓑ Ⓒ Ⓓ
43. Ⓐ Ⓑ Ⓒ Ⓓ
44. Ⓐ Ⓑ Ⓒ Ⓓ
45. Ⓐ Ⓑ Ⓒ Ⓓ
46. Ⓐ Ⓑ Ⓒ Ⓓ
47. Ⓐ Ⓑ Ⓒ Ⓓ
48. Ⓐ Ⓑ Ⓒ Ⓓ
49. Ⓐ Ⓑ Ⓒ Ⓓ
50. Ⓐ Ⓑ Ⓒ Ⓓ
51. Ⓐ Ⓑ Ⓒ Ⓓ
52. Ⓐ Ⓑ Ⓒ Ⓓ
53. Ⓐ Ⓑ Ⓒ Ⓓ
54. Ⓐ Ⓑ Ⓒ Ⓓ
55. Ⓐ Ⓑ Ⓒ Ⓓ
56. Ⓐ Ⓑ Ⓒ Ⓓ
57. Ⓐ Ⓑ Ⓒ Ⓓ
58. Ⓐ Ⓑ Ⓒ Ⓓ
59. Ⓐ Ⓑ Ⓒ Ⓓ
60. Ⓐ Ⓑ Ⓒ Ⓓ
61. Ⓐ Ⓑ Ⓒ Ⓓ
62. Ⓐ Ⓑ Ⓒ Ⓓ
63. Ⓐ Ⓑ Ⓒ Ⓓ
64. Ⓐ Ⓑ Ⓒ Ⓓ
65. Ⓐ Ⓑ Ⓒ Ⓓ
66. Ⓐ Ⓑ Ⓒ Ⓓ
67. Ⓐ Ⓑ Ⓒ Ⓓ
68. Ⓐ Ⓑ Ⓒ Ⓓ
69. Ⓐ Ⓑ Ⓒ Ⓓ
70. Ⓐ Ⓑ Ⓒ Ⓓ
71. Ⓐ Ⓑ Ⓒ Ⓓ
72. Ⓐ Ⓑ Ⓒ Ⓓ
73. Ⓐ Ⓑ Ⓒ Ⓓ
74. Ⓐ Ⓑ Ⓒ Ⓓ
75. Ⓐ Ⓑ Ⓒ Ⓓ

The contents of this worksheet are copyrighted. Disclosure or reproduction of any portion herein without express written consent of Track 2 Success © is strictly prohibited.

Track 2 Success
"Dedicated to helping you stay on the track to academic success"

Worksheet #4

The questions contained in this worksheet cover all the topics presented in the study guide section and are ordered by increasing difficulty from Arithmetic through College Level Math.

Since calculators are not generally allowed while taking the Postsecondary Education Readiness Test, use of calculators should be limited while completing this worksheet so as to replicate and properly prepare for exam conditions.

1. $\dfrac{2}{3} + \dfrac{4}{5} =$

   A. $\dfrac{6}{15}$ B. $1\dfrac{7}{15}$ C. $1\dfrac{1}{4}$ D. $2\dfrac{1}{4}$

2. $2 - \dfrac{7}{8} =$

   A. $1\dfrac{1}{8}$ B. $-\dfrac{5}{8}$ C. $\dfrac{3}{8}$ D. $1\dfrac{3}{4}$

3. $-\dfrac{1}{4} \times \dfrac{2}{5} =$

   A. $\dfrac{1}{20}$ B. $-\dfrac{1}{5}$ C. $-\dfrac{1}{10}$ D. $-\dfrac{5}{8}$

4. $3\dfrac{1}{3} \div 4\dfrac{1}{5} =$

   A. $\dfrac{50}{63}$ B. $2\dfrac{2}{3}$ C. 6   D. 14

5. What is 110% of 50?
   A. 55   B. 10   C. 60   D. 65

6. 35 is what percent of 28?
   A. 80%   B. 95%   C. 125%   D. 140%

7. 36 is 45% of what number?
   A. 80   B. 75   C. 60   D. 108

The contents of this worksheet are copyrighted. Disclosure or reproduction of any portion herein without express written consent of Track 2 Success © is strictly prohibited.

8. The Fab Boutique decided to increase the price of their skirts by 25%. If their skirts cost $32.60 now, how much will they cost after the increase?

A. $8.15   **B.** $10.50   **C.** $40.75   **D.** $43.10

9. The price of a $40 camera is increased by 30% and then later decreased by 30% during a sale. What is the price of the camera during the sale?

A. $28.00   **B.** $36.40   **C.** $40.00   **D.** $52.00

10. $1.375 =$

A. 1.375% **B.** 13.75% **C.** 137.5% **D.** $\dfrac{1.375}{100}$

11. $4.621 + 3.104 =$

A. 6.725 **B.** 7.705 **C.** 7.725 **D.** 7.735

12. $4.147 - (-1.503) =$

A. 2.644 **B.** 5.65 **C.** -2.644 **D.** -5.65

13. $5.17 \times 0.06 =$

A. 0.03102 **B.** 0.3102 **C.** 3.102 **D.** 31.02

14. $12.75 \div (-2.5) =$

A. 5.1 **B.** -5.1 **C.** 51 **D.** -0.51

15. During a field trip, I spent $1.92 on a hotdog, $2.35 on a soda, $5.42 on popcorn and $4.75 on a ride. What is a reasonable estimate of the amount of money spent during the field trip?

A. $12   **B.** $13   **C.** $14   **D.** $15

16. Mary has $60 to spend on her daughter's birthday party. If she spends ⅕ of the budget on decorations, and ⅔ on food and drinks, how much money does she have remaining for entertainment?

A. $8   **B.** $12   **C.** $40   **D.** $52

17.

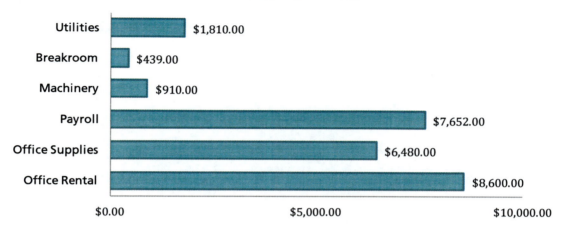

A company is currently operating at a loss each month due to the fact that they are spending more on expenses than they are earning in revenue. They decide to cut back on the amount of office supplies purchased each month. Based on the chart above, which shows the amount spent each month, by what percent should they reduce the total spent on office supplies in order to meet their monthly budget of $25,000?

A. 10.36% B. 13.75% C. 22.36% D. 25.92%

18. What is the mode of the numbers 20, 21, 27, 31, 21, 25 and 27?

A. 21 B. 25 C. There are two modes. D. There is no mode.

19. The perimeter of a rectangular garden is 54 feet. If the length of the garden is 18 feet, what is the width?

A. 6 B. 9 C. 12 D. 18

20. Calculate the value of the following: $\dfrac{9.2 \times 0.00073}{0.021}$

A. 0.0032 B. 0.032 C. 0.32 D. 3.2

21. A boutique sold their last three blouses yesterday. They sold the first for $19.60, the second for $22.10 and the last for $18.90. If the boutique bought the blouses for $15.00 a piece, what is the average profit per blouse?

A. $5.20 B. $6.70 C. $15.60 D. $20.20

22. A recipe for pancakes that feeds 6 people requires 3¼ cups of flour. How many cups of flour would be needed to feed 24 people pancakes?

A. 4 B. 13 C. 52 D. 78

23. Susan saved $1623.73 over the past eight months. If she deposits $572.91 today and withdraws $818.23 tomorrow what will be her new account balance?

A. $805.50 **B.** $1,378.41 **C.** $1,869.05 **D.** $2,196.64

---

24. Which of the following numbers is the smallest?

A. $\left(-\frac{1}{2}\right)(6)$ **B.** -5 **C.** $-\frac{18}{5}$ **D.** $|-11+7|$

---

25. Which of the following is NOT a true statement?

A. $12^2 = 144$

B. $4.210 \geq 4.2010$

C. $\frac{3}{2} \times \frac{4}{9} = \frac{27}{8}$

D. $\sqrt{0.04} = 0.2$

---

26. Ms. Daniel's class wants to buy candy for their class Halloween party. If one bag of candy costs $4.78, how many bags can the class buy if they saved $44.95?

A. 8 **B.** 9 **C.** 10 **D.** 11

---

27. Identify the place value of the underlined digit in the number 32.01$\underline{9}$2

A. Ones **B.** Tenths **C.** Hundredths **D.** Thousandths

---

28. Find the area of a circle whose radius is ⅘ inches.

A. $\frac{4}{5}\pi$ Sq. inches

B. $\frac{8}{5}\pi$ Sq. inches

C. $\frac{16}{25}\pi$ Sq. inches

D. $\frac{16}{25}\pi^2$ Sq. inches

---

29. Evaluate: $|-(-3)^2 - 14|$

A. -5 **B.** 5 **C.** -23 **D.** 23

---

30. Factor: $12x^2 + 4x$

A. $4x(3x)$ **B.** $16x^2$ **C.** $4x(3x + 1)$ **D.** $4x(3x + x)$

---

31. $(x^3 + 6x^2 - 4x + 7) - (3x^3 + 2x - 4) =$

A. $4x^3 + 6x^2 - 2x + 3$

B. $-2x^3 + 6x^2 - 6x + 11$

C. $4x^3 + 6x^2 - 6x + 11$

D. $-2x^3 + 4x^2 - 4x + 3$

32. Simplify: $-3(5x-4)^2$

A. $-75x^2 + 120x - 48$
B. $-75x^2 + 120x + 48$
C. $75x^2 - 48$
D. $225x^2 - 180x + 144$

33. Simplify: $\left(\dfrac{c^4 d^3}{cd^2}\right)\left(\dfrac{d^2}{c^3}\right)^3$

A. c^3
B. $c^3 d^4$
C. $\dfrac{d^7}{c^6}$
D. $\dfrac{d^6}{c^3}$

34. Simplify: $\dfrac{x^2 - 2x}{2x^2 - x - 6}$

A. $\dfrac{1}{5}$
B. $\dfrac{1}{2x+3}$
C. $\dfrac{x}{x-3}$
D. $\dfrac{x}{2x+3}$

35. If $x^2 + 2 = 2(x+5)$, then $x =$

A. -2 and 4
B. -2 and $-\dfrac{5}{2}$
C. -5
D. 2 and -4

36. What is the value of the expression $2a + b(1 - a^2)$ when $a = 5$ and $b = -3$?

A. -46
B. 37
C. 82
D. -68

37. Evaluate: $3i^2(i^7 - 4) - 2i^5$

A. $12 + i$
B. $-12 - i$
C. $9 - 2i$
D. $-15 - 2i$

38. Choose the equation that is equivalent to the statement below:
"The difference between a number n and 7 more than three times the number is 18."

A. $(7n + 3) + n = 18$
B. $n - (3n + 7) = 18$
C. $3n + 7 = 18$
D. $n(3n - 7) = 18$

39. Given that LM || OR, find the value of $\angle b$ in the diagram below.

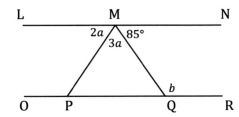

A. $75°$
B. $85°$
C. $95°$
D. Cannot be determined.

40. If $\frac{2}{5}c - 7 = 9$, then $c =$

A. 5   B. 16   C. 25   D. 40

41. If $4x + 6 \geq 14 - 2x$, then

A. $x \geq \frac{4}{3}$ B. $x \geq \frac{7}{3}$ C. $x \geq \frac{10}{3}$ D. $x \geq 10$

42. Solve the following for b: $ab - d = cb$

A. $\frac{d}{a-c}$ B. $\frac{d}{ac}$ C. $\frac{d}{a+c}$ D. $\frac{a-c}{d}$

43. If $-3(x+5) > 14$, then

A. $x > -3$ B. $x < \frac{1}{3}$ C. $x > -\frac{19}{3}$ D. $x < -\frac{29}{3}$

44. Divide: $\frac{-21m^3 + 18m^7 - 15m^2}{-3m^2}$

A. $-7m + 6m^5 - 5$ C. $7m^3 - 6m^7 + 5m^2$

B. $7m - 6m^5 + 5$ D. $7m - 6m^5$

45. The window of a house resembles the figure below. How many square feet of glass are required if the length of the rectangle is x feet and the width is 2 feet?

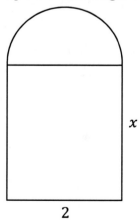

A. $2x^2$ B. $2x + \frac{\pi}{2}$ C. $2x + 2\pi$ D. $2(x + \pi r^2)$

46. If $f(x)$ is a linear function such that $f(6) = -2$ and $f(-2) = 10$, what is the value of a if $f(a) = -8$?

A. -4 B. 0 C. 10 D. 19

47. The sum of twice a number and 9 is five less than three times the number. Write an expression that can be used to find the number x.

A. $18x = 3x - 5$ C. $2x + 9 = 3x + 5$

B. $2x + 9 = 5 - 3x$ D. $2x + 9 = 3x - 5$

48. Graph the line: $y = -4$

A.
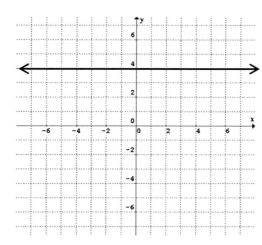

C.
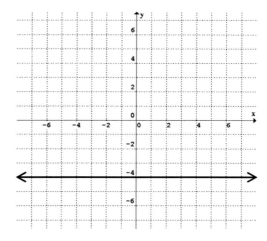

B.
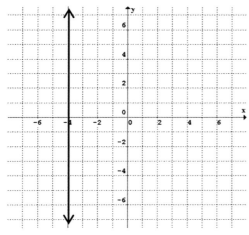

D.
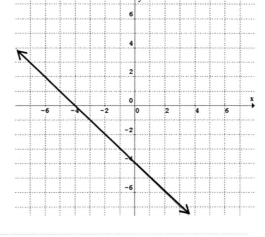

49. The figure below contains two squares. What is the area of the shaded region?

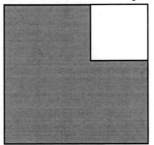

2x

5y

- **A.** $(5y - 2x)(5y - 2x)$
- **B.** $(5y - 2x)^2$
- **C.** $(5y - 2x)^2 - 4x^2$
- **D.** $(5y + 2x)(5y - 2x)$

50. Simplify: $5\sqrt{72a^5b^6c^7}$

- **A.** $20a^4b^6c^6\sqrt{18ac}$
- **B.** $9a^2b^3c^3\sqrt{18ac}$
- **C.** $30a^2b^3c^3\sqrt{2ac}$
- **D.** $11a^2b^3c^3\sqrt{2ac}$

51. Convert to scientific notation: 0.00000567

- **A.** 567×10^6
- **B.** 567×10^{-6}
- **C.** 5.67×10^6
- **D.** 5.67×10^{-6}

52. Find the x − intercept of the equation: $y = \frac{4}{3}x - 8$

- **A.** $(-8, 0)$
- **B.** $(0, -8)$
- **C.** $(0, 6)$
- **D.** $(6, 0)$

53. If $3^m = 8$ and $3^n = 2$, then $9^{m-n} =$

- **A.** 4
- **B.** 12
- **C.** 16
- **D.** 32

54. Simplify: $\dfrac{\sqrt[4]{81x^{10}y^4}}{\sqrt[4]{3xy^7}}$

- **A.** $x^2\sqrt[4]{27xy^3}$
- **B.** $3x^2y\sqrt[4]{x}$
- **C.** $\dfrac{x^2\sqrt[4]{27xy}}{y}$
- **D.** $\dfrac{3x^2}{\sqrt[4]{y^4}}$

55. $\dfrac{\sqrt{2}}{\sqrt{5} - 2} =$

- **A.** $\sqrt{7} + 2\sqrt{2}$
- **B.** $\sqrt{10} - 2\sqrt{2}$
- **C.** $\sqrt{10} + 2\sqrt{2}$
- **D.** $\sqrt{7} - 2\sqrt{2}$

56. Simplify: $\dfrac{x^3 + x^2 - 2x}{2x^2 + 10x + 12} \div \dfrac{x^2 + x}{4x + 12}$

A. $\dfrac{x - 1}{2(x + 1)}$ **B.** $\dfrac{x + 1}{2(x - 1)}$ **C.** $\dfrac{2(x + 1)}{x - 1}$ **D.** $\dfrac{2(x - 1)}{x + 1}$

---

57. Simplify: $\dfrac{4}{x(x - 2)} + \dfrac{2}{x} - \dfrac{5}{3(x - 2)}$

A. $\dfrac{6 - 2x}{3x(x - 2)}$ **B.** $\dfrac{x + 10}{3x(x - 2)}$ **C.** $\dfrac{1}{3(x - 2)}$ **D.** $\dfrac{x + 24}{3(x - 2)}$

---

58. Factor: $m^4 + 2m^3 - 16m^2 - 32m$

A. $m(m + 2)(m^2 + 16)$ **C.** $m(m + 2)(m + 4)(m - 4)$

B. $m(m + 16)(m^2 + 2)$ **D.** $m(m + 2)(m - 4)(m - 4)$

---

59. Which of the following is a factor of the equation $x^3 - 5x^2 - 9x + 45$?

A. $x + 5$ **C.** $x + 9$

B. $x - 5$ **D.** $x^2 + 9$

---

60. Solve the equation: $|3x - 4| = -12$

A. $-\dfrac{8}{3}$ only **B.** $-\dfrac{8}{3}$ and $\dfrac{16}{3}$ **C.** $\dfrac{16}{3}$ only **D.** No Solution

---

61. If $2(3x - 1) - 2x = 5x - 3$, then $x =$

A. 1 **B.** -1 **C.** -2 **D.** $-\dfrac{5}{9}$

---

62. If $-9x \le 8 - 3x$, then

A. $x \ge -\dfrac{4}{3}$ **B.** $x \le -\dfrac{4}{3}$ **C.** $x > -\dfrac{4}{3}$ **D.** $x \ge -\dfrac{2}{3}$

---

63. If $x^2 = -12(x + 3)$, then $x =$

A. -6 only **B.** -6 and 6 **C.** -4 and -9 **D.** 4 and 9

---

64. If $2m - 2 = -m^2$, then $m =$

A. -2 and 0 **B.** $-1 \pm 2\sqrt{3}$ **C.** $-1 \pm \sqrt{5}$ **D.** $-1 \pm \sqrt{3}$

65. Which of the following represents the solution of $\frac{1}{2}|5 - x| > 2$?

A.

```
   -2   -1   0   1   2   3   4   5   6   7   8   9   10   11   12
—————————————○···············○————————————————
             1                9
```

B.

```
   -2   -1   0   1   2   3   4   5   6   7   8   9   10   11   12
·············●———————————————●·····························
             1                9
```

C.

```
   -2   -1   0   1   2   3   4   5   6   7   8   9   10   11   12
—————————————●···············●————————————————
             1                9
```

D.

```
   -2   -1   0   1   2   3   4   5   6   7   8   9   10   11   12
·············○———————————————○·····························
             1                9
```

66. Solve the system of equations:

$2y - 3x = 13$
$6x - 4y = -26$

A. $(-13, 13)$

B. $\left(\dfrac{-13}{5}, \dfrac{13}{5}\right)$

C. The graphs are dependent

D. No Solution

67. Solve the equation: $3(2x + 1)^2 = 48$

A. $x = \dfrac{3}{2}$

B. $x = \dfrac{7}{2}$

C. $x = -\dfrac{9}{2}$ and $x = \dfrac{7}{2}$

D. $x = -\dfrac{5}{2}$ and $x = \dfrac{3}{2}$

68. Which of the following linear equations matches the data given in the table below?

x	-4	-2	-1	0	3
y	-7	-1	2	5	14

A. $5y - 2x = 10$

B. $y - 3x = 5$

C. $3y - x = 5$

D. $2y + 10 = 6x$

69. Which of the following represents a line perpendicular to the graph $y = -3x + 17$?

A. $x = 3y + 15$

B. $y + 3x = 2$

C. $y - 3x = 10$

D. $3y + x = -1$

70. Find the vertex of the equation $y = 4x^2 - 12x + 23$

A. $\left(\frac{3}{2}, 14\right)$ B. $\left(-\frac{3}{2}, 50\right)$ C. $(-3, 95)$ D. $\left(\frac{1}{2}, 18\right)$

71. If $f_2 = f_1 - 1$ and $f_3 = 5f_2 + 3$, then $f_1 + f_2 + f_3 =$

A. $6f_1 + 2$ B. $6f_1 - 3$ C. $7f_1 - 3$ D. $7f_1 - 2$

72. If $h^{-1}(x) = \dfrac{x+3}{5}$ and $h^{-1}(x)$ is the inverse of $h(x)$, then $h(2) =$

A. -1 B. 1 C. 4 D. 7

73. For what real numbers is $-x^2 - x + 6$ negative?

A. $x < -3 \text{ or } x > 2$ C. $-3 < x < 2$
B. $x \leq -3 \text{ or } x \geq 2$ D. $-3 \leq x < 2$

74. The cost of laminate flooring is directly proportional to its area. If a rectangular piece of laminate measuring 15 inches by 20 inches costs $75.00, how much would a piece measuring 3 inches by 5 inches cost?

A. $3.75 B. $5.00 C. $15.00 D. $18.00

75.

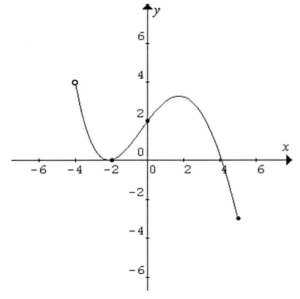

What is the domain of the function above?

A. $(-3, 4]$ B. $[-3, 4)$ C. $(-4, 5]$ D. $[-4, 5)$

◆············ END WORKSHEET ············◆

Postsecondary Education Readiness Test (PERT) – Worksheet #4 Answer Sheet

1. B	26. B	51. D
2. A	27. D	52. D
3. C	28. C	53. C
4. A	29. D	54. C
5. A	30. C	55. C
6. C	31. B	56. D
7. A	32. A	57. C
8. C	33. C	58. C
9. B	34. D	59. B
10. C	35. A	60. D
11. C	36. C	61. A
12. B	37. A	62. A
13. B	38. B	63. A
14. B	39. C	64. D
15. C	40. D	65. A
16. A	41. A	66. C
17. B	42. A	67. D
18. C	43. D	68. B
19. B	44. B	69. A
20. C	45. B	70. A
21. A	46. C	71. C
22. B	47. D	72. D
23. B	48. C	73. A
24. B	49. D	74. A
25. C	50. C	75. C

The contents of this worksheet are copyrighted. Disclosure or reproduction of any portion herein without express written consent of Track 2 Success © is strictly prohibited.

Workbook Exam #4 Solutions

1. **B**

    $\dfrac{2}{3} + \dfrac{4}{5} =$

    LCD: 15

    $\dfrac{(5)2}{(5)3} + \dfrac{4(3)}{5(3)} = \dfrac{10}{15} + \dfrac{12}{15} = \dfrac{22}{15} = 1\dfrac{7}{15}$

2. **A**

    $2 - \dfrac{7}{8} =$

    Place the whole number over 1 and use the LCD to combine the two fractions.

    $\dfrac{(8)2}{(8)1} - \dfrac{7}{8} = \dfrac{16}{8} - \dfrac{7}{8} = \dfrac{9}{8} = 1\dfrac{1}{8}$

3. **C**

    $-\dfrac{1}{4} \times \dfrac{2}{5} = -\dfrac{2}{20} = -\dfrac{1}{10}$

4. **A**

    $3\dfrac{1}{3} \div 4\dfrac{1}{5} =$

    Change to the mixed numbers to improper fractions

    $\dfrac{10}{3} \div \dfrac{21}{5}$

    Take the reciprocal of the second fraction and multiply

    $\dfrac{10}{3} \times \dfrac{5}{21} = \dfrac{50}{63}$

5. **A**

    What is 110% of 50?
    This is the simplest percentage problem. Multiply the number by the decimal representation of the percentage.
    110% of 50 = 1.10 × 50 = 55

6. **C**

    35 is what percent of 28?

    $\dfrac{is}{of} = \dfrac{\%}{100}$

The contents of this worksheet are copyrighted. Disclosure or reproduction of any portion herein without express written consent of Track 2 Success © is strictly prohibited.

$$\frac{35}{28} = \frac{x}{100}$$

Cross multiply and solve

$28x = (35)(100)$

$28x = 3500$

$x = \dfrac{3500}{28} = 125\%$

7. **A**

36 is 45% of what number?

$$\frac{is}{of} = \frac{\%}{100}$$

$$\frac{36}{x} = \frac{45}{100}$$

Cross multiply and solve

$45x = (36)(100)$

$45x = 3600$

$x = \dfrac{3600}{45} = 80$

8. **C**

The Fab Boutique decided to increase the price of their skirts by 25%. If their skirts cost $32.60 now, how much will they cost after the increase?

An increase of 25% can be represented by 125% of $32.60

125% of 32.60 = 1.25 × 32.6

Multiply 125 by 326

125 × 326 = 40,750

There were 3 decimal places in the original two decimals so move the decimal point 3 places to the left. Therefore, the new cost is $40.75

9. **B**

The price of a $40 camera is increased by 30% and then later decreased by 30% during a sale. What is the price of the camera during the sale?

An increase of 30% can be represented by 130%

130% of 40 = 1.3 × 40 = $52.00

Price after increase: $52

Next calculate the price after the 30% decrease which can be represented by 70%.

70% of 52 = 0.7 × 52 = $36.40

Therefore, the price of camera during the sale is $36.40

10. **C**

1.375 =

Convert to percentage by moving the decimal point two places to the right

137.5%

11. C

$4.621 + 3.104 =$

Line up the decimal points and add

$$
\begin{array}{r}
4.621 \\
+\ \ 3.104 \\
\hline
7.725
\end{array}
$$

12. B

$4.147 - (-1.503) =$

$4.147 + 1.503$

Line up the decimal points and add

$$
\begin{array}{r}
4.147 \\
+\ \ 1.503 \\
\hline
5.650
\end{array}
$$

13. B

$5.17 \times 0.06 =$

Multiply 517 by 6

$517 \times 6 = 3102$

There were 4 decimal places in the two original decimals combined so move the decimal point in the answer 4 places to the left

0.3102

14. B

$12.75 \div (-2.5) =$

Move the decimal point two places to the right in both decimals

$1275 \div -250$

$$= -\frac{1275}{250} = -\frac{51}{10}$$

Division by 10 can be accomplished by moving the decimal point one place to the left.

$= -5.1$

15. C

During a field trip, I spent $1.92 on a hotdog, $2.35 on a soda, $5.42 on popcorn and $4.75 on a ride. What is a reasonable estimate of the amount of money spent during the field trip? Round all the numbers to the nearest dollar.

$\$2 + \$2 + \$5 + \$5 = \$14$

16. A

Mary has $60 to spend on her daughter's birthday party. If she spends ⅕ of the budget on decorations, and ⅔ on food and drinks, how much money does she have remaining for entertainment?

$$\frac{1}{5} + \frac{2}{3} = \frac{(3)1}{(3)5} + \frac{(5)2}{(5)3} = \frac{3}{15} + \frac{10}{15} = \frac{13}{15}$$

Fraction left:

$$1 - \frac{13}{15} = \frac{15}{15} - \frac{13}{15} = \frac{2}{15}$$

Money left:

$$\frac{2}{15} \times \$60 = \frac{120}{15} = \$8$$

17. B

A company is currently operating at a loss each month due to the fact that they are spending more on expenses than they are earning in revenue. They decide to cut back on the amount of office supplies purchased each month. Based on the chart above, which shows the amount spent each month, by what percent should they reduce the total spent on office supplies in order to meet their monthly budget of $25,000?

First add all their expenses to determine the amount by which they are currently exceeding the budget.

$1,810 + \$439 + \$910 + \$7,652 + \$6,480 + \$8,600 = \$25,891$

Since their total budget is $25, 000$ per month, they need to decrease office supplies purchased by $891.

$$percentage\ decrease = \frac{decrease}{current\ amount\ spent} \times 100\%$$

$$= \frac{891}{6480} \times 100\% = \frac{89100}{6480}\% = \frac{8910}{648}\% = 13.75\%$$

18. C

What is the mode of the numbers 20, 21, 27, 31, 21, 25 and 27?

Ordering the numbers from least to greatest: 20, 21, 21, 25, 27, 27 and 31

The mode is the number that appears the most in the list. 21 and 27 both appear twice so the set is bimodal (having two modes)

19. B

The perimeter of a rectangular garden is 54 feet. If the length of the garden is 18 feet, what is the width?

$P = 2l + 2w$

$54 = 2(18) + 2w$

$54 = 36 + 2w$

$2w = 54 - 36$

$2w = 18$

$w = \dfrac{18}{2} = 9\ ft$

20. C

Calculate the value of the following: $\dfrac{9.2 \times 0.00073}{0.021}$

Convert all the terms to scientific notation.

$$\frac{9.2 \times 7.3 \times 10^{-4}}{2.1 \times 10^{-2}}$$

If we round the coefficients it would be easier to calculate. Use law #5 of exponents to combine the powers of 10.

$$\frac{9 \times 7}{2} \times 10^{-2} = \frac{63}{2} \times 10^{-2} = 31.5 \times 10^{-2} = 0.315 \approx 0.32$$

21. A

A boutique sold their last three blouses yesterday. They sold the first for $19.60, the second for $22.10 and the last for $18.90. If the boutique bought the blouses for $15.00 a piece, what is the average profit per blouse?

$19.60 - $15.00 = $4.60

$22.10 - $15.00 = $7.10

$18.90 - $15.00 = $3.90

$$\text{Average profit} = \frac{\text{total profit}}{\text{\# blouses}} = \frac{4.60 + 7.10 + 3.90}{3}$$
$$= \frac{\$15.60}{3} = \$5.20$$

22. B

A recipe for pancakes that feeds 6 people requires 3¼ cups of flour. How many cups of flour would be needed to feed 24 people pancakes?

A ratio would be perfect to solve this problem.

$$\frac{6\ people}{3\frac{1}{4}\ cups} = \frac{24\ people}{x\ cups}$$

Cross multiply and solve

$$6x = (24)\left(3\frac{1}{4}\right)$$
$$6x = (24)\left(\frac{13}{4}\right)$$
$$6x = (6)(13)$$
$$x = \frac{78}{6} = 13\ cups$$

23. B

Susan saved $1623.73 over the past eight months. If she deposits $572.91 today and withdraws $818.23 tomorrow what will be her new account balance?

$1623.73 + 572.91 - 818.23 = 1378.41$

24. B

Which of the following numbers is the smallest?

$$\left(-\frac{1}{2}\right)(6),\quad -5,\quad -\frac{18}{5},\quad |-11 + 7|$$

$$-\frac{1}{2} \times 6 = -\frac{6}{2} = -3$$

$$-\frac{18}{5} = -3\frac{3}{5}$$

$$|-11 + 7| = |-4| = 4$$

Putting all the number on a number line

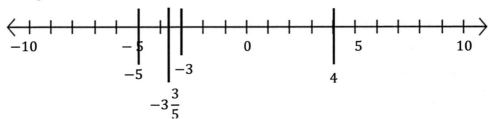

Therefore the smallest number is: -5

25. **C**

    Which of the following is NOT a true statement?
    $12^2 = 144$
    $4.210 \geq 4.2010$
    $\frac{3}{2} \times \frac{4}{9} = \frac{27}{8}$
    $\sqrt{0.04} = 0.2$

    Choice A is correct.
    We can tell that 4.210 is greater than 4.2010 by comparing their hundredths place value so choice B is also correct.
    Choice C is incorrect because we do not flip the second fraction when multiplying fractions.
    $\frac{3}{2} \times \frac{4}{9} = \frac{12}{18} \neq \frac{27}{8}$

26. **B**

    Ms. Daniel's class wants to buy candy for their class Halloween party. If one bag of candy costs $4.78, how many bags can the class buy if they saved $44.95?
    Round the two then divide to quickly estimate the number of bags that can be bought.
    $44.95 can be rounded to $45 and $4.78 can be rounded to $5.
    $$\frac{\$45}{\$5} = 9 \ bags$$
    Notice that $44.98 ÷ $4.78 = 9.40 bags which rounds down to 9 bags.

27. **D**

    Identify the place value of the underlined digit in the number 32.01<u>9</u>2
    The third decimal place to the right represents the thousandths position.

28. **C**

    Find the area of a circle whose radius is ⅘ inches.

$$A = \pi r^2$$

$$A = \pi \left(\frac{4}{5}\right)^2 = \frac{16}{25}\pi \text{ square inches}$$

29. D

Evaluate: $\quad |-(-3)^2 - 14|$

$= |-9 - 14|$

$= |-23| = 23$

30. C

Factor: $\quad 12x^2 + 4x$

$12x^2 + 4x = 4x(3x + 1)$

31. B

$(x^3 + 6x^2 - 4x + 7) - (3x^3 + 2x - 4) =$

Combine like terms

$= x^3 + 6x^2 - 4x + 7 - 3x^3 - 2x + 4$

$= x^3 - 3x^3 + 6x^2 - 4x - 2x + 7 + 4$

$= -2x^3 + 6x^2 - 6x + 11$

32. A

Simplify: $\quad -3(5x - 4)^2$

$-3(5x - 4)(5x - 4)$

FOIL

$= -3(25x^2 - 20x - 20x + 16)$

$= -3(25x^2 - 40x + 16)$

$= -75x^2 + 120x - 48$

33. C

Simplify: $\quad \left(\frac{c^4 d^3}{cd^2}\right)\left(\frac{d^2}{c^3}\right)^3$

Using law #1 of exponents

$= \left(\frac{c^4 d^3}{cd^2}\right)\left(\frac{d^6}{c^9}\right)$

Using law #4 of exponents

$\frac{c^4 d^{3+6}}{c^{1+9}d^2} = \frac{c^4 d^9}{c^{10}d^2}$

Using law #5 of exponents

$c^{4-10}d^{9-2} = c^{-6}d^7$

Using law #6 of exponents

$= \frac{d^7}{c^6}$

34. D

Simplify: $\dfrac{x^2 - 2x}{2x^2 - x - 6}$

Factor then eliminate factors common to both the numerator and denominator

$\dfrac{x(x - 2)}{(2x + 3)(x - 2)}$

$\dfrac{x}{2x + 3}$

35. A

If $x^2 + 2 = 2(x + 5)$, then $x =$

$x^2 + 2 = 2x + 10$

$x^2 - 2x + 2 - 10 = 0$

$x^2 - 2x - 8 = 0$

$(x - 4)(x + 2) = 0$

$x - 4 = 0 \qquad x + 2 = 0$

$x = 4 \qquad\quad x = -2$

36. C

What is the value of the expression $2a + b(1 - a^2)$ when $a = 5$ and $b = -3$?

Substitute the values of a and b.

$2a + b(1 - a^2)$

$= 2(5) + (-3)(1 - 5^2)$

$= 10 + (-3)(1 - 25) = 10 + (-3)(-24) = 10 + 72 = 82$

37. A

Evaluate: $3i^2(i^7 - 4) - 2i^5$

Distribute and then simplify using the rules of exponents and $i^2 = -1$

$3i^2(i^7 - 4) - 2i^5 = 3i^2 i^7 - 12i^2 - 2i^5 = 3i^9 - 12i^2 - 2i^5$

In order to use $i^2 = -1$:

If the exponent is odd, factor an i so that the remaining exponent is even.

$3i^9 - 12i^2 - 2i^5 = 3(i^8)i - 12i^2 - 2(i^4)i$

Next divide the even powers by 2 and replace i^2 by -1 as shown.

$3(i^8)i - 12i^2 - 2(i^4)i$

$= 3(i^2)^4 i - 12i^2 - 2(i^2)^2 i$

$= 3(-1)^4 i - 12(-1) - 2(-1)^2 i$

$= 3(1)i + 12 - 2(1)i = 3i + 12 - 2i = 12 + i$

38. B

Choose the equation that is equivalent to the statement below:

"The difference between a number n and 7 more than three times the number is 18."

7 more than three times the number $= 3n + 7$

Therefore: $n - (3n + 7) = 18$

39. C

Given that LM || OR, find the value of ∠b in the diagram below.

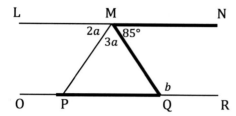

There are two ways to solve this problem but one method is far shorter than the other. The long way would be to first find the value of a, then find the values of the angles in triangle MPQ and finally use the value of angle MQP to find the value of b.

The short way would be to recognize that angles ∠NMQ and ∠MQP are transverse angles. Note the Z shape in the diagram above even though it is backwards.

∠MQP = ∠NMQ = 85°

∠b and ∠MQP are supplementary angles so:

∠b = 180° − ∠MQP = 180° − 85° = 95°

40. D

If $\frac{2}{5}c - 7 = 9$, then $c =$

$\frac{2}{5}c - 7 = 9$

$\frac{2}{5}c = 9 + 7$

$\frac{2}{5}c = 16$

$\left(\frac{5}{2}\right)\frac{2}{5}c = 16\left(\frac{5}{2}\right)$

$c = 8 \times 5 = 40$

41. A

If $4x + 6 \geq 14 - 2x$, then

$4x + 6 \geq 14 - 2x$

$4x + 2x \geq 14 - 6$

$6x \geq 8$

$x \geq \frac{8}{6}$

$x \geq \frac{4}{3}$

42. A

Solve for b: $ab - d = cb$

$ab - d = cb$

$$ab - cb = d$$
$$b(a - c) = d$$
$$b = \frac{d}{a - c}$$

43. D

If $-3(x + 5) > 14$, then
$-3x - 15 > 14$
$-3x > 14 + 15$
$-3x > 29$
Remember to switch the direction of the inequality sign when dividing or multiplying by a negative number.
$$x < -\frac{29}{3}$$

44. B

Divide: $\dfrac{-21m^3 + 18m^7 - 15m^2}{-3m^2}$

$-\dfrac{21m^3}{-3m^2} + \dfrac{18m^7}{-3m^2} - \dfrac{15m^2}{-3m^2}$

$7m^{3-2} - 6m^{7-2} + 5m^{2-2}$

$7m - 6m^5 + 5$

45. B

The window of a house resembles the figure below. How many square feet of glass are required if the length of the rectangle is x feet and the width is 2 feet?

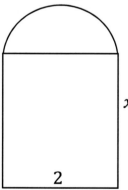

From the diagram, the diameter of the circle is 2 and so the radius is 1 (half the diameter)
Area of semicircle:
$$A = \frac{\pi r^2}{2} = \frac{\pi(1)^2}{2} = \frac{\pi}{2}$$
Area of rectangle:
$$A = l \times w = x \times 2 = 2x$$
$$Total\ area = 2x + \frac{\pi}{2}$$

46. **C**

Using the given information we can extract two points on the line and use the points to find the equation of the line.

The two points are: $(6, -2)$ and $(-2, 10)$

Label the points

$(x_1, y_1) \quad (x_2, y_2)$
$(6, -2) \quad (-2, 10)$

Find the slope

$$m = \frac{10 - (-2)}{-2 - (6)} = \frac{10 + 2}{-2 - 6} = \frac{12}{-8} = -\frac{3}{2}$$

Use the point-slope equation

$$y - y_1 = m(x - x_1)$$
$$y - (-2) = -\frac{3}{2}(x - 6)$$
$$y + 2 = -\frac{3}{2}x + 9$$
$$y = -\frac{3}{2}x + 7$$

Now set $y = -8$ and $x = a$ then solve for a

$$-\frac{3}{2}(a) + 7 = -8$$
$$-\frac{3}{2}a = -15$$
$$-3a = -30$$
$$a = \frac{-30}{-3} = 10$$

47. **D**

The sum of twice a number and 9 is five less than three times the number. Write an expression that can be used to find the number x.

Sum of twice a number and $9 = 2x + 9$
Five less than three times the number $= 3x - 5$
$2x + 9 = 3x - 5$

48. **C**

Graph the line: $y = -4$
The graph of the line $y = -4$ is a horizontal line passing through -4 on the y–axis.

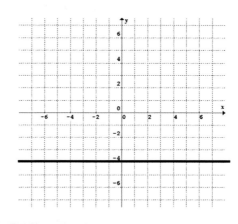

49. D

The figure below contains two squares. What is the area of the shaded region?

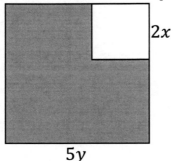

Area of larger square:
$A = side^2$
$A = (5y)^2 = 25y^2$
Area of smaller square:
$A = side^2$
$A = (2x)^2 = 4x^2$
Area of shaded region = area of large square – area of smaller square
$$= 25y^2 - 4x^2 = (5y + 2x)(5y - 2x)$$

50. C

Simplify: $\quad 5\sqrt{72a^5b^6c^7}$
$\sqrt{72} = \sqrt{2 \times 2 \times 2 \times 3 \times 3} = 2 \times 3\sqrt{2} = 6\sqrt{2}$

With exponents, to take the square root, divide the exponent by 2. The quotient becomes the exponent outside the radical and the remainder becomes the exponent inside the radical.

$\sqrt{a^5} = a^2\sqrt{a} \quad$ because $5 \div 2 = 2\ R1$
$\sqrt{b^6} = b^3 \quad$ because $6 \div 2 = 3\ R0$
$\sqrt{c^7} = c^3\sqrt{c} \quad$ because $7 \div 2 = 3\ R1$
$5\sqrt{72a^5b^6c^7} = 5(6a^2b^3c^3\sqrt{2ac}) = 30a^2b^3c^3\sqrt{2ac}$

51. D

Convert to scientific notation: $\quad$ 0.00000567

Move the decimal point to the right of the first non-zero digit and count the number of places moved. The number of places moved becomes the exponent of 10 and the sign is negative because the decimal point was moved to the right.
5.67×10^{-6}

52. D

Find the x – intercept of the equation: $\quad y = \dfrac{4}{3}x - 8$

Substitute $y = 0$

$$0 = \frac{4}{3}x - 8$$

$$\frac{4}{3}x = 8$$

$$\left(\frac{3}{4}\right)\frac{4}{3}x = 8\left(\frac{3}{4}\right)$$

$$x = \frac{24}{4} = 6$$

$$(6, 0)$$

53. **C**

We first need to change base 9 to base 3.

$9 = 3^2$ therefore:

$$9^{m-n} = 3^{2(m-n)} = 3^{2m-2n}$$

Using rule #5 of exponents:

$$3^{2m-2n} = \frac{3^{2m}}{3^{2n}}$$

Using rule #1 of exponents

$$\frac{3^{2m}}{3^{2n}} = \frac{(3^m)^2}{(3^n)^2}$$

Substituting the values of 3^m and 3^n

$$\frac{(3^m)^2}{(3^n)^2} = \frac{8^2}{2^2} = \frac{64}{4} = 16$$

54. **C**

Simplify: $\dfrac{\sqrt[4]{81x^{10}y^4}}{\sqrt[4]{3xy^7}}$

We can rationalize the denominator first then simplify or we can simply then rationalize.

$$\frac{\sqrt[4]{81x^{10}y^4}}{\sqrt[4]{3xy^7}} = \sqrt[4]{\frac{81x^{10}y^4}{3xy^7}}$$

Using law #5 of exponents

$$= \sqrt[4]{27x^{10-1}y^{4-7}} = \sqrt[4]{27x^9y^{-3}}$$

Using law #6 of exponents

$$= \sqrt[4]{\frac{27x^9}{y^3}}$$

Now we can rationalize. Since we need the 4th root, we need to make the expression under the radical in the denominator a power of 4. So we multiply by $\sqrt[4]{y}$

$$= \frac{\sqrt[4]{27x^9}}{\sqrt[4]{y^3}} \cdot \frac{\sqrt[4]{y}}{\sqrt[4]{y}} = \frac{\sqrt[4]{27x^9y}}{\sqrt[4]{y^4}}$$

Now simplify the radicals in the numerator and denominator

$$= \frac{x^2 \sqrt[4]{27xy}}{y}$$

55. C

$$\frac{\sqrt{2}}{\sqrt{5}-2} =$$

Multiply by the conjugate of the denominator

$$\frac{\sqrt{2}}{\sqrt{5}-2} \cdot \frac{\sqrt{5}+2}{\sqrt{5}+2}$$

FOIL

$$\frac{\sqrt{10}+2\sqrt{2}}{5+2\sqrt{5}-2\sqrt{5}-4}$$

$$\frac{\sqrt{10}+2\sqrt{2}}{5-4} = \sqrt{10}+2\sqrt{2}$$

56. D

Simplify: $\dfrac{x^3+x^2-2x}{2x^2+10x+12} \div \dfrac{x^2+x}{4x+12}$

Factor

$$\frac{x(x^2+x-2)}{2(x^2+5x+6)} \div \frac{x(x+1)}{4(x+3)}$$

$$\frac{x(x+2)(x-1)}{2(x+2)(x+3)} \div \frac{x(x+1)}{4(x+3)}$$

Take the reciprocal of the second fraction and multiply

$$\frac{x(x+2)(x-1)}{2(x+2)(x+3)} \cdot \frac{4(x+3)}{x(x+1)}$$

$$\frac{4x(x+2)(x-1)(x+3)}{2x(x+2)(x+3)(x+1)}$$

Eliminate common factors

$$\frac{2(x-1)}{x+1}$$

57. C

Simplify: $\dfrac{4}{x(x-2)} + \dfrac{2}{x} - \dfrac{5}{3(x-2)}$

LCD: $3x(x-2)$

$$\frac{(3)4}{(3)x(x-2)} + \frac{(3(x-2))2}{(3(x-2))x} - \frac{(x)5}{(x)3(x-2)}$$

$$\frac{12}{3x(x-2)} + \frac{6(x-2)}{3x(x-2)} - \frac{5x}{3x(x-2)}$$

$$\frac{12+6(x-2)-5x}{3x(x-2)}$$

$$\frac{12 + 6x - 12 - 5x}{3x(x-2)}$$

$$\frac{x}{3x(x-2)} = \frac{1}{3(x-2)}$$

58. C

Factor: $m^4 + 2m^3 - 16m^2 - 32m$

$m(m^3 + 2m^2 - 16m - 32)$

Factor by grouping

$m[m^2(m+2) - 16(m+2)]$

$m[(m^2 - 16)(m+2)] = m(m+4)(m-4)(m+2)$

59. B

Which of the following is a factor of the equation $x^3 - 5x^2 - 9x + 45$?

Let's factor by grouping

$x^3 - 5x^2 - 9x + 45$

$= x^2(x-5) - 9(x-5)$

$= (x^2 - 9)(x-5)$

Using difference of squares on the first factor

$= (x-3)(x+3)(x-5)$

60. D

Solve the equation: $|3x - 4| = -12$

The absolute value of an expression can never equal a negative number so no solution.

61. A

If $2(3x - 1) - 2x = 5x - 3$, then $x =$

$2(3x - 1) - 2x = 5x - 3$

$6x - 2 - 2x = 5x - 3$

$4x - 2 = 5x - 3$

$4x - 5x = -3 + 2$

$-x = -1$

$x = 1$

62. A

If $-9x \leq 8 - 3x$, then

$-9x \leq 8 - 3x$

$-9x + 3x \leq 8$

$-6x \leq 8$

Reverse the direction of the inequality sign when dividing or multiplying by a negative.

$x \geq \frac{8}{-6}$

$x \geq -\frac{4}{3}$

63. **A**

If $x^2 = -12(x+3)$, then $x =$
$x^2 = -12x - 36$
$x^2 + 12x + 36 = 0$
$(x+6)(x+6) = 0$
$x + 6 = 0$
$x = -6$

64. **D**

If $2m - 2 = -m^2$, then $m =$
$m^2 + 2m - 2 = 0$
This is not factorable so use the quadratic formula
$a = 1, \ b = 2, \ c = -2$
$m = \dfrac{-b \pm \sqrt{b^2 - 4ac}}{2a}$
$m = \dfrac{-2 \pm \sqrt{2^2 - 4(1)(-2)}}{2(1)}$
$m = \dfrac{-2 \pm \sqrt{4 + 8}}{2}$
$m = \dfrac{-2 \pm \sqrt{12}}{2}$
$m = \dfrac{-2 \pm 2\sqrt{3}}{2}$
$m = -1 \pm \sqrt{3}$

65. **A**

Which of the following represents the solution of $\dfrac{1}{2}|5 - x| > 2$?
$|5 - x| > 4$

$5 - x > 4$ $\qquad\qquad\qquad$ $5 - x < -4$
$-x > 4 - 5$ $\qquad\qquad\qquad$ $-x < -4 - 5$
$-x > -1$ $\qquad\qquad\qquad\quad$ $-x < -9$

Reverse the direction of the inequality sign when dividing or multiplying by a negative.
$x < 1$ $\qquad\qquad\qquad\qquad\qquad$ $x > 9$

66. **C**

Solve the system of equations:
$2y - 3x = 13$
$6x - 4y = -26$
Multiply the top equation by 2
$2(2y - 3x = 13)$

$$4y - 6x = 26$$
$$-6x + 4y = 26$$
Add the two equations
$$-6x + 4y = 26$$
$$6x - 4y = -26$$
$$0 = 0$$
This is a true statement without variables so the lines are consistent and dependent.

67. D

Solve the equation: $3(2x + 1)^2 = 48$

$$(2x + 1)^2 = \frac{48}{3}$$
$$(2x + 1)^2 = 16$$
$$2x + 1 = \pm\sqrt{16}$$
$$2x = -1 \pm 4$$

$$2x = -1 + 4 \qquad\qquad 2x = -1 - 4$$
$$2x = 3 \qquad\qquad\qquad 2x = -5$$
$$x = \frac{3}{2} \qquad\qquad\qquad x = -\frac{5}{2}$$

68. B

Which of the following linear equations matches the data given in the table below?

x	-4	-2	-1	0	3
y	-7	-1	2	5	14

Choose any of the points in the table.
Using $(0, 5)$ and $(3, 14)$
$$m = \frac{y_2 - y_1}{x_2 - x_1} = \frac{14 - 5}{3 - 0} = \frac{9}{3} = 3$$
Since the y-intercept $b = 5$, the equation of the line is:
$$y = mx + b$$
$$y = 3x + 5$$
Therefore the answer is B.

69. A

Which of the following represents a line perpendicular to the graph $y = -3x + 17$?
Slope of this graph is -3
To find the slope of the perpendicular line, take the negative reciprocal of the current slope
$$m = -\left(-\frac{1}{3}\right) = \frac{1}{3}$$
Solve all the equations for y to determine which has a slope of ⅓
$$x = 3y + 15$$
$$3y = x - 15$$

$$y = \frac{1}{3}x - \frac{15}{3}$$

$$y = \frac{1}{3}x - 5$$

This line has a slope of ⅓ and is therefore perpendicular to the line $y = -3x + 17$

70. A

Find the vertex of the equation $y = 4x^2 - 12x + 23$

$$x = -\frac{b}{2a} = -\frac{(-12)}{2(4)} = \frac{12}{8} = \frac{3}{2}$$

Substitute for x in the original equation to find the y–coordinate.

$$y = 4\left(\frac{3}{2}\right)^2 - 12\left(\frac{3}{2}\right) + 23$$

$$y = 4\left(\frac{9}{4}\right) - \frac{36}{2} + 23$$

$$y = 9 - 18 + 23 = 14$$

Vertex: $\left(\frac{3}{2}, 14\right)$

71. C

If $f_2 = f_1 - 1$ and $f_3 = 5f_2 + 3$, then $f_1 + f_2 + f_3 =$
Notice the all the answer choices are written using f_1

$$
\begin{aligned}
f_1 + f_2 + f_3 &= f_1 + (f_1 - 1) + (5f_2 + 3) \\
&= f_1 + f_1 - 1 + 5(f_1 - 1) + 3 \\
&= 2f_1 - 1 + 5f_1 - 5 + 3 \\
&= 7f_1 - 3
\end{aligned}
$$

72. D

If $h^{-1}(x) = \frac{x+3}{5}$ and $h^{-1}(x)$ is the inverse of $h(x)$, then $h(2) =$
By the rule of inverses:

$$\frac{x+3}{5} = 2$$

Cross multiply and solve

$$\frac{x+3}{5} = \frac{2}{1}$$

$$x + 3 = 10$$

$$x = 7$$

73. A

For what real number is $-x^2 - x + 6$ negative?

$$-x^2 - x + 6 < 0$$

$$-(x^2 + x - 6) < 0$$

$$-(x + 3)(x - 2) < 0$$

Set each factor equal to zero and solve for the critical points

$x + 3 = 0 \qquad x - 2 = 0$

$x = -3 \qquad x = 2$

Place the critical values on a number line and choose test points in the 3 regions formed.

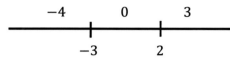

Substitute the test points into the original inequality to see which satisfy the inequality.

When $x = -4$	When $x = 0$	When $x = 3$
$-(-4)^2 - (-4) + 6 < 0$	$-(0)^2 - (0) + 6 < 0$	$-(3)^2 - (3) + 6 < 0$
$-16 + 4 + 6 < 0$	$0 + 0 + 6 < 0$	$-9 - 3 + 6 < 0$
$-6 < 0$	$6 < 0$	$-6 < 0$
True	False	True

Therefore the solution set is:

$x < -3 \text{ or } x > 2$

74. **A**

The cost of laminate flooring is directly proportional to its area. If a rectangular piece of laminate measuring 15 inches by 20 inches costs $75.00, how much would a piece measuring 3 inches by 5 inches cost?

Area of the first piece:

$A = l \times w$

$A = 15 \times 20 = 300$

Area of the second piece:

$A = l \times w$

$A = 3 \times 5 = 15$

Now set up a proportion to solve for the cost.

$\dfrac{\text{area}}{\text{cost}} = \dfrac{300}{75} = \dfrac{15}{x}$

Cross multiply and solve.

$300x = (15)(75)$

$300x = 1125$

$x = \dfrac{1125}{300} = \$3.75$

75. **C**

The domain is comprised of the x-values for which the function is defined.

The graph is bound between $x = -4$ and $x = 5$.

The open circle at $x = -4$ indicates that the end point is not included in the domain so we use the symbol "(".

The closed circle at $x = 5$ indicates that the end point is included in the domain so we use the symbol "]".

Therefore the domain is $(-4, 5]$.

Track 2 Success
"Dedicated to helping you stay on the track to academic success"

Postsecondary Education Readiness Test (PERT) – Worksheet #5 Bubble Sheet

1. Ⓐ Ⓑ Ⓒ Ⓓ
2. Ⓐ Ⓑ Ⓒ Ⓓ
3. Ⓐ Ⓑ Ⓒ Ⓓ
4. Ⓐ Ⓑ Ⓒ Ⓓ
5. Ⓐ Ⓑ Ⓒ Ⓓ
6. Ⓐ Ⓑ Ⓒ Ⓓ
7. Ⓐ Ⓑ Ⓒ Ⓓ
8. Ⓐ Ⓑ Ⓒ Ⓓ
9. Ⓐ Ⓑ Ⓒ Ⓓ
10. Ⓐ Ⓑ Ⓒ Ⓓ
11. Ⓐ Ⓑ Ⓒ Ⓓ
12. Ⓐ Ⓑ Ⓒ Ⓓ
13. Ⓐ Ⓑ Ⓒ Ⓓ
14. Ⓐ Ⓑ Ⓒ Ⓓ
15. Ⓐ Ⓑ Ⓒ Ⓓ
16. Ⓐ Ⓑ Ⓒ Ⓓ
17. Ⓐ Ⓑ Ⓒ Ⓓ
18. Ⓐ Ⓑ Ⓒ Ⓓ
19. Ⓐ Ⓑ Ⓒ Ⓓ
20. Ⓐ Ⓑ Ⓒ Ⓓ
21. Ⓐ Ⓑ Ⓒ Ⓓ
22. Ⓐ Ⓑ Ⓒ Ⓓ
23. Ⓐ Ⓑ Ⓒ Ⓓ
24. Ⓐ Ⓑ Ⓒ Ⓓ
25. Ⓐ Ⓑ Ⓒ Ⓓ

26. Ⓐ Ⓑ Ⓒ Ⓓ
27. Ⓐ Ⓑ Ⓒ Ⓓ
28. Ⓐ Ⓑ Ⓒ Ⓓ
29. Ⓐ Ⓑ Ⓒ Ⓓ
30. Ⓐ Ⓑ Ⓒ Ⓓ
31. Ⓐ Ⓑ Ⓒ Ⓓ
32. Ⓐ Ⓑ Ⓒ Ⓓ
33. Ⓐ Ⓑ Ⓒ Ⓓ
34. Ⓐ Ⓑ Ⓒ Ⓓ
35. Ⓐ Ⓑ Ⓒ Ⓓ
36. Ⓐ Ⓑ Ⓒ Ⓓ
37. Ⓐ Ⓑ Ⓒ Ⓓ
38. Ⓐ Ⓑ Ⓒ Ⓓ
39. Ⓐ Ⓑ Ⓒ Ⓓ
40. Ⓐ Ⓑ Ⓒ Ⓓ
41. Ⓐ Ⓑ Ⓒ Ⓓ
42. Ⓐ Ⓑ Ⓒ Ⓓ
43. Ⓐ Ⓑ Ⓒ Ⓓ
44. Ⓐ Ⓑ Ⓒ Ⓓ
45. Ⓐ Ⓑ Ⓒ Ⓓ
46. Ⓐ Ⓑ Ⓒ Ⓓ
47. Ⓐ Ⓑ Ⓒ Ⓓ
48. Ⓐ Ⓑ Ⓒ Ⓓ
49. Ⓐ Ⓑ Ⓒ Ⓓ
50. Ⓐ Ⓑ Ⓒ Ⓓ

51. Ⓐ Ⓑ Ⓒ Ⓓ
52. Ⓐ Ⓑ Ⓒ Ⓓ
53. Ⓐ Ⓑ Ⓒ Ⓓ
54. Ⓐ Ⓑ Ⓒ Ⓓ
55. Ⓐ Ⓑ Ⓒ Ⓓ
56. Ⓐ Ⓑ Ⓒ Ⓓ
57. Ⓐ Ⓑ Ⓒ Ⓓ
58. Ⓐ Ⓑ Ⓒ Ⓓ
59. Ⓐ Ⓑ Ⓒ Ⓓ
60. Ⓐ Ⓑ Ⓒ Ⓓ
61. Ⓐ Ⓑ Ⓒ Ⓓ
62. Ⓐ Ⓑ Ⓒ Ⓓ
63. Ⓐ Ⓑ Ⓒ Ⓓ
64. Ⓐ Ⓑ Ⓒ Ⓓ
65. Ⓐ Ⓑ Ⓒ Ⓓ
66. Ⓐ Ⓑ Ⓒ Ⓓ
67. Ⓐ Ⓑ Ⓒ Ⓓ
68. Ⓐ Ⓑ Ⓒ Ⓓ
69. Ⓐ Ⓑ Ⓒ Ⓓ
70. Ⓐ Ⓑ Ⓒ Ⓓ
71. Ⓐ Ⓑ Ⓒ Ⓓ
72. Ⓐ Ⓑ Ⓒ Ⓓ
73. Ⓐ Ⓑ Ⓒ Ⓓ
74. Ⓐ Ⓑ Ⓒ Ⓓ
75. Ⓐ Ⓑ Ⓒ Ⓓ

The contents of this worksheet are copyrighted. Disclosure or reproduction of any portion herein without express written consent of Track 2 Success © is strictly prohibited.

Worksheet #5

The questions contained in this worksheet cover all the topics presented in the study guide section and are ordered by increasing difficulty from Arithmetic through College Level Math.

Since calculators are not generally allowed while taking the Postsecondary Education Readiness Test, use of calculators should be limited while completing this worksheet so as to replicate and properly prepare for exam conditions.

1. $\dfrac{7}{8} + 1\dfrac{3}{4} =$

   A. $1\dfrac{5}{8}$ B. $1\dfrac{1}{4}$ C. $2\dfrac{5}{8}$ D. $2\dfrac{1}{4}$

2. $-\dfrac{5}{9} - \dfrac{1}{4} =$

   A. $\dfrac{11}{36}$ B. $-\dfrac{4}{5}$ C. $-\dfrac{29}{36}$ D. $\dfrac{29}{36}$

3. $-\dfrac{8}{11} \times \dfrac{3}{4} =$

   A. $-\dfrac{11}{15}$ B. $-\dfrac{6}{11}$ C. $-\dfrac{32}{33}$ D. $-1\dfrac{1}{32}$

4. $-5\dfrac{1}{4} \div \left(-\dfrac{7}{10}\right) =$

   A. $3\dfrac{3}{14}$ B. $3\dfrac{27}{40}$ C. $-7\dfrac{1}{2}$ D. $7\dfrac{1}{2}$

5. $17 - [4 - 2(3^2 + 5 - 7)] \div 2 * \dfrac{3}{5} =$

   A. 3 B. $8\dfrac{1}{10}$ C. $10\dfrac{1}{2}$ D. 20

6. 14 is what percent of 280?

   A. 5%   B. 20%   C. 10%   D. 25%

The contents of this worksheet are copyrighted. Disclosure or reproduction of any portion herein without express written consent of Track 2 Success © is strictly prohibited.

7. Sammy spent 20% of his allowance last week. If he spent $16, how much money did he get for allowance?

A. $70   **B.** $90   **C.** $80   **D.** $60

8. The cost of Fun Park tickets is decreased by 18%. How much money would you save if a ticket originally cost $120?

A. $6.67   **B.** $21.60   **C.** $98.40   **D.** $141.60

9. $\dfrac{9}{20} =$

A. 48%   **B.** 4.5   **C.** 4.8   **D.** 0.45

10. The Kwik Stop grocery store charges 8% tax. If Susan buys 2 gallons of milk at $3.75 each, one dozen eggs for $1.20 and 3 bags of cookies at $4.50 each, what is Susan's total grocery bill?

A. $10.45   **B.** $19.93   **C.** $22.20   **D.** $23.98

11. $0.213 + 3.001 + 1.52 =$

A. 3.366   **B.** 3.734   **C.** 4.734   **D.** 6.551

12. The Johnson family is going on vacation. Their 8 suitcases weigh an average of 58 lbs each which is over the allowed limit of 55 lbs each. In an attempt to meet the weight requirement, they decide to leave their heaviest suitcases weighing 76, 59 and 64 lbs. What is the mean weight of the remaining suitcases?

A. 33 *lbs*   **B.** 53 *lbs*   **C.** 55 *lbs*   **D.** 66 *lbs*

13. $-2.81 \times 1.5 =$

A. −2.815   **B.** 2.815   **C.** 4.215   **D.** −4.215

14. $9.63 \div (-0.3) =$

A. −3.21   **B.** −32.1   **C.** 32.1   **D.** −321

15. Which of the following is greatest?

A. $0.\overline{45}$ **B.** $0.4\overline{5}$ **C.** 0.045   **D.** 0.448

16. The Tennis Club decided to sell boxes of cookies to raise money. The table below shows the number of boxes sold by each member. What is a reasonable estimate of the average number of boxes sold by each member?

Member	Boxes of cookies sold
Brianna	31
Sean	25
Lacy	29
Cory	42
Bryan	23
John	17
Charlotte	8
Katie	37
Sam	29
William	16
Nicky	33

A. 26   B. 27   C. 29   D. 30

17. Write the result of the following in scientific notation: $0.0000031 \times 27{,}000$

A. 8.37×10^{-2} B. 8.37×10^{-1} C. 8.37 D. 8.37×10^{1}

18. A mother splits a bag of candy equally between her two children. Her son decides to give ⅖ of his candy to his best friend. What fraction of the candy does her son now have?

A. $\dfrac{1}{5}$ B. $\dfrac{2}{5}$ C. $\dfrac{3}{5}$ D. $\dfrac{3}{10}$

19. Water flows from one faucet at a rate of 2 gallons per hour and from another faucet at a rate of 5 gallons per hour. How long will it take the faucets to fill a 63 gallon tub together?

A. 7 hrs   B. 8 hrs   C. 9 hrs   D. 10 hrs

20. Find the arithmetic mean of the following: $\dfrac{2}{3}, \dfrac{4}{5},$ and $\dfrac{14}{15}$

A. $\dfrac{4}{5}$ B. $\dfrac{4}{9}$ C. $\dfrac{12}{5}$ D. 4

21. A couple buys a new house and needs to purchase new kitchen appliances. A refrigerator costs $1290.31, a dishwasher costs $489.50, a stove costs $876.32 and a microwave costs $192.89. Estimate the amount of money the couple would need to save to buy the new appliances to the nearest hundred dollars.

A. $2,800   B. $2,900   C. $3,000   D. $3,100

22. A sixth grade class paid $63.50 to buy candy for their Halloween party. If they bought 10 lbs of jelly beans for $1.25 per pound, 12 lbs of candy corn for $1.10/lb and chocolate bars at $0.90 each, how many chocolate bars did they buy?
    A. 38
    B. 42
    C. 45
    D. 48

23. Find the area of a circle whose radius is 3.5 inches.
    A. 6.13π
    B. $6.13\pi^2$
    C. 12.25π
    D. $12.25\pi^2$

24.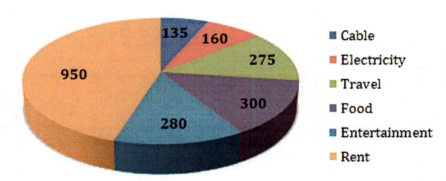

    The above graph shows the allocated budget for the Smith household in dollars. Approximately what percent of the budget is allocated to food?
    A. 12%
    B. 14%
    C. 18%
    D. 20%

25. How many ounces of milk are necessary to triple a recipe that calls for 13.4 ounces?
    A. 16.8 oz
    B. 40.2 oz
    C. 41.4 oz
    D. 23.6 oz

26. Shelby joined the Pound Shedders fitness club. She weighed 189 pounds when she started, 152 pounds after 3 months and then lost another 21 pounds two months later. What was her average monthly weight loss?
    A. 7.4 lbs
    B. 10.5 lbs
    C. 11.6 lbs
    D. 19.3 lbs

27. Given that $(a^x)^3 = a^{15}$ and $b^y b^6 = b^{12}$, what is the value of $2x - y$?
    A. -1
    B. 4
    C. 6
    D. 22

28. Evaluate: $-\dfrac{2}{5}\left[\dfrac{1}{8} - 3\left(\dfrac{2}{3} - \dfrac{3}{4}\right)\right]$
    A. $1\dfrac{1}{20}$
    B. $-\dfrac{3}{20}$
    C. $\dfrac{9}{20}$
    D. $\dfrac{23}{240}$

29. Use the table below to solve the problem:

Approximately how many times larger is the largest lake as compared to the smallest lake?

City	Population (in thousands)
San Juan	152
Key Largo	88
St. Joseph	395
Thomasville	415
Littlerock	107
Winter Park	385

A. 3   **B.** 4   **C.** 5   **D.** 6

---

30. Which of the following is a linear factor of $5x^2 + 13x - 6$?

A. $x - 3$ **B.** $5x - 2$ **C.** $5x + 2$ **D.** $x - 2$

---

31. Simplify: $4(x^2 - 3x + 5) - 3(x^2 - 2x + 1)$

A. $7x^2 - 18x + 23$ **C.** $x^2 - 6x + 17$

B. $7x^2 + 18x + 2$ **D.** $x^2 + 6x + 23$

---

32. Simplify: $(2x - 5)(x^2 - x + 1)$

A. $x^2 + x - 4$ **C.** $2x^3 - 2x^2 + 2x - 5$

B. $2x^3 - 7x^2 + 7x - 5$ **D.** $2x^3 - 7x^2 + 5x + 1$

---

33. Simplify: $(4x^6 y^8)^{3/2}$

A. $8x^9 y^{12}$ **B.** $3x^9 y^{12}$ **C.** $8x^{15/2} y^{19/2}$ **D.** $3x^{15/2} y^{19/2}$

---

34. Simplify: $\dfrac{x^2 + 2x + 1}{x^2 - 2x + 3}$

A. $\dfrac{x + 1}{x - 3}$ **B.** $\dfrac{x - 1}{x - 3}$ **C.** 3 **D.** $x - 3$

---

35. If $2(x + 2)^2 = 50$, then $x =$

A. 3   **B.** -7 and 3 **C.** 8   **D.** 3 and 8

36. Which of the following represents the solution of $5 - 3x \leq -19$?

A.

B.

C.

D.

37. The height $h(t)$ of a ball thrown into the air is given by the equation $h(t) = 40t - 16t^2$ where t is time in seconds. What is the height of the ball after 2 seconds?

A. 16 B. 24 C. 32 D. 40

38. $2\sqrt{3} + 4\sqrt{48} - \sqrt{3} =$

A. $5\sqrt{3}$ B. $20\sqrt{3}$ C. $17\sqrt{3}$ D. $12\sqrt{3}$

39. The product of two consecutive even integers is 108. Which equation should be used to find the numbers if one of the numbers is represented by n?

A. $n(2n) = 108$ C. $n(n+2) = 108$
B. $n + (2n) = 108$ D. $n + (n+2) = 108$

40. If $4(3x+2) - 11 = 3(3x-2)$, then $x =$

A. -4 B. -1 C. 2 D. 5

41. If $1 + \dfrac{4}{5}x < x + 6$, then

A. $x < -1$ B. $x > -1$ C. $x > -25$ D. $x > 25$

42. Solve for x: $\dfrac{x+4}{a} = \dfrac{y}{b}$

A. $ay - 4b$ B. $\dfrac{ay - 4b}{b}$ C. $\dfrac{y}{b} - \dfrac{4}{a}$ D. $\dfrac{ay - 4b}{ab}$

43. Given $g(x) = 2 + 4x - x^2 - 3x^3$, find $g\left(-\dfrac{2}{3}\right)$

A. $-\dfrac{2}{9}$ B. $\dfrac{2}{3}$ C. $\dfrac{10}{3}$ D. $\dfrac{38}{9}$

44. A can of paint costs $9.75 and covers 150 square ft. What will be the cost of painting the walls of a 6 ft by 8 ft room if the walls are 10 ft high?

A. $9.75   B. $19.50   C. $29.25   D. $39.00

45. In the figure below, a circle is inscribed in a square and the radius of the circle is x inches. Write an expression that represents the area of the shaded region.

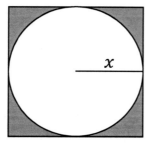

A. $x^2(4 - \pi)$ B. $x^2(2 - \pi)$ C. $4x - \pi x^2$ D. $16\pi x^2$

46. Which of the following is a solution of the equation $x^2 - 7x + 12$?

A. $x = -2$ B. $x = -3$ C. $x = 3$ D. $x = -4$

47. If the equation of the linear function in the figure is $y = mx + b$, then $m =$

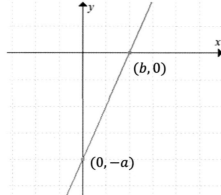

A. $\dfrac{a}{b}$ B. $-\dfrac{a}{b}$ C. $-\dfrac{b}{a}$ D. $\dfrac{b}{a}$

48. Simplify: $18m - 3n - 4m + 5 + 11n$

A. $14m + 8n + 5$ C. $15m + 7n + 5$
B. $22m + 14n + 5$ D. $11m + 11n + 5$

49. Simplify: $\dfrac{3}{4a^2} + \dfrac{5}{12a}$

A. $\dfrac{8}{12a}$ B. $\dfrac{3a+5}{12a}$ C. $\dfrac{9+5a}{12a^2}$ D. $\dfrac{8}{12a^2}$

50. The product of a number and four less than three times the number is 60. Write an expression that can be used to find the number x.

A. $x(4 - 3x) = 60$
B. $x(3x - 4) = 60$
C. $x + (3x - 4) = 60$
D. $3x(x - 4) = 60$

51. In the triangle below, what is the value of x?

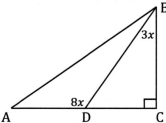

A. 8 B. 12 C. 18 D. Cannot be determined.

52. What is the greatest common factor of the expression: $18x^3y - 6x^2y^4$

A. $3xy$ B. $6x^2y$ C. $6xy^2$ D. $3x^2y$

53. Find the slope of the graph below:

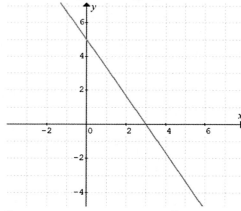

A. $\dfrac{5}{3}$ B. $-\dfrac{5}{3}$ C. $\dfrac{3}{5}$ D. $-\dfrac{3}{5}$

54. Convert to standard form: 3.73×10^6

A. 0.00000373
B. 0.000000373
C. 373,000,000
D. 3,730,000

55. Simplify: $\left(\dfrac{25}{64}\right)^{-3/2}$

A. $\dfrac{125}{512}$ B. $\dfrac{25}{96}$ C. $\dfrac{512}{125}$ D. $-\dfrac{15}{24}$

---

56. Simplify: $(81)^{-0.5}$

A. -9 B. $\dfrac{1}{9}$ C. $\dfrac{1}{27}$ D. -121.5

---

57. $\dfrac{\sqrt{3}+\sqrt{2}}{\sqrt{3}-\sqrt{2}}=$

A. $13+2\sqrt{6}$ B. $13-2\sqrt{6}$ C. $5-2\sqrt{6}$ D. $5+2\sqrt{6}$

---

58. Simplify: $\dfrac{3x}{x^2+5x-14}+\dfrac{5}{x^2+7x}$

A. $\dfrac{3x-1}{x(x+7)}$ C. $\dfrac{3x+8}{x(x+7)(x-2)}$

B. $\dfrac{3x^2+5x-10}{x(x+7)(x-2)}$ D. $\dfrac{3x+1}{x-7}$

---

59. Factor: x^3-27y^3

A. $(x+3y)(x^2-3xy+9y^2)$ C. $(x-3y)(x^2+3xy+9y^2)$

B. $(x+3)(x+3y)(x-3y)$ D. Cannot be factored

---

60. Expand: $a(a-b)^2$

A. a^3-ab^2 C. $a^3-2a^2b-ab^2$

B. a^3+ab^2 D. $a^3-2a^2b+ab^2$

---

61. If $\dfrac{-4}{y-12}=\dfrac{2}{5y-3}$, then $y=$

A. $-\dfrac{2}{3}$ B. $-\dfrac{6}{11}$ C. $\dfrac{18}{11}$ D. $-\dfrac{18}{11}$

---

62. Fred orders a 3-course meal at the Yummy Restaurant which charges 8% tax. He pays $8.95 for the appetizer, $12.50 for the main course and $3.25 for dessert. If he adds a 15% tip to the total bill after tax, what is the cost of Fred's meal?

A. $24.70   B. $26.68   C. $28.41   D. $30.68

63. Find the area of the trapezoid below.

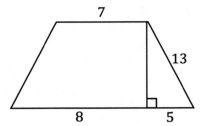

- **A.** 90 $units^2$
- **B.** 120 $units^2$
- **C.** 130 $units^2$
- **D.** 156 $units^2$

64. Solve the inequality: $|2x - 3| \geq 15$

- **A.** $-6 \leq x \leq 9$
- **B.** $x \leq -6$ or $x \geq 9$
- **C.** $x \geq -6$
- **D.** No Solution

65. Solve the inequality: $\dfrac{3x + 2}{5} < 2x$

- **A.** $x < \dfrac{2}{7}$
- **B.** $x > \dfrac{2}{7}$
- **C.** $x < -7$
- **D.** $x > 7$

66. Which of the following is a linear factor of $x^2 - 2x - 8$?

- **A.** $x + 2$
- **B.** $x + 4$
- **C.** $x - 2$
- **D.** $x - 8$

67. Find all real solutions of the equation: $3y^2 + 12y - 18 = 2y^2 + 9$

- **A.** -9 and -3
- **B.** $6 \pm 3\sqrt{7}$
- **C.** $-6 \pm 3\sqrt{7}$
- **D.** $-6 \pm 6\sqrt{7}$

68. Solve the system of equations:
$y = x^2 - 3x - 10$
$y - 2x = -4$

- **A.** $(6, 8)$ and $(-1, -6)$
- **B.** $(3, 2)$ and $(-2, -8)$
- **C.** $(6, -6)$ and $(-1, 8)$
- **D.** No Solution

69. Write the following expression in the form $a + bi$: $\dfrac{-3}{4 - i}$

- **A.** $-\dfrac{12}{17} - \dfrac{3}{17}i$
- **B.** $-\dfrac{4}{5} - \dfrac{1}{5}i$
- **C.** $-\dfrac{7}{17} - \dfrac{3}{17}i$
- **D.** $-\dfrac{4}{5} + \dfrac{1}{5}i$

70. If $f(x) = \sqrt[3]{x + 9}$ then $f^{-1}(x) = ?$

- **A.** $\dfrac{1}{x + 9}$
- **B.** $x^3 - 9$
- **C.** $x^3 + 9$
- **D.** $\dfrac{1}{x^3 - 9}$

71. Which of the following represents the graph of $6y - 3x = -12$?

A.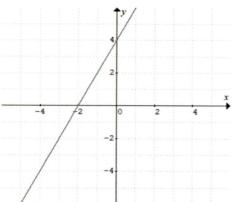

C.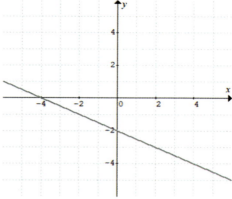

B.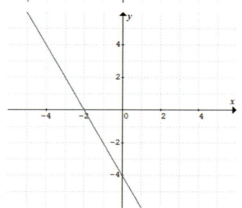

D.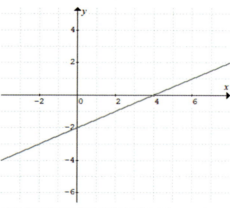

72. Which of the following is the equation of the line joining $(3, 7)$ and $(9, -5)$?

A. $y + 2x = 13$
B. $y + x = 10$
C. $y + 2x = 23$
D. $y + x = 14$

73. Which of the following is the equation of the graph below?

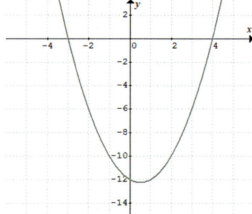

A. $g(x) = x^2 + x - 12$
B. $g(x) = x^2 - x - 12$
C. $g(x) = -x^2 + x + 12$
D. $g(x) = -x^2 - x + 12$

74. Which of the following graphs represents the relationship $f(-2) = 6$?

A.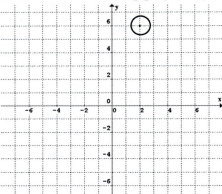

C.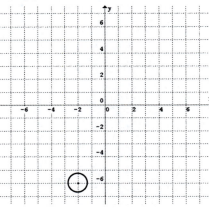

B.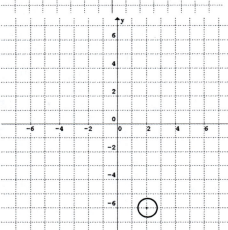

D.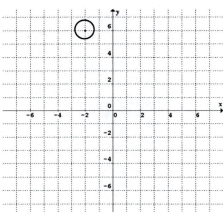

75. From the graph below find the value of $h\big(g(f(-2))\big)$

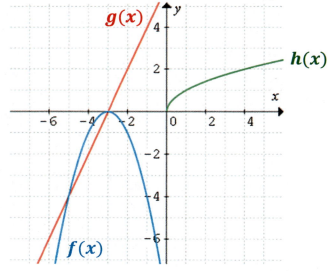

A. -1 B. 2 C. 4 D. 8

END WORKSHEET

Postsecondary Education Readiness Test (PERT) – Worksheet #5 Answer Sheet

1. C	26. C	51. C
2. C	27. B	52. B
3. B	28. B	53. B
4. D	29. C	54. D
5. D	30. B	55. C
6. A	31. C	56. B
7. C	32. B	57. D
8. B	33. A	58. B
9. D	34. A	59. C
10. D	35. B	60. D
11. C	36. B	61. C
12. B	37. A	62. D
13. D	38. C	63. B
14. B	39. C	64. B
15. B	40. B	65. B
16. D	41. C	66. A
17. A	42. B	67. C
18. D	43. A	68. A
19. C	44. B	69. A
20. A	45. A	70. B
21. B	46. C	71. D
22. B	47. A	72. A
23. C	48. A	73. B
24. B	49. C	74. D
25. B	50. B	75. B

The contents of this worksheet are copyrighted. Disclosure or reproduction of any portion herein without express written consent of Track 2 Success © is strictly prohibited.

Worksheet #5 Solutions

1. **C**

   $$\frac{7}{8} + 1\frac{3}{4} =$$

   Change the mixed number to an improper fraction

   $$\frac{7}{8} + \frac{7}{4}$$

   LCD: 8

   $$\frac{7}{8} + \frac{7(2)}{4(2)} = \frac{7}{8} + \frac{14}{8} = \frac{21}{8} = 2\frac{5}{8}$$

2. **C**

   $$-\frac{5}{9} - \frac{1}{4} =$$

   The signs are the same so add the fractions and keep the sign

   LCD: 36

   $$-\frac{(4)5}{(4)9} - \frac{1(9)}{4(9)} = -\frac{20}{36} - \frac{9}{36} = -\frac{29}{36}$$

3. **B**

   $$-\frac{8}{11} \times \frac{3}{4} =$$

   Multiply straight across and reduce

   $$-\frac{24}{44} = -\frac{6}{11}$$

4. **D**

   $$-5\frac{1}{4} \div \left(-\frac{7}{10}\right) =$$

   Convert the mixed number to an improper fraction

   $$-\frac{21}{4} \div -\frac{7}{10}$$

   Take the reciprocal of the second fraction and multiply

   $$-\frac{21}{4} \times -\frac{10}{7} = \frac{210}{28} = \frac{15}{2} = 7\frac{1}{2}$$

5. **D**

   $$17 - [4 - 2(3^2 + 5 - 7)] \div 2 * \frac{3}{5} =$$

The contents of this worksheet are copyrighted. Disclosure or reproduction of any portion herein without express written consent of Track 2 Success © is strictly prohibited.

Begin by solving the expression within the innermost set of parentheses following the order of GEMDAS. Simplify the exponent first then complete the addition and subtraction from left to right.

$$= 17 - [4 - 2(\mathbf{9 + 5 - 7})] \div 2 * \frac{3}{5}$$

$$= 17 - [4 - 2(\mathbf{14 - 7})] \div 2 * \frac{3}{5}$$

$$= 17 - [4 - 2 * \mathbf{7}] \div 2 * \frac{3}{5}$$

Now simplify the expression within the brackets, again following GEMDAS.

$$= 17 - [4 - \mathbf{14}] \div 2 * \frac{3}{5}$$

$$= 17 - (\mathbf{-10}) \div 2 * \frac{3}{5}$$

$$= 17 + \mathbf{10} \div 2 * \frac{3}{5}$$

Simplify the multiplication and division from left to right.

$$= 17 + \mathbf{5} * \frac{3}{5}$$

$$= 17 + \frac{5}{1} * \frac{3}{5}$$

$$= 17 + \mathbf{3}$$

Complete any addition or subtraction from left to right.

$$= 20$$

6. **A**

   14 is what percent of 280?

   $$\frac{is}{of} = \frac{\%}{100}$$

   $$\frac{14}{280} = \frac{x}{100}$$

   $$280x = (14)(100)$$

   $$280x = 1400$$

   $$x = \frac{1400}{280} = 5\%$$

7. **C**

   Sammy spent 20% of his allowance last week. If he spent $16, how much money did he get for allowance?

   The question is asking, 20% of what number is 16?

   $$\frac{is}{of} = \frac{\%}{100}$$

   $$\frac{16}{x} = \frac{20}{100}$$

Cross multiply and solve

$20x = (16)(100)$

$x = \dfrac{1600}{20} = \$80$

8. **B**

The cost of Fun Park tickets is decreased by 18%. How much money would you save if a ticket originally cost $120?

Find 18% of $120

$= 0.18 \times 120 = \$21.60$

9. **D**

$\dfrac{9}{20} =$

First convert to a decimal by using long division or the table of common fractions.

$\dfrac{9}{20} = 9 \times \dfrac{1}{20} = 9 \times 0.05 = 0.45$

This is one of the answers so we stop there.

10. **D**

The Kwik Stop grocery store charges 8% tax. If Susan buys 2 gallons of milk at $3.75 each, one dozen eggs for $1.20 and 3 bags of cookies at $4.50 each, what is Susan's total grocery bill?

Find the total before tax:

$2(\$3.75) + \$1.20 + 3(\$4.50)$

$= \$7.50 + \$1.20 + \$13.50$

$= \$22.20$

Next add the 8% tax. This is an increase of 8% which can be represented by 108% of the total bill.

$\$22.20 \times 1.08 = \$23.98$

11. **C**

$0.213 + 3.001 + 1.52 =$

Line up the decimal points, add extra zeroes as necessary and add

$$
\begin{array}{r}
0.213 \\
+ \quad 3.001 \\
1.520 \\
\hline
4.734 \\
\hline
\end{array}
$$

12. **B**

The Johnson family is going on vacation. Their 8 suitcases weigh an average of 58 lbs each which is over the allowed limit of 55 lbs each. In an attempt to meet the weight requirement, they decide to leave their heaviest suitcases weighing 76, 59 and 64 lbs. What is the mean weight of the remaining suitcases?

For the original 8 bags:

$$mean = \frac{total\ weight}{\#\ bags}$$

$$58 = \frac{total\ weight}{8}$$

$$total\ weight = 58 \times 8 = 464\ lbs$$

Subtracting the weights of the 3 removed bags:

$$Weight\ of\ remaining\ 5\ bags = 464 - (76 + 59 + 64) = 464 - 199 = 265\ lbs$$

$$new\ mean\ weight = \frac{new\ total\ weight}{\#\ bags} = \frac{265}{5} = 53\ lbs$$

13. D

$-2.81 \times 1.5 =$

Multiply -281 by 15

$-281 \times 15 = -4215$

There were 3 decimal places in the two original decimals combined so move the decimal point 3 places to the left in the answer

-4.215

14. B

$9.63 \div (-0.3) =$

Move the decimal point two places to the right in both decimals then divide

$963 \div -30$

$$-\frac{963}{30} = -\frac{321}{10} = -32.1$$

15. B

Which of the following is greatest?

$0.\overline{45},\ 0.4\overline{5},\ 0.045,\ 0.448$

Write out the recurring decimals to make comparing easier

$0.454545 \ldots$

$0.45555 \ldots$

0.045

0.448

Eliminate answer choice C after comparing the tenths position

Eliminate answer choice D after comparing the hundredths position

Eliminate answer choice A after comparing the thousandths position

Therefore the correct answer is B

16. D

The Tennis Club decided to sell boxes of cookies to raise money. The table below shows the number of boxes sold by each member. What is a reasonable estimate of the average number of boxes sold by each member?

Member	Boxes of cookies sold
Brianna	31
Sean	25
Lacy	29
Cory	42
Bryan	23
John	17
Charlotte	8
Katie	37
Sam	29
William	16
Nicky	33

Round the number of boxes to the nearest ten, add and then divide by 10 instead of 11 (11 rounded to the nearest ten is 10)

$$\frac{30 + 30 + 30 + 40 + 20 + 20 + 10 + 40 + 30 + 20 + 30}{10} = \frac{300}{10} = 30$$

17. A

Write the result of the following in scientific notation: $0.0000031 \times 27,000$

Convert each decimal to scientific notation

$3.1 \times 10^{-6} \times 2.7 \times 10^{4}$

$= 3.1 \times 2.7 \times 10^{-6} \times 10^{4}$

$= 8.37 \times 10^{-2}$

18. D

A mother splits a bag of candy equally between her two children. Her son decides to give ⅖ of his candy to his best friend. What fraction of the candy does her son now have?

If he gave away $\frac{2}{5}$ then he still has $\frac{3}{5}$ left

$\frac{3}{5}$ of $\frac{1}{2} = \frac{3}{5} \times \frac{1}{2} = \frac{3}{10}$

19. C

Water flows from one faucet at a rate of 2 gallons per hour and from another faucet at a rate of 5 gallons per hour. How long will it take the faucets to fill a 63 gallon tub together?

Together the pipes fill the tub at 7 gallons per hour.

Gallons = rate × time

$63 = 7 \times t$

$t = \frac{63}{7} = 9$ hours

Track 2 Success © 2014   All Rights Reserved

20. A

Find the arithmetic mean of the following: $\dfrac{2}{3}, \dfrac{4}{5}$, and $\dfrac{14}{15}$

$$\dfrac{\dfrac{2}{3} + \dfrac{4}{5} + \dfrac{14}{15}}{3}$$

LCD of numerator: 15

$$\dfrac{\dfrac{(5)2}{(5)3} + \dfrac{(3)4}{(3)5} + \dfrac{14}{15}}{3}$$

$$\dfrac{\dfrac{10}{15} + \dfrac{12}{15} + \dfrac{14}{15}}{3}$$

$$\dfrac{\dfrac{36}{15}}{\dfrac{3}{1}} = \dfrac{36}{15} \cdot \dfrac{1}{3} = \dfrac{36}{45} = \dfrac{4}{5}$$

21. B

A couple buys a new house and needs to purchase new kitchen appliances. A refrigerator costs $1290.31, a dishwasher costs $489.50, a stove costs $876.32 and a microwave costs $192.89. Estimate the amount of money the couple would need to save to buy the new appliances to the nearest hundred dollars.

Round the cost of each appliance to the nearest hundred and add.

$1300 + $500 + $900 + $200 = $2900

22. B

A sixth grade class paid $63.50 to buy candy for their Halloween party. If they bought 10 lbs of jelly beans for $1.25 per pound, 12 lbs of candy corn for $1.10/lb and chocolate bars at $0.90 each, how many chocolate bars did they buy?

Jelly beans = $1.25 × 10 = $12.50

Candy corn = $1.10 × 12 = $ 13.20

$12.50 + $13.50 = $25.70

Money left for chocolate bars = $63.50 – $25.70 = $37.80

To find out how many chocolate bars were bought, divide the money left by the cost per chocolate bar.

$37.80 ÷ $0.90

Move the decimal point two places to the right in both numbers to convert to whole numbers and make division easier.

$$\dfrac{3780}{90} = \dfrac{378}{9} = 42 \; chocolate \; bars$$

23. C

Find the area of a circle whose radius is 3.5 inches.

$A = \pi r^2 = \pi(3.5)^2 = 12.25\pi$ square inches

24. **B**

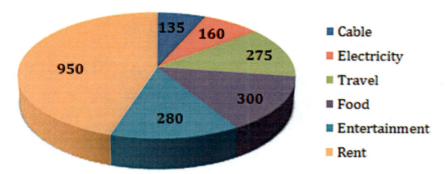

The above graph shows the allocated budget for the Smith household in dollars.
Approximately what percent of the budget is allocated to food?
First find the total amount of money in the Smith's budget.
$275 + 160 + 135 + 300 + 280 + 950 = 2100$
$300 is allocated to food so the percentage allocated to food is:
$$\frac{300}{2100} \times 100 = \frac{1}{7} \times 100 = \frac{100}{7} = 14.29 \approx 14\%$$

25. **B**

How many ounces of milk are necessary to triple a recipe that calls for 13.4 ounces of milk?
$13.4 \times 3 = 40.2 \ ounces$

26. **C**

Shelby joined the Pound Shedders fitness club. She weighed 189 pounds when she started, 152 pounds after 3 months and then lost another 21 pounds two months later. What was her average monthly weight loss?
After 3 months she had lost:
189 − 152 = 37
Since she lost another 21 pounds in the next two months, her total weight loss was:
37 + 21 = 58 pounds
$$Average \ weight \ loss = \frac{total \ weight \ loss}{total \ \# \ months} = \frac{58}{5} = 11.6 \ lbs$$

27. **B**

$(a^x)^3 = a^{15}$ $b^y b^6 = b^{12}$
$a^{3x} = a^{15}$ $b^{y+6} = b^{12}$
$3x = 15$ $y + 6 = 12$
$x = \frac{15}{3} = 5$ $y = 12 - 6 = 6$
Therefore:
$2x - y = 2(5) - 6 = 10 - 6 = 4$

28. B

Evaluate: $\quad -\dfrac{2}{5}\left[\dfrac{1}{8} - 3\left(\dfrac{2}{3} - \dfrac{3}{4}\right)\right]$

Using GEMDAS, simplify the innermost parentheses first

$-\dfrac{2}{5}\left[\dfrac{1}{8} - 3\left(\dfrac{(4)2}{(4)3} - \dfrac{3(3)}{4(3)}\right)\right]$

$-\dfrac{2}{5}\left[\dfrac{1}{8} - 3\left(\dfrac{8}{12} - \dfrac{9}{12}\right)\right]$

$-\dfrac{2}{5}\left[\dfrac{1}{8} - 3\left(-\dfrac{1}{12}\right)\right]$

Next complete the innermost multiplication

$-\dfrac{2}{5}\left[\dfrac{1}{8} - \dfrac{3}{1}\left(-\dfrac{1}{12}\right)\right]$

$-\dfrac{2}{5}\left[\dfrac{1}{8} + \dfrac{3}{12}\right]$

$-\dfrac{2}{5}\left[\dfrac{1}{8} + \dfrac{1}{4}\right]$

Simplify the expression within the brackets

$-\dfrac{2}{5}\left[\dfrac{1}{8} + \dfrac{1(2)}{4(2)}\right] = -\dfrac{2}{5}\left[\dfrac{1}{8} + \dfrac{2}{8}\right] = -\dfrac{2}{5}\left[\dfrac{3}{8}\right]$

Finally multiply and reduce

$-\dfrac{6}{40} = -\dfrac{3}{20}$

29. C

Use the table below to solve the problem:

Approximately how many times larger is the largest lake as compared to the smallest lake?

City	Population (in thousands)
San Juan	152
Key Largo	88
St. Joseph	395
Thomasville	415
Littlerock	107
Winter Park	385

The largest lake has a population of 415 thousand. The smallest lake has a population 88 thousand.

$\dfrac{415}{88} = 4\dfrac{63}{88}$

Since 63 is more than half of 88 we can round up. Therefore the largest lake is approximately 5 times as large as the smallest lake.

30. B

Which of the following is a linear factor of $5x^2 + 13x - 6$?

Factor the quadratic

$(5x - 2)(x + 3)$

Compare to the answer choices to see which of the two factors above appear. The correct choice is B.

31. C

Simplify: $4(x^2 - 3x + 5) - 3(x^2 - 2x + 1)$

Distribute and combine like terms

$4x^2 - 12x + 20 - 3x^2 + 6x - 3$

$4x^2 - 3x^2 - 12x + 6x + 20 - 3$

$x^2 - 6x + 17$

32. B

Simplify: $(2x - 5)(x^2 - x + 1)$

$2x(x^2 - x + 1) - 5(x^2 - x + 1)$

$2x^3 - 2x^2 + 2x - 5x^2 + 5x - 5$

$2x^3 - 7x^2 + 7x - 5$

33. A

Simplify: $(4x^6y^8)^{3/2}$

$4^{3/2}(x^6)^{3/2}(y^8)^{3/2}$

$\left(\sqrt{4}\right)^3 x^{6\left(\frac{3}{2}\right)} y^{8\left(\frac{3}{2}\right)}$

$2^3 x^9 y^{12}$

$8x^9 y^{12}$

34. A

Simplify: $\dfrac{x^2 + 2x + 1}{x^2 - 2x + 3}$

Factor the numerator and denominator then eliminate any common factors

$\dfrac{(x + 1)(x + 1)}{(x - 3)(x + 1)} = \dfrac{x + 1}{x - 3}$

35. B

If $2(x + 2)^2 = 50$, then $x =$

$2(x + 2)^2 = 50$

$(x + 2)^2 = \dfrac{50}{2}$

$(x + 2)^2 = 25$

$x + 2 = \pm\sqrt{25}$

$x + 2 = \pm 5$

$x = -2 \pm 5$

$x = -2 - 5 \qquad\qquad x = -2 + 5$

$x = -7 \qquad\qquad\quad x = 3$

36. **B**

Which of the following represents the solution of $5 - 3x \leq -19$?

$5 - 3x \leq -19$

$-3x \leq -24$

Remember that the direction of the inequality sign switches when dividing or multiplying by a negative number

$\dfrac{-3x}{-3} \geq \dfrac{-24}{-3}$

$x \geq 8$

The circle will be closed because of the greater than or equal to symbol.

37. **A**

The height $h(t)$ of a ball thrown into the air is given by the equation $h(t) = 40t - 16t^2$ where t is time in seconds. What is the height of the ball after 2 seconds?

$h(2) = 40(2) - 16(2)^2$

$h(2) = 80 - 16(4) = 80 - 64 = 16$

38. **C**

$2\sqrt{3} + 4\sqrt{48} - \sqrt{3} =$

Simplify the radicands if possible

$2\sqrt{3} + 4\sqrt{4 \times 4 \times 3} - \sqrt{3}$

$2\sqrt{3} + 4(4\sqrt{3}) - \sqrt{3}$

$2\sqrt{3} + 16\sqrt{3} - \sqrt{3} = 17\sqrt{3}$

39. **C**

The product of two consecutive even integers is 108. Which equation should be used to find the numbers if one of the numbers is represented by n?

If the integers are even and consecutive, we need to add 2 to the first integer to get to the second integer.

First integer $= n$

Second integer $= n + 2$

The product of the two integers is 108 therefore:

$n(n + 2) = 108$

40. **B**

If $4(3x + 2) - 11 = 3(3x - 2)$, then $x =$

$12x + 8 - 11 = 9x - 6$

$12x - 9x = -6 - 8 + 11$

$3x = -3$

$x = \dfrac{-3}{3} = -1$

41. C

If $1 + \dfrac{4}{5}x < x + 6$, then

$1 + \dfrac{4}{5}x < x + 6$

$\dfrac{4}{5}x - x < 6 - 1$

$\dfrac{4}{5}x - \dfrac{5}{5}x < 5$

$-\dfrac{1}{5}x < 5$

Remember to reverse the direction of the inequality sign when multiplying or dividing by a negative number

$\left(-\dfrac{5}{1}\right) - \dfrac{1}{5}x > 5\left(-\dfrac{5}{1}\right)$

$x > -25$

42. B

Solve for x : $\quad \dfrac{x + 4}{a} = \dfrac{y}{b}$

Cross multiply

$b(x + 4) = ay$

$bx + 4b = ay$

$bx = ay - 4b$

$x = \dfrac{ay - 4b}{b}$

43. A

Given $g(x) = 2 + 4x - x^2 - 3x^3$, find $g\left(-\dfrac{2}{3}\right)$

$g\left(-\dfrac{2}{3}\right) = 2 + 4\left(-\dfrac{2}{3}\right) - \left(-\dfrac{2}{3}\right)^2 - 3\left(-\dfrac{2}{3}\right)^3$

$g\left(-\dfrac{2}{3}\right) = 2 - \dfrac{8}{3} - \dfrac{4}{9} - 3\left(-\dfrac{8}{27}\right)$

$g\left(-\dfrac{2}{3}\right) = 2 - \dfrac{8}{3} - \dfrac{4}{9} + \dfrac{8}{9}$

$g\left(-\dfrac{2}{3}\right) = \dfrac{(9)2}{(9)1} - \dfrac{(3)8}{(3)3} - \dfrac{4}{9} + \dfrac{8}{9}$

$g\left(-\dfrac{2}{3}\right) = \dfrac{18}{9} - \dfrac{24}{9} - \dfrac{4}{9} + \dfrac{8}{9} = -\dfrac{2}{9}$

44. B

A can of paint costs $9.75 and covers 150 square ft. What will be the cost of painting the walls of a 6 ft by 8 ft room if the walls are 10 ft high?

Imagine that you are in the room now. You will have two walls that are 6ft by 10ft high and two walls that are 8ft by 10ft high.

Total area of walls:
2(6 × 10) + 2(8 × 10)
2(60) + 2(80)
120 + 160 = 280 *square ft*
To find out how many cans are needed, divide the total area of the walls by the area that each can of paint can cover.
$\frac{280}{150} = 1\frac{13}{15}$
Since we cannot purchase a fraction of a can, we need to buy 2 cans of paint.
$9.75 × 2 = $19.50

45. **A**
In the figure below, a circle is inscribed in a square and the radius of the circle is x inches. Write an expression that represents the area of the shaded region.

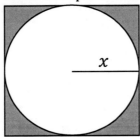

Area of shaded area = area of square – area of circle
Area of circle:
$A = \pi r^2$
$A = \pi x^2$
Area of square:
$A = side^2$
From the diagram, notice that side = $2x$
$A = (2x)^2 = 4x^2$
Shaded Area = $4x^2 - \pi x^2 = x^2(4 - \pi)$

46. **C**
Which of the following is a solution of the equation $x^2 - 7x + 12$?
Factor the quadratic, set each factor equal to zero and solve
$x^2 - 7x + 12 = 0$
$(x - 3)(x - 4) = 0$
$x - 3 = 0$ $x - 4 = 0$
$x = 3$ $x = 4$
Choose the solution that appears as one of the answer choices

47. **A**
If the equation of the linear function in the figure is $y = mx + b$, then $m =$

Prep for Success: Florida's PERT Math Study Guide   **367**

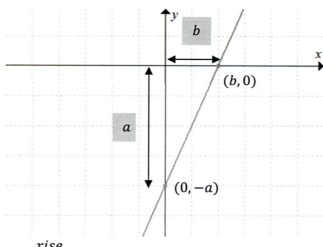

$$m = \frac{rise}{run}$$

Starting at the y-intercept and working towards the x-intercept, the y-coordinate of the y-intercept gives rise and the x-coordinate of the x-intercept gives run.

$$m = \frac{rise}{run} = \frac{a}{b}$$

Notice we take the positive of each because we moved in the positive direction for both rise and run.

48. A

Simplify: $18m - 3n - 4m + 5 + 11n$

Combine like terms

$18m - 4m - 3n + 11n + 5$

$14m + 8n + 5$

49. C

Simplify: $\dfrac{3}{4a^2} + \dfrac{5}{12a}$

LCD: $12a^2$

$\dfrac{(3)3}{(3)4a^2} + \dfrac{5(a)}{12a(a)}$

$= \dfrac{9}{12a^2} + \dfrac{5a}{12a^2} = \dfrac{9 + 5a}{12a^2}$

50. B

The product of a number and four less than three times the number is 60. Write an expression that can be used to find the number x.

First number $= x$

Second number:

Four less than three times the first number $= 3x - 4$

The product of the two numbers is 60 therefore:

$x(3x - 4) = 60$

51. **C**

In the triangle below, what is the value of x?

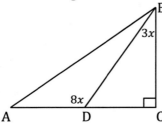

Focusing on $\triangle BCD$, $\angle D = 180° - 8x$
The sum of the angles in a triangle is $180°$
$\angle B + \angle C + \angle D = 180°$
$3x + 90° + 180° - 8x = 180°$
$-5x = -90°$
$x = \dfrac{-90°}{-5} = 18°$

52. **B**

What is the greatest common factor of the expression: $18x^3y - 6x^2y^4$
Factor the instance of each variable with the lowest exponent therefore the greatest common factor is:
$6x^2y$

53. **B**

Find the slope of the graph below:

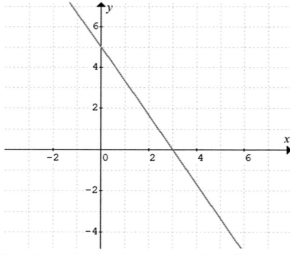

Starting at the $y - intercept$ and working towards the $x - intercept$:
Rise $= -5$
Run $= 3$
Slope $m = \dfrac{rise}{run} = -\dfrac{5}{3}$

54. **D**

Convert to standard form: 3.73×10^6

Since the exponent is positive, move the decimal point 6 places to the right

3,730,000

55. **C**

Simplify: $\left(\dfrac{25}{64}\right)^{-3/2}$

The negative exponent can be made positive by taking the reciprocal of the fraction inside the parentheses.

$\left(\dfrac{64}{25}\right)^{3/2}$

$\left(\sqrt{\dfrac{64}{25}}\right)^3 = \left(\dfrac{8}{5}\right)^3 = \dfrac{512}{125}$

56. **B**

Simplify: $(81)^{-0.5}$

The negative exponent can be made positive by taking the reciprocal of the number inside the parentheses.

$\left(\dfrac{1}{81}\right)^{0.5}$

$\left(\dfrac{1}{81}\right)^{1/2}$

Using law #7 of exponents, we convert the fractional exponent to a root.

$= \sqrt{\dfrac{1}{81}} = \dfrac{1}{9}$

57. **D**

$\dfrac{\sqrt{3} + \sqrt{2}}{\sqrt{3} - \sqrt{2}} =$

Multiply by the conjugate of the denominator

$\dfrac{\sqrt{3} + \sqrt{2}}{\sqrt{3} - \sqrt{2}} \cdot \dfrac{\sqrt{3} + \sqrt{2}}{\sqrt{3} + \sqrt{2}}$

FOIL

$\dfrac{3 + \sqrt{3}\sqrt{2} + \sqrt{2}\sqrt{3} + 2}{3 + \sqrt{3}\sqrt{2} - \sqrt{2}\sqrt{3} - 2}$

$\dfrac{3 + \sqrt{6} + \sqrt{6} + 2}{3 - 2} = 5 + 2\sqrt{6}$

58. B

Simplify: $\dfrac{3x}{x^2 + 5x - 14} + \dfrac{5}{x^2 + 7x}$

Factor the denominators of both

$\dfrac{3x}{(x+7)(x-2)} + \dfrac{5}{x(x+7)}$

LCD: $x(x+7)(x-2)$

$\dfrac{(x)3x}{(x)(x+7)(x-2)} + \dfrac{5(x-2)}{x(x+7)(x-2)}$

$\dfrac{3x^2 + 5(x-2)}{x(x+7)(x-2)}$

$\dfrac{3x^2 + 5x - 10}{x(x+7)(x-2)}$

59. C

Factor: $x^3 - 27y^3$

$(x)^3 - (3y)^3$

Use the difference of cubes formula

$a^3 - b^3 = (a-b)(a^2 + ab + b^2)$

$(x)^3 - (3y)^3 = (x-3y)((x)^2 + (x)(3y) + (3y)^2)$

$\qquad\qquad = (x-3y)(x^2 + 3xy + 9y^2)$

60. D

Expand: $a(a-b)^2$

$a(a-b)(a-b)$

FOIL the second and third factors

$a(a^2 - ab - ab + b^2)$

$a(a^2 - 2ab + b^2) = a^3 - 2a^2b + ab^2$

61. C

If $\dfrac{-4}{y-12} = \dfrac{2}{5y-3}$, then $y =$

Cross multiply and solve

$-4(5y-3) = 2(y-12)$

$-20y + 12 = 2y - 24$

$-20y - 2y = -24 - 12$

$-22y = -36$

$y = \dfrac{-36}{-22} = \dfrac{18}{11}$

62. D

Fred orders a 3-course meal at the Yummy Restaurant which charges 8% tax. He pays $8.95 for the appetizer, $12.50 for the main course and $3.25 for dessert. If he adds a 15% tip to the total bill after tax, what is the cost of Fred's meal?

Find the pre-tax total of his bill
$8.95 + $12.50 + $3.25 = 24.70
He adds the tip to the total after tax so calculate the total after tax. Adding tax makes the bill 108% of the total.
108% of $24.7 = 1.08 \times 24.7 = 26.68$
Now add 15% to this total.
115% of $26.68 = 1.15 \times 26.68 = 30.68$
Fred's meal costs $30.68.

63. **B**

Find the area of the trapezoid below.

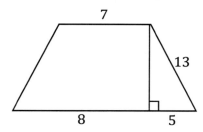

Find the height using the Pythagorean Theorem
$5^2 + h^2 = 13^2$
$25 + h^2 = 169$
$h^2 = 169 - 25 = 144$
$h = \sqrt{144} = 12$
$B = 8 + 5 = 13$
Area of trapezoid $= \frac{1}{2}h(B + b) = \frac{1}{2}(12)(13 + 7) = \frac{1}{2}(12)(20) = 120\ units^2$

64. **B**

Solve the inequality: $|2x - 3| \geq 15$

$2x - 3 \geq 15$ $2x - 3 \leq -15$
$2x \geq 18$ $2x \leq -12$
$x \geq \frac{18}{2}$ $x \leq -\frac{12}{2}$
$x \geq 9$ $x \leq -6$

Therefore: $x \leq -6\ or\ x \geq 9$

65. **B**

Solve the inequality: $\frac{3x + 2}{5} < 2x$

Remember never cross multiply when dealing with inequalities

$\frac{3x + 2}{5} - 2x < 0$

$\frac{3x + 2}{5} - \frac{2x(5)}{1(5)} < 0$

$$\frac{3x+2-10x}{5} < 0$$
$$\frac{-7x+2}{5} < 0$$

Set the numerator equal to zero to find the critical point (there is no variable in the denominator so ignore it)

$-7x + 2 = 0$
$-7x = -2$
$x = \frac{2}{7}$

Place the critical value on a number line and choose test points to the left and right.

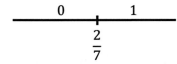

Substitute the test points into original inequality to see which satisfy the inequality.

When $x = 0$
$$\frac{3(0)+2}{5} < 2(0)$$
$$\frac{0+2}{5} < 2$$
$$\frac{2}{5} < 0$$
False

When $x = 1$
$$\frac{3(1)+2}{5} < 2(1)$$
$$\frac{3+2}{5} < 2$$
$$\frac{5}{5} < 2$$
$1 < 2$
True

The solution exists where the inequality is true therefore the solution is:
$$x > \frac{2}{7}$$

66. **A**

Which of the following is a linear factor of $x^2 - 2x - 8$?
Factor the quadratic
$(x-4)(x+2)$
Choose the factor that appears as one of the answer choices

67. **C**

Find all real solutions of the equation: $3y^2 + 12y - 18 = 2y^2 + 9$
Move all terms to one side of the equation and combine like terms
$3y^2 + 12y - 18 - 2y^2 - 9 = 0$
$3y^2 - 2y^2 + 12y - 18 - 9 = 0$
$y^2 + 12y - 27 = 0$
This cannot be factored so use the quadratic formula with $a = 1$, $b = 12$, $c = -27$
$$y = \frac{-b \pm \sqrt{b^2 - 4ac}}{2a}$$

$$y = \frac{-12 \pm \sqrt{12^2 - 4(1)(-27)}}{2(1)}$$

$$y = \frac{-12 \pm \sqrt{144 + 108}}{2}$$

$$y = \frac{-12 \pm \sqrt{144 + 108}}{2}$$

$$y = \frac{-12 \pm \sqrt{252}}{2}$$

$$y = \frac{-12 \pm \sqrt{6 \times 6 \times 7}}{2}$$

$$y = \frac{-12 \pm 6\sqrt{7}}{2}$$

$$y = -6 \pm 3\sqrt{7}$$

68. A

Solve the system of equations:

$y = x^2 - 3x - 10$

$y - 2x = -4$

Since the first equation is already solved for y, substitution would be ideal in this situation

$(x^2 - 3x - 10) - 2x = -4$

$x^2 - 5x - 10 + 4 = 0$

$x^2 - 5x - 6 = 0$

$(x - 6)(x + 1) = 0$

$x - 6 = 0 \qquad\qquad x + 1 = 0$

$x = 6 \qquad\qquad\quad x = -1$

Substitute the values of x into either of the two original equations to find the corresponding y values. The second equation is simpler so we use that one.

$y - 2x = -4$

When $x = 6$ When $x = -1$

$y - 2(6) = -4$ $y - 2(-1) = -4$

$y - 12 = -4$ $y + 2 = -4$

$y = 8$ $y = -6$

$(6, 8)$ $(-1, -6)$

69. A

Write the following expression in the form $a + bi$: $\dfrac{-3}{4 - i}$

Multiply by the conjugate of the denominator

$$\frac{-3}{4 - i} \cdot \frac{4 + i}{4 + i}$$

FOIL

$$\frac{-12 - 3i}{16 + 4i - 4i - i^2}$$

$$\frac{-12-3i}{16-(-1)} = -\frac{12}{17} - \frac{3}{17}i$$

70. B

If $f(x) = \sqrt[3]{x+9}$ then $f^{-1}(x) =$?
Let $f(x) = y$
$y = \sqrt[3]{x+9}$
Interchange x and y
$x = \sqrt[3]{y+9}$
Solve for y
$x^3 = \left(\sqrt[3]{y+9}\right)^3$
$x^3 = y + 9$
$y = x^3 - 9$
Replace y with $f^{-1}(x)$
$f^{-1}(x) = x^3 - 9$

71. D

Which of the following represents the graph of $6y - 3x = -12$?
Find the x-intercept by setting $y = 0$
$6(0) - 3x = -12$
$-3x = -12$
$x = \frac{-12}{-3} = 4$
$(4, 0)$
Find the y-intercept by setting $x = 0$
$6y - 3(0) = -12$
$6y = -12$
$y = \frac{-12}{6} = -2$
$(0, -2)$
Choose the graph that has the correct x and y intercepts

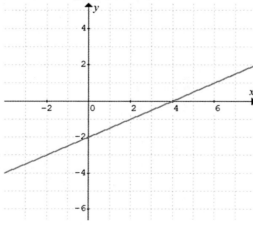

Prep for Success: Florida's PERT Math Study Guide   375

72. **A**

Use the two points to first find the slope of the line
$$m = \frac{y_2 - y_1}{x_2 - x_1} = \frac{-5 - 7}{9 - 3} = \frac{-12}{6} = -2$$
Next use the slope and one of the points in the point–slope formula
$$y - y_1 = m(x - x_1)$$
$$y - 7 = -2(x - 3)$$
$$y - 7 = -2x + 6$$
$$y = -2x + 13$$
$$y + 2x = 13$$

73. **B**

Which of the following is the equation of the graph below?

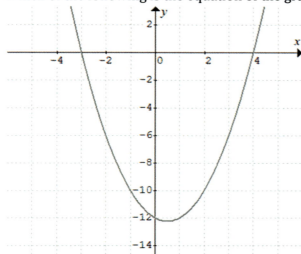

The two x-intercepts occur at $x = -3$ and $x = 4$ therefore the factors of the quadratic equation are $(x + 3)$ and $(x - 4)$. Remember that we set the factors equal to zero to find the x-intercepts hence we use the opposite sign when going from x-intercepts to factor form.
$$(x + 3)(x - 4)$$
$$x^2 - 4x + 3x - 12$$
$$x^2 - x - 12$$
Note that the graph is u-shaped and so the leading coefficient is positive. If the graph were n-shaped, multiply the entire quadratic by –1 to get the proper equation.
Therefore the equation of the quadratic shown above is:
$$g(x) = x^2 - x - 12$$

74. **D**

Which of the following graphs represents the relationship $f(-2) = 6$?
The point represented by the relationship is $(-2, 6)$

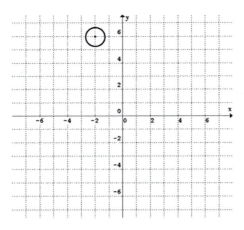

75. B

From the graph below find the value of $h\big(g(f(-2))\big)$

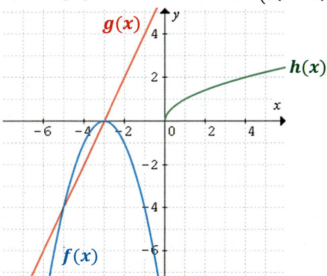

First find the value of $f(-2)$ from the graph by reading the y value corresponding to $x = -2$ on the graph of $f(x)$

$f(-2) = -1$

The problem now becomes $h(g(-1))$

Find $g(-1)$ from the graph by reading the y value corresponding to $x = -1$ on the graph of $g(x)$

$g(-1) = 4$

The problem now becomes $h(4)$

Find $h(4)$ from the graph by reading the y value corresponding to $x = 4$ on the graph of $h(x)$

$h(4) = 2$

Therefore:

$h\big(g(f(-2))\big) = 2$

Track 2 Success
"Dedicated to helping you stay on the track to academic success"

Postsecondary Education Readiness Test (PERT) – Extra Bubble Sheet

1. A B C D
2. A B C D
3. A B C D
4. A B C D
5. A B C D
6. A B C D
7. A B C D
8. A B C D
9. A B C D
10. A B C D
11. A B C D
12. A B C D
13. A B C D
14. A B C D
15. A B C D
16. A B C D
17. A B C D
18. A B C D
19. A B C D
20. A B C D
21. A B C D
22. A B C D
23. A B C D
24. A B C D
25. A B C D

26. A B C D
27. A B C D
28. A B C D
29. A B C D
30. A B C D
31. A B C D
32. A B C D
33. A B C D
34. A B C D
35. A B C D
36. A B C D
37. A B C D
38. A B C D
39. A B C D
40. A B C D
41. A B C D
42. A B C D
43. A B C D
44. A B C D
45. A B C D
46. A B C D
47. A B C D
48. A B C D
49. A B C D
50. A B C D

51. A B C D
52. A B C D
53. A B C D
54. A B C D
55. A B C D
56. A B C D
57. A B C D
58. A B C D
59. A B C D
60. A B C D
61. A B C D
62. A B C D
63. A B C D
64. A B C D
65. A B C D
66. A B C D
67. A B C D
68. A B C D
69. A B C D
70. A B C D
71. A B C D
72. A B C D
73. A B C D
74. A B C D
75. A B C D

The contents of this worksheet are copyrighted. Disclosure or reproduction of any portion herein without express written consent of Track 2 Success © is strictly prohibited.

378 Prep for Success: Florida's PERT Math Study Guide

Track 2 Success

"Dedicated to helping you stay on the track to academic success"

Postsecondary Education Readiness Test (PERT) – Extra Bubble Sheet

1. Ⓐ Ⓑ Ⓒ Ⓓ
2. Ⓐ Ⓑ Ⓒ Ⓓ
3. Ⓐ Ⓑ Ⓒ Ⓓ
4. Ⓐ Ⓑ Ⓒ Ⓓ
5. Ⓐ Ⓑ Ⓒ Ⓓ
6. Ⓐ Ⓑ Ⓒ Ⓓ
7. Ⓐ Ⓑ Ⓒ Ⓓ
8. Ⓐ Ⓑ Ⓒ Ⓓ
9. Ⓐ Ⓑ Ⓒ Ⓓ
10. Ⓐ Ⓑ Ⓒ Ⓓ
11. Ⓐ Ⓑ Ⓒ Ⓓ
12. Ⓐ Ⓑ Ⓒ Ⓓ
13. Ⓐ Ⓑ Ⓒ Ⓓ
14. Ⓐ Ⓑ Ⓒ Ⓓ
15. Ⓐ Ⓑ Ⓒ Ⓓ
16. Ⓐ Ⓑ Ⓒ Ⓓ
17. Ⓐ Ⓑ Ⓒ Ⓓ
18. Ⓐ Ⓑ Ⓒ Ⓓ
19. Ⓐ Ⓑ Ⓒ Ⓓ
20. Ⓐ Ⓑ Ⓒ Ⓓ
21. Ⓐ Ⓑ Ⓒ Ⓓ
22. Ⓐ Ⓑ Ⓒ Ⓓ
23. Ⓐ Ⓑ Ⓒ Ⓓ
24. Ⓐ Ⓑ Ⓒ Ⓓ
25. Ⓐ Ⓑ Ⓒ Ⓓ
26. Ⓐ Ⓑ Ⓒ Ⓓ
27. Ⓐ Ⓑ Ⓒ Ⓓ
28. Ⓐ Ⓑ Ⓒ Ⓓ
29. Ⓐ Ⓑ Ⓒ Ⓓ
30. Ⓐ Ⓑ Ⓒ Ⓓ
31. Ⓐ Ⓑ Ⓒ Ⓓ
32. Ⓐ Ⓑ Ⓒ Ⓓ
33. Ⓐ Ⓑ Ⓒ Ⓓ
34. Ⓐ Ⓑ Ⓒ Ⓓ
35. Ⓐ Ⓑ Ⓒ Ⓓ
36. Ⓐ Ⓑ Ⓒ Ⓓ
37. Ⓐ Ⓑ Ⓒ Ⓓ
38. Ⓐ Ⓑ Ⓒ Ⓓ
39. Ⓐ Ⓑ Ⓒ Ⓓ
40. Ⓐ Ⓑ Ⓒ Ⓓ
41. Ⓐ Ⓑ Ⓒ Ⓓ
42. Ⓐ Ⓑ Ⓒ Ⓓ
43. Ⓐ Ⓑ Ⓒ Ⓓ
44. Ⓐ Ⓑ Ⓒ Ⓓ
45. Ⓐ Ⓑ Ⓒ Ⓓ
46. Ⓐ Ⓑ Ⓒ Ⓓ
47. Ⓐ Ⓑ Ⓒ Ⓓ
48. Ⓐ Ⓑ Ⓒ Ⓓ
49. Ⓐ Ⓑ Ⓒ Ⓓ
50. Ⓐ Ⓑ Ⓒ Ⓓ
51. Ⓐ Ⓑ Ⓒ Ⓓ
52. Ⓐ Ⓑ Ⓒ Ⓓ
53. Ⓐ Ⓑ Ⓒ Ⓓ
54. Ⓐ Ⓑ Ⓒ Ⓓ
55. Ⓐ Ⓑ Ⓒ Ⓓ
56. Ⓐ Ⓑ Ⓒ Ⓓ
57. Ⓐ Ⓑ Ⓒ Ⓓ
58. Ⓐ Ⓑ Ⓒ Ⓓ
59. Ⓐ Ⓑ Ⓒ Ⓓ
60. Ⓐ Ⓑ Ⓒ Ⓓ
61. Ⓐ Ⓑ Ⓒ Ⓓ
62. Ⓐ Ⓑ Ⓒ Ⓓ
63. Ⓐ Ⓑ Ⓒ Ⓓ
64. Ⓐ Ⓑ Ⓒ Ⓓ
65. Ⓐ Ⓑ Ⓒ Ⓓ
66. Ⓐ Ⓑ Ⓒ Ⓓ
67. Ⓐ Ⓑ Ⓒ Ⓓ
68. Ⓐ Ⓑ Ⓒ Ⓓ
69. Ⓐ Ⓑ Ⓒ Ⓓ
70. Ⓐ Ⓑ Ⓒ Ⓓ
71. Ⓐ Ⓑ Ⓒ Ⓓ
72. Ⓐ Ⓑ Ⓒ Ⓓ
73. Ⓐ Ⓑ Ⓒ Ⓓ
74. Ⓐ Ⓑ Ⓒ Ⓓ
75. Ⓐ Ⓑ Ⓒ Ⓓ

The contents of this worksheet are copyrighted. Disclosure or reproduction of any portion herein without express written consent of Track 2 Success © is strictly prohibited.

Prep for Success: Florida's PERT Math Study Guide

Track 2 Success
"Dedicated to helping you stay on the track to academic success"

Postsecondary Education Readiness Test (PERT) – Extra Bubble Sheet

1. Ⓐ Ⓑ Ⓒ Ⓓ
2. Ⓐ Ⓑ Ⓒ Ⓓ
3. Ⓐ Ⓑ Ⓒ Ⓓ
4. Ⓐ Ⓑ Ⓒ Ⓓ
5. Ⓐ Ⓑ Ⓒ Ⓓ
6. Ⓐ Ⓑ Ⓒ Ⓓ
7. Ⓐ Ⓑ Ⓒ Ⓓ
8. Ⓐ Ⓑ Ⓒ Ⓓ
9. Ⓐ Ⓑ Ⓒ Ⓓ
10. Ⓐ Ⓑ Ⓒ Ⓓ
11. Ⓐ Ⓑ Ⓒ Ⓓ
12. Ⓐ Ⓑ Ⓒ Ⓓ
13. Ⓐ Ⓑ Ⓒ Ⓓ
14. Ⓐ Ⓑ Ⓒ Ⓓ
15. Ⓐ Ⓑ Ⓒ Ⓓ
16. Ⓐ Ⓑ Ⓒ Ⓓ
17. Ⓐ Ⓑ Ⓒ Ⓓ
18. Ⓐ Ⓑ Ⓒ Ⓓ
19. Ⓐ Ⓑ Ⓒ Ⓓ
20. Ⓐ Ⓑ Ⓒ Ⓓ
21. Ⓐ Ⓑ Ⓒ Ⓓ
22. Ⓐ Ⓑ Ⓒ Ⓓ
23. Ⓐ Ⓑ Ⓒ Ⓓ
24. Ⓐ Ⓑ Ⓒ Ⓓ
25. Ⓐ Ⓑ Ⓒ Ⓓ

26. Ⓐ Ⓑ Ⓒ Ⓓ
27. Ⓐ Ⓑ Ⓒ Ⓓ
28. Ⓐ Ⓑ Ⓒ Ⓓ
29. Ⓐ Ⓑ Ⓒ Ⓓ
30. Ⓐ Ⓑ Ⓒ Ⓓ
31. Ⓐ Ⓑ Ⓒ Ⓓ
32. Ⓐ Ⓑ Ⓒ Ⓓ
33. Ⓐ Ⓑ Ⓒ Ⓓ
34. Ⓐ Ⓑ Ⓒ Ⓓ
35. Ⓐ Ⓑ Ⓒ Ⓓ
36. Ⓐ Ⓑ Ⓒ Ⓓ
37. Ⓐ Ⓑ Ⓒ Ⓓ
38. Ⓐ Ⓑ Ⓒ Ⓓ
39. Ⓐ Ⓑ Ⓒ Ⓓ
40. Ⓐ Ⓑ Ⓒ Ⓓ
41. Ⓐ Ⓑ Ⓒ Ⓓ
42. Ⓐ Ⓑ Ⓒ Ⓓ
43. Ⓐ Ⓑ Ⓒ Ⓓ
44. Ⓐ Ⓑ Ⓒ Ⓓ
45. Ⓐ Ⓑ Ⓒ Ⓓ
46. Ⓐ Ⓑ Ⓒ Ⓓ
47. Ⓐ Ⓑ Ⓒ Ⓓ
48. Ⓐ Ⓑ Ⓒ Ⓓ
49. Ⓐ Ⓑ Ⓒ Ⓓ
50. Ⓐ Ⓑ Ⓒ Ⓓ

51. Ⓐ Ⓑ Ⓒ Ⓓ
52. Ⓐ Ⓑ Ⓒ Ⓓ
53. Ⓐ Ⓑ Ⓒ Ⓓ
54. Ⓐ Ⓑ Ⓒ Ⓓ
55. Ⓐ Ⓑ Ⓒ Ⓓ
56. Ⓐ Ⓑ Ⓒ Ⓓ
57. Ⓐ Ⓑ Ⓒ Ⓓ
58. Ⓐ Ⓑ Ⓒ Ⓓ
59. Ⓐ Ⓑ Ⓒ Ⓓ
60. Ⓐ Ⓑ Ⓒ Ⓓ
61. Ⓐ Ⓑ Ⓒ Ⓓ
62. Ⓐ Ⓑ Ⓒ Ⓓ
63. Ⓐ Ⓑ Ⓒ Ⓓ
64. Ⓐ Ⓑ Ⓒ Ⓓ
65. Ⓐ Ⓑ Ⓒ Ⓓ
66. Ⓐ Ⓑ Ⓒ Ⓓ
67. Ⓐ Ⓑ Ⓒ Ⓓ
68. Ⓐ Ⓑ Ⓒ Ⓓ
69. Ⓐ Ⓑ Ⓒ Ⓓ
70. Ⓐ Ⓑ Ⓒ Ⓓ
71. Ⓐ Ⓑ Ⓒ Ⓓ
72. Ⓐ Ⓑ Ⓒ Ⓓ
73. Ⓐ Ⓑ Ⓒ Ⓓ
74. Ⓐ Ⓑ Ⓒ Ⓓ
75. Ⓐ Ⓑ Ⓒ Ⓓ

The contents of this worksheet are copyrighted. Disclosure or reproduction of any portion herein without express written consent of Track 2 Success © is strictly prohibited.

382   Prep for Success: Florida's PERT Math Study Guide

Appendix A – Formulas

Percentages

$$\frac{is}{of} = \frac{\%}{100}$$

Place Values

Millions	Hundred Thousands	Ten Thousands	Thousands	Hundreds	Tens	Ones	.	Tenths	Hundredths	Thousandths	Ten Thousandths	Hundred Thousandths	Millionths
10^6	10^5	10^4	10^3	10^2	10^1	10^0	.	$\frac{1}{10^1}$	$\frac{1}{10^2}$	$\frac{1}{10^3}$	$\frac{1}{10^4}$	$\frac{1}{10^5}$	$\frac{1}{10^6}$

Important conversions

$1\ foot = 12\ inches$
$3\ feet = 1\ yard$
$1\ mile = 5,280\ feet$

$60\ seconds = 1\ minute$
$60\ minutes = 1\ hour$
$24\ hours = 1\ day$
$7\ days = 1\ week$
$52\ weeks = 1\ year$

$1\ pound\ (lb) = 16\ ounces\ (oz)$

$1\ cup = 8\ fluid\ ounces\ (fl\ oz)$
$1\ pint\ (pt) = 2\ cups$
$1\ quart\ (qt) = 2\ pints$
$1\ gallon\ (gal) = 4\ quarts$

$1\ gal = 4\ qt = 8\ pt = 16\ cups = 128\ fl\ oz$

Distance, Rate and Time

$$\text{rate} = \frac{distance}{time} \qquad distance = rate \times time \qquad Average\ speed = \frac{total\ distance}{total\ time}$$

Distance between two points

Given any two points (x_1, y_1) and (x_2, y_2), the distance between them can be found using the distance formula:

$$d = \sqrt{(x_2 - x_1)^2 + (y_2 - y_1)^2}$$

Midpoint of two points

$$M = \left(\frac{x_1 + x_2}{2}, \frac{y_1 + y_2}{2}\right)$$

Geometry

	Area	Perimeter

Square

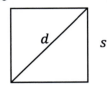

$A = s^2$
$A = \frac{1}{2}d^2$

$P = 4s$

Rectangle

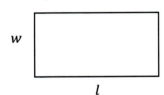

$A = l \times w$

$P = 2l + 2w$

Triangle

$A = \frac{1}{2}bh$

Add lengths of sides

Circle

$A = \pi r^2$

$C = 2\pi r$
$C = \pi d$

Other Geometric Shapes

Trapezoid

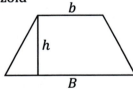

A trapezoid is a shape with exactly 1 pair of parallel sides called bases (B and b).
The area is $A = \frac{1}{2}h(B + b)$

Parallelogram

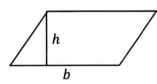

A parallelogram is a four sided figure with two pairs of parallel sides. Like any other four-sided figure, the sum of the interior angles is 360°.
The area is $A = bh$ and its diagonals bisect each other (divide each other in half)

Rhombus

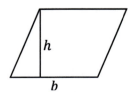

A rhombus is a parallelogram with four equal sides. The diagonals are perpendicular and bisect each other.
Area $= bh = \frac{1}{2}$ (product of diagonals)

diameter of a circle = 2r (twice the radius)
diagonal of a square = $s\sqrt{2}$
The sum of the angles in a triangle is 180°.
The sum of any two sides of a triangle is greater than the third.
The sum of the exterior angles in a polygon equals 360°
The sum of the interior angles is 180°(n − 2) where n is the number of sides of the polygon.
If the polygon is a regular polygon (all sides equal), it has the additional properties:

$$\text{Value of an exterior angle} = \frac{360°}{n}$$

$$\text{Value of an interior angle} = \frac{180°(n-2)}{n}$$

Types of Triangles

Triangle	Property
Isosceles 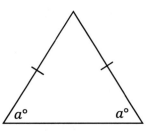	Two congruent (equal) sides. The base angles are equal.
Equilateral 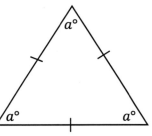	All three sides congruent. All three angles equal.
Scalene 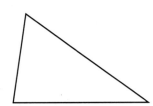	No sides equal. No angles equal.

Pythagorean Theorem

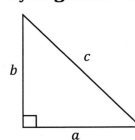

Right-angled triangles use a special formula called the **Pythagorean Theorem**.
In a right-angled triangle where the hypotenuse (longest side) is c:
$a^2 + b^2 = c^2$

Here are a few right-angled triangles that can be memorized to shorten your calculation time.

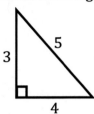 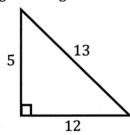 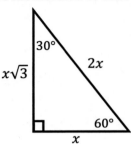 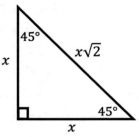

Types of Angles

Angle	Property	Angle	Property
Acute Angles	Less than 90°	Supplementary Angles 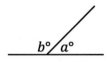	Straight Angle divided in 2 therefore supplementary angles add to 180° $\angle a + \angle b = 180°$
Right Angles 	Equal to 90°	Complementary Angles	Right angle divided in 2 therefore complementary angles add to 90° $\angle a + \angle b = 90°$
Obtuse Angles	Greater than 90°	Vertical Angles 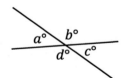	When two lines cross to form an X, the angles opposite each other are equal. $\angle a = \angle c$ and $\angle b = \angle d$
Straight Angles	Sit on a straight line and is equal to 180°	Transverse Angles 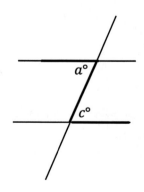	Formed when a line intersects two parallel lines. The intersecting line creates vertical angles with each parallel line. It also creates some other equal angles. The easiest way to identify these transverse angles is to picture the Z. $\angle a = \angle c$

Circles
The standard form of a circle is $(x - h)^2 + (y - k)^2 = r^2$ where the radius $= r$ and center is (h, k).

Factoring
$a^2 - b^2 = (a + b)(a - b)$
$a^3 - b^3 = (a - b)(a^2 + ab + b^2)$
$a^3 + b^3 = (a + b)(a^2 - ab + b^2)$

The Laws of Exponents:
1. $(x^a)^b = x^{ab}$
2. $(xy)^a = x^a y^a$
3. $\left(\dfrac{x}{y}\right)^a = \dfrac{x^a}{y^a}$
4. $x^a x^b = x^{a+b}$
5. $\dfrac{x^a}{x^b} = x^{a-b}$
6. $x^{-a} = \dfrac{1}{x^a}$ and $\dfrac{1}{x^{-a}} = x^a$
7. $x^{1/a} = \sqrt[a]{x}$

Note that $x^a + x^b \neq x^{a+b}$

Linear Equations
Given (x_1, y_1) and (x_2, y_2)

$$slope = \frac{vertical\ distance\ moved}{horizontal\ distance\ moved} = \frac{rise}{run} = \frac{y_2 - y_1}{x_2 - x_1}$$

Point – slope formula:
$$y - y_1 = m(x - x_1)$$

Slope – intercept formula:
$$y = mx + b \quad \text{where } m \text{ is the slope and } b \text{ is the } y\text{–intercept}$$

Quadratic Equations
$$y = ax^2 + bx + c \quad or \quad y = a(x - h)^2 + k$$
Quadratic Formula:
$$x = \frac{-b \pm \sqrt{b^2 - 4ac}}{2a}$$

Vertex (h, k):
$$x = h = -\frac{b}{2a}$$
$$y = k = f\left(-\frac{b}{2a}\right)$$

388 Prep for Success: Florida's PERT Math Study Guide

Index

A

Absolute Values	5
Additive Inverse	1
Angles	189
Area	63
Arithmetic Mean	42

C

Circles	64
Coefficient	107
Comparisons	58
Complex Numbers	141
Conjugate	141
Composite Functions	179
Composite Number	7
Congruent	64
Constant	107
Conversions	17, 46
Coordinate Plane	77
Cubes	10

D

Decimals	
To Percentage	18
To Fraction	19
Adding	28
Dividing	29
Subtracting	28
Multiplying	29
Diameter	64
Difference of Cubes	121
Distance Formula	85
Dividend	19, 125
Divisor	19, 125
Domain	148

E

Equilateral Triangles	65, 187
Estimation	55

Evaluating Equations	108
Exponents	5, 129
Exponential Equations	110
Expression	107

F

Factoring	111, 118
Factors	107
Prime	7
FOIL	122
Fractions	
Adding	26
Subtracting	26
Multiplying	27
Dividing	27
Improper	18
To Decimal	19
To Percentage	20
Reducing	17
First-Degree Equation	67

G

GCF	9
Geometric Reasoning	185

H

Horizontal Lines	83
Horizontal Line Test	148

I

Imaginary Numbers	141
Inequalities	93
Graphs	95
Absolute Values	99
Rational Expressions	60
Integers	1
Inverse Functions	177
Irrational Numbers	2

Isosceles Triangles	65, 187

L

LCD	25
LCM	8
Like Terms	117
Linear Equations	77
Graphs	80
Literal Equations	107

M

Median	43
Metric Conversions	46
Midpoint	86
Mixed Numbers	18
Mode	43

N

Natural Numbers	1

O

Order of Operations	6, 109
One-to-One Function	148

P

Parallel Lines	84
Parallelogram	64
Percentage	37
Error	39
To Fraction	20
To Decimal	21
Perimeter	63
Perpendicular Lines	85
Place Value	53
Point-Slope Formula	78
Polygons	190
Polynomials	107, 117
Adding	117

Dividing	124	Dividing	12	Elimination	165
Expanding	126	Subtracting	11	Substitution	168
Factoring	118	Multiplying	12		
Multiplying	123	Range	43, 148	**T**	
Subtracting	117	Rates	46		
Prime Factors	7	Rational Expressions	155	Trapezoid	64
Properties of Real Numbers		Add/Subtract	155	Triangles	65, 187
Associative	2	Dividing	157		
Commutative	2	Multiplying	156	**V**	
Distributive	2	Solving Equations	158		
Inverse	2	Rational Numbers	1	Vertex	147
Proportions	44	Rationalize	133	Vertical Asymptote	149
Direct	44	Ratios	44	Vertical Lines	83
Indirect	44	Remainder	125	Vertical Line Test	148
Pythagorean Theorem	188	Rectangles	63		
		Rhombus	64		
		Rounding	54	**W**	

Q

Quadratic Equations	141			Whole Numbers	1
Solving	141	**S**			
Factoring	118	Scalene Triangle	65, 187		
Graphing	147	Scientific Notation	56	**X**	
Quadratic Formula	144	Signed Numbers	1		
Quotient	125	Simple Interest	40	X-intercepts	79
		Slope	78		
		Slope-Intercept Form	78	**Y**	

R

		Squares	63	Y-intercepts	79
Radicals and Roots	9, 132	Square Root	10		
Adding	11	Sum of Cubes	121		
		Systems of Equations	165		